BIOSURFACTANTS
Research Trends and Applications

BIOSURFACTANTS
Research Trends and Applications

Edited by
Catherine N. Mulligan
Sanjay K. Sharma
Ackmez Mudhoo

CRC Press
Taylor & Francis Group
Boca Raton London New York

CRC Press is an imprint of the
Taylor & Francis Group, an **informa** business

CRC Press
Taylor & Francis Group
6000 Broken Sound Parkway NW, Suite 300
Boca Raton, FL 33487-2742

First issued in paperback 2019

ISBN-13: 978-1-4665-1823-0 (hbk)
ISBN-13: 978-0-367-37889-9 (pbk)

Visit the Taylor & Francis Web site at
http://www.taylorandfrancis.com

and the CRC Press Web site at
http://www.crcpress.com

In remembrance of my father-in-law, late O.P. Sharma.

Sanjay K. Sharma

For Camilla

Ackmez Mudhoo

Contents

Preface

Microorganisms produce a variety of surface active compounds mostly on cell surfaces or excreted extracellularly. These microbially derived surfactants, called biosurfactants, have different chemical structures and hence diverse chemical properties. Every biosurfactant has its natural role in the life cycle of the microorganism that produces it. Biosurfactants are broadly classified into several different groups based on their chemical composition and microbial origin. These include glycolipids, lipopeptides, phospholipids, neutral lipids, fatty acids, and polymeric and particulate biosurfactants. Synthetic surfactants have been shown to remove nonpolar compounds from surfaces, but problems can be associated with their use, such as reduced availability of compounds sequestered into micelles, their toxicity, and their ultimate resistance to biodegradation, leading to increased pollution. The use of microbially produced surfactants, that is, biosurfactants, is an alternative with immense potential. The distinct advantage of biosurfactants over whole cells or exopolymers is their small size (generally biosurfactant molecular weights are less than 1500). Another advantage is that biosurfactants have a wide variety of chemical structures that may show different metal selectivities and, thus, metal removal efficiencies. Also, biosurfactant production is potentially less expensive than synthetic surfactants and can be easily achieved in situ at contaminated sites from inexpensive raw materials. Several biosurfactants have found applications in environmental remediation, such as the acceleration of the biodegradation of hydrophobic hydrocarbons in oil-contaminated beach soils and soil slurried in bioreactors. Results to date are promising, but consolidated knowledge of biosurfactants is still required to have a reliable predictive framework that would further the knowledge on the scientific and technological components of applications of biosurfactants.

The writing of this book has been undertaken because of the need for an updated manual covering the latest research and development findings of the most relevant aspects linked to biological, biochemical, chemical, and physical processes and applications of biosurfactants. Novel research findings and applications of biosurfactants have been presented, covering the following areas: medical applications, drug and gene delivery, nanotechnology, food technology, synthetic applications and organic chemistry, pollutant degradation and remediation, polymer chemistry, industrial syntheses and processes, reactor design, electrochemical systems, and combined ultrasound–microwave technologies.

Chapter 1 reviews the current knowledge and the latest advances in biosurfactant applications as well as the biotechnological strategies being developed for improving production processes. This chapter also aims at providing an up-to-date overview of biosurfactant roles, applications, and future potential keeping in view the goals of green chemistry and sustainable development. Chapter 2 critically reviews the current status of biosynthetic and genetic regulation mechanisms for microbial surfactants and their various classifications and characteristics and also suggests future directions toward achieving higher yields and new applications. Chapter 3 focuses on research that has been done on rhamnolipids, a promising biosurfactant with high surface activities, and the major advances in production processes. Sophorolipids are surface-active compounds synthesized by a select number of yeast species. They offer an environment-friendly alternative for petrochemical-based surfactants. Chapter 4 provides an overview and updates of basic as well as applied research on sophorolipids. Chapter 5 describes the sources and characteristics of the different types of marine microbial biosurfactants. Chapter 6 details the factors influencing biosurfactant production and potential applications of lipopeptides. Chapter 7 addresses several aspects related to the application of biosurfactants in the food industry.

Trehalose lipids are surface-active compounds synthesized by most species belonging to the mycolates group like *Mycobacterium*, *Rhodoccoccus*, *Arthrobacter*, *Nocardia*, and *Gordonia*. Several structural types of trehalose lipids have been reported, differing in the size and structure of their hydrophobic moiety. Apart from their environmental and industrial use, trehalose lipids can be used as biologically active compounds in the biomedical field. Despite their high potential for application, industrial use of trehalose lipids, like other biosurfactants, is still limited because of low yields and high costs involved in the production process, particularly in the recovery step. Chapter 8 reviews the latter concerns with regard to trehalose lipids.

Eco-friendly synthesis of nanoparticles has many more advantages than conventional synthesis, which is perennially associated with environmental hazards caused by the use of chemicals and solvents. Due to biodegradability, low toxicity, and capability of forming micelles, biosurfactants are considered a soft template for the green synthesis of nanoparticles. Glycolipids and lipopeptides have been found to be very effective as capping and reducing agents in the synthesis of metal nanoparticles. They also ensure the stability of suspension for suitable applications. Though achieving nanoparticles of uniform shape and size using biosurfactants with nonlinear structural conformations still remains a major challenge, it is the green process coupled with environmental sustainability that will drive future research on nanoparticles. Chapter 9 presents a number of the latter aspects related to biosurfactants and nanoscience. Chapter 10 presents the applications of biosurfactants in environmental industries as a result of their biodegradability, low toxicity, and effectiveness in enhancing biodegradation and solubilization of low solubility compounds. Chapter 11 describes known interactions between metals and different classes of biosurfactants. It examines the molecular nature of these interactions where possible and highlights potential biosurfactant-based approaches to metal remediation technologies, including soil washing, phytoremediation, co-contaminant removal, micellar-enhanced ultrafiltration, and ion flotation. Chapter 12 presents a summary of the prospects in relation to the sustainable development of biosurfactants.

We sincerely hope that this book contributes substantially to the pool of knowledge on the applications of biosurfactants. We extend our gratitude to all the contributors, experts who have voluntarily commented on the manuscripts, and the production staff of Taylor & Francis Group, CRC Press. They have all helped us immensely in their respective roles in producing this unique work of science and technology.

Catherine N. Mulligan
Sanjay K. Sharma
Ackmez Mudhoo

Acknowledgments

The compilation of this book has provided us with a unique opportunity to renew old friendships and weave new ones. We thank all the contributors for having graciously responded to deadlines. We appreciate their constructive criticisms, which have improved the content of this work. We hope that the final result does justice to their efforts.

Professor Sanjay K. Sharma expresses his heartfelt gratitude to his respected parents, Dr. M.P. Sharma and Smt. Parmeshwari Devi. He also extends his regards to Brig. (Dr.) S.S. Pabla (president, JECRC University, Jaipur) and to Amit Agarwal and Arpit Agarwal (directors, JECRC University, Jaipur) for their encouraging words and motivation. He would also like to thank his wife Dr. Pratima and his children Kunal and Kritika for their never-ending support and love, without which this book would not have been possible. Ackmez Mudhoo expresses his appreciation for the faith his parents, sister-in-law, brother, and lovely nieces Yanna and Camilla have placed in him throughout the writing and compilation of this handbook. He is also grateful to Professor Romeela Mohee, Dr. Geeta D. Somaroo, Dr. Gianeshwar Ramsawock, Dr. Vinod K. Garg, Dr. Sunil Kumar, Darshini D. Narain, Prof. Mohammad Taherzadeh, Dr. Naraindra Kistamah, Dr. Reena Amatya Shrestha, and Prof. Mika Sillanpää for their support and encouragement. Catherine Mulligan wishes to acknowledge the discussions and contributions of collaborators, research students, and other professionals.

Editors

Professor Catherine N. Mulligan earned her BEng and MEng in chemical engineering and her PhD in geoenvironmental engineering from McGill University, Montreal, Quebec, Canada. She has more than 25 years of research experience in government, industrial, and academic environments. She worked at the Biotechnology Research Institute of the National Research Council and SNC Research Corp., a subsidiary of SNC-Lavalin, Montreal, Quebec, Canada. She then joined the Department of Building, Civil and Environmental Engineering at Concordia University, Montreal, Quebec, Canada, in 1999. She has taught courses in site remediation, environmental engineering, fate and transport of contaminants, and geoenvironmental engineering and conducts research in remediation of contaminated soils, sediments, and water. Dr. Mulligan holds a Concordia Research Chair in Geoenvironmental Sustainability (Tier I) and is a full professor and the associate dean of research and graduate studies of the Faculty of Engineering and Computer Science. She has authored more than 80 refereed papers in various journals, holds 3 patents, and has supervised to completion more than 40 graduate students. She serves as the director of the Concordia Institute of Water, Energy and Sustainable Systems, which trains students in sustainable development practices and promotes research into new systems, technologies, and solutions for water, energy, and resource conservation.

Professor (Dr.) Sanjay K. Sharma is a well-known author and editor of many books, research journals, and hundreds of articles over the last 20 years. His recent publications include *Waste Water Reuse and Management* (Springer, UK), *Advances in Water Treatment and Pollution Prevention* (Springer, UK), *Green Corrosion Chemistry and Engineering* (Wiley, Germany), *Handbook of Applied Biopolymer Technology: Synthesis, Degradation and Applications* (Royal Society of Chemistry, UK), *Handbook on Applications of Ultrasound: Sonochemistry and Sustainability* (CRC Press, Taylor & Francis Group, Boca Raton, Florida), and *Green Chemistry for Environmental Sustainability* (CRC Press, Taylor & Francis Group, Boca Raton, Florida). He is a member of the American Chemical Society (United States) and the Royal Society of Chemistry (UK) and is also a life member of various international professional societies, including the International Society of Analytical Scientists, the Indian Council of Chemists, the International Congress of Chemistry and Environment, and the Indian Chemical Society. His research interests include green chemistry, heavy metals removal by biosorption, biopolymers, and green corrosion inhibition. Dr. Sharma has published 13 books on chemistry distributed by national and international publishers and has over 52 research papers of national and international repute to his credit. He has been appointed as series editor for Springer UK's prestigious book series *Green Chemistry for Sustainability*. He has also had a long association with this publisher and has edited 12 titles for them. Dr. Sharma currently serves as a professor and head in the Department of Chemistry, JECRC University, Jaipur, India, where he teaches engineering chemistry and environmental chemistry to BTech students and organic chemistry to MS students. He also pursues his research interest in the domain of green chemistry. Dr. Sharma serves as editor-in-chief for three international research journals: *RASAYAN Journal of Chemistry*; *International Journal of Chemical, Environmental and Pharmaceutical Research*; and *Water: Research & Development*. He is also a reviewer for many other international journals, including the prestigious *Green Chemistry*, *Green Chemistry Letters and Reviews*, and *Ultrasonics Sonochemistry*.

Ackmez Mudhoo earned his BEng (Hons) in chemical and environmental engineering and his MPhil in chemical engineering from the University of Mauritius, Réduit, Mauritius. His research interests encompass the bioremediation of solid wastes and wastewaters. He has over 60 publications in highly reputed journals, conference papers, and book chapters. Mudhoo serves as a peer reviewer for *Waste Management*, *International Journal of Environment and Waste Management, Journal of Hazardous Materials, Journal of Environmental Informatics, Environmental Engineering Science, Ecological Engineering, Green Chemistry Letters and Reviews, Chemical Engineering Journal*, and *Water Research*. He is an associate editor on the editorial board of *Environmental Chemistry Letters* (Springer). He is also the coeditor of several books on sustainability, green technology, bioremediation, and anaerobic digestion published by Taylor & Francis Group, Scrivener Publishing—Wiley, and Springer. He has worked as a consultant chemical process engineer with China International Water & Electric Corp. (CWE, Mauritius; February 2006–March 2008) in the construction and operation and maintenance of the Montagne Jacquot Sewerage Plant (Mauritius). Mudhoo currently serves as a lecturer in the Department of Chemical and Environmental Engineering at the University of Mauritius and is pursuing his PhD in environmental engineering under the supervision of Professor Romeela Mohee and Associate Professor Dr. Bhola R. Gurjar (Department of Civil Engineering, Indian Institute of Technology Roorkee, Roorkee, India). His aim is to contribute to the development of the chemical and environmental engineering disciplines through mature teaching, motivational instruction, and interdisciplinary fundamental and applied research.

Contributors

Fereshteh Arab
Department of Building, Civil and
  Environmental Engineering
Concordia University
Montreal, Quebec, Canada

Ibrahim M. Banat
Faculty of Life and Health Sciences
School of Biomedical Sciences
University of Ulster
Coleraine, Northern Ireland

Nelly Christova
Institute of Microbiology
Bulgarian Academy of Sciences
Sofia, Bulgaria

Siddhartha G.V.A.O. Costa
Secretariat for Policies and Programs on
  Research and Development-SEPED
Ministry of Science, Technology and
  Innovation-MCTI
Brasília, Brazil

Gunaseelan Dhanarajan
Department of Biotechnology
Indian Institute of Technology
Kharagpur, India

David E. Hogan
Department of Soil, Water, and
  Environmental Science
University of Arizona
Tucson, Arizona

Eun-Ki Kim
Department of Biological Engineering
Inha University
Incheon, Republic of South Korea

Raina M. Maier
Department of Soil, Water, and
  Environmental Science
University of Arizona
Tucson, Arizona

Snigdha Majumder
Department of Biotechnology
Indian Institute of Technology
Kharagpur, India

Komal Makhijani
Green Chemistry & Sustainability
  Research Group
Department of Chemistry
JECRC University
Jaipur, India

Vivek K. Morya
Department of Biological Engineering
Inha University
Incheon, Republic of South Korea

Ackmez Mudhoo
Department of Chemical and
  Environmental Engineering
Faculty of Engineering
University of Mauritius
Réduit, Mauritius

Catherine N. Mulligan
Department of Building, Civil and
  Environmental Engineering
Concordia University
Montreal, Quebec, Canada

Marcia Nitschke
Institute of Chemistry of São Carlos
University of São Paulo
São Carlos, Brazil

Jeanne E. Pemberton
Department of Chemistry and Biochemistry
University of Arizona
Tucson, Arizona

Vivek Rangarajan
Department of Biotechnology
Indian Institute of Technology
Kharagpur, India

Ramkrishna Sen
Department of Biotechnology
Indian Institute of Technology
Kharagpur, India

Sanjay K. Sharma
Green Chemistry & Sustainability
    Research Group
Department of Chemistry
JECRC University
Jaipur, India

Ivanka Stoineva
Institute of Organic
    Chemistry with Centre
    of Phytochemistry
Bulgarian Academy of Sciences
Sofia, Bulgaria

Rengathavasi Thavasi
Jeneil Biotech Inc.
Saukville, Wisconsin

Tracey A. Veres-Schalnat
Department of Chemistry and
    Biochemistry
University of Arizona
Tucson, Arizona

1 Green Chemistry and Biosurfactant Research

Catherine N. Mulligan, Sanjay K. Sharma,
Ackmez Mudhoo, and Komal Makhijani

CONTENTS

INTRODUCTION

There is no doubt that our lives have been enhanced by chemistry. That is something chemists and students need to celebrate. The chemical enterprise generates critical living needs such as food for the world's population, achieves various medical wonders that save millions of lives and improve people's health, and produces materials essential

1

to the present and future needs of mankind. "This role of chemistry is not generally recognized by government or the public. In fact, chemicals, chemistry, and chemists are actually seen by many as the cause of the problems" (Clark, 1999). Indeed, the chemical industry releases more hazardous wastes to the environment than any other industry sector and more in total than that is released by the next nine sectors combined (Anastas and Warner, 1998). It is true that in the production of fine chemicals and pharmaceuticals many individual steps are usually required, and a majority of these are stoichiometric, thereby consuming large amounts of reagents. Most of these materials consume excessive energy in their manufacture. They also entail the use of aggressive, corrosive, and sometimes explosive reagents, and much waste is generated, which currently is disposed of in landfills. The need to devise new methods of manufacture that minimize consumption of energy and materials (including volatile solvents) and that also minimize the liberation of harmful gases (typified by the notorious greenhouse gas, N_2O) is pressing. An emerging approach to this grand challenge seeks to embed the diverse set of environmental perspectives and interests in the everyday practice of the people most responsible for using and creating new materials—chemists. The approach, which has come to be known as green chemistry, intends to eliminate the intrinsic hazard itself, rather than focusing on reducing risk by minimizing exposure.

Hydrophobic pollutants present in petroleum hydrocarbons and soil and water environment require solubilization before being degraded by microbial cells. Mineralization is governed by desorption of hydrocarbons from soil. Surfactants can increase the surface area of hydrophobic materials, such as pesticides in soil and water environment, thereby increasing their water solubility. Hence, the presence of surfactants may increase microbial degradation of pollutants. Use of biosurfactants (BSs) for degradation of pesticides in soil and water environment has gained importance only recently. BSs are attracting much interest due to their potential advantages over their synthetic counterparts in many fields spanning environmental, food, biomedical, and other industrial applications.

GREEN CHEMISTRY: AN INNOVATIVE TECHNOLOGY

The term green chemistry was first used in 1991 by Paul T. Anastas in a special program launched by the U.S. Environmental Protection Agency (USEPA) to implement sustainable development in chemistry and chemical technology by industry, academia, and government.

DEFINITION OF GREEN CHEMISTRY

The term green chemistry, as adopted by the Working Party, is defined as "The invention, design and application of chemical products and processes to reduce or to eliminate the use and generation of hazardous substances" (Anastas and Warner, 1998). Green chemistry is the use of chemistry for pollution prevention. More specifically, it is the design of chemical products and processes that are environmentally benign. Advances in green chemistry address both obvious hazards and those associated with such global issues as climate change, energy production, availability of a safe and adequate water supply, food production, and the presence of toxic substances in the environment (Anastas and Kirchoff, 2002). The design of environmentally benign products and processes may be guided by the 12 principles of green chemistry (Anastas and Warner, 1998). These principles are a categorization of the fundamental approaches taken to achieve the green chemistry goals of benign products and processes and have been used as guidelines and design criteria by molecular scientists.

TWELVE PRINCIPLES OF GREEN CHEMISTRY

1. Prevention: It is better to prevent waste than to treat or clean up waste after it has been created.
2. Atom Economy: Synthetic methods should be designed to maximize the incorporation of all materials used in the process into the final product.
3. Less Hazardous Chemical Syntheses: Wherever practicable, synthetic methods should be designed to use and generate substances that possess little or no toxicity to human health and the environment.
4. Designing Safer Chemicals: Chemical products should be designed to affect their desired function while minimizing their toxicity.
5. Safer Solvents and Auxiliaries: The use of auxiliary substances (solvents, separation agents, etc.) should be made unnecessary wherever possible and innocuous when used.
6. Design for Energy Efficiency: Energy requirements of chemical processes should be recognized for their environmental and economic impacts and should be minimized. If possible, synthetic methods should be conducted at ambient temperature and pressure.
7. Use of Renewable Feedstocks: A raw material or feedstock should be renewable rather than depleting whenever technically and economically practicable.
8. Reduce Derivatives: Unnecessary derivatization (use of blocking groups, protection/deprotection, and temporary modification of physical/chemical processes) should be minimized or avoided if possible, because such steps require additional reagents and can generate waste.
9. Catalysis: Catalytic reagents (as selective as possible) are superior to stoichiometric reagents.
10. Design for Degradation: Chemical products should be designed so that at the end of their function, they break down into innocuous degradation products and do not persist in the environment.
11. Real-Time Analysis for Pollution Prevention: Analytical methodologies need to be further developed to allow for real-time in-process monitoring and control prior to the formation of hazardous substances.
12. Inherently Safer Chemistry for Accident Prevention: Substances and the form of a substance used in a chemical process should be chosen to minimize the potential for chemical accidents, including releases, explosions, and fires.

ILLUSTRATIVE CASE IN GREEN CHEMISTRY

Argonne National Laboratory (ANL) Energy Systems Division won the 1998 presidential Green Chemistry Challenge Award in the Alternative Solvents/Reaction category. Potential substitution of green/bio-based solvents is foreseen for up to 80% of the 30 billion pounds (metric?) of environmentally damaging solvents employed in the world. It is interesting to note that this effort did not win for creating a new solvent. Their product is ethyl lactate (Coupland, 2003) known for years as a technically effective alternative solvent and approved by the Food and Drug Administration for use in food. Until this innovation, ethyl lactate has been too expensive to employ as an alternative solvent. ANL's breakthrough transformed the economics of producing ethyl lactate, which allows it to compete with existing solvents. The ANL team's primary breakthrough was developing a cost-cutting manufacturing process based on new membrane technology that enabled more cost-effective separation and purification techniques. ANL did it in a fashion that is more environmentally benign, and this case is judged to illustrate most of the 12 principles of green chemistry.

1. The process eliminates salt waste (gypsum) and undesirable by-products achieving Principle 1 Prevent Waste.
2. The process innovation demonstrates Principle 2 Atom Economy in that undesirable by-products are avoided and much more of the input materials are incorporated in the final product.
3. ANL's synthetic process minimizes the use of ethanol and the need to distill it and keeps the alcohol in the reaction system reducing the risk of fire or explosion—fulfilling Principle 3 Less Hazardous Synthesis.
4. Ethyl lactate is nontoxic and therefore a good example of Principle 4 Safer Chemicals.
5. Their innovation provides an option for users to address Principle 5 Alternative Solvents.
6. The use of catalysis and the membrane separation technology enables the reaction to consume 90% less energy than traditional processes fulfilling Principle 6 Energy Efficiency.
7. This process for ethyl lactate is carbohydrate based rather than petrochemical based, using corn, for example. It therefore demonstrates Principle 7 Renewable Feedstocks.
8. The process illustrates Principle 9 in its use of catalysis for cracking the carbohydrates.
9. Ethyl lactate is biodegradable. It hydrolyzes into ethyl alcohol and lactic acid—common constituents of food. Thus, the end product builds in Principle 10 Design for Degradation.
10. The handling of the end product is less dangerous than most other solvents. Therefore, it illustrates Principle 12 Accident Prevention.

HISTORICAL BACKGROUND

The term "green chemistry" was first coined by the USEPA in the early 1990s, and major interest in green chemistry in the United States began in earnest with the passage of the "Pollution Prevention Act" of 1990. Thus, green chemistry is becoming a formal focus of the EPA in 1991 (Kidwai and Mohan, 2005). Passage of this act signaled a move away from the "command and control" response to environmental issues and toward pollution prevention as a more effective strategy that focused on preventing waste from being formed in the first place (Anastas and Kirchoff, 2002). In 1991, the Office of Pollution Prevention and Toxics in the USEPA launched the first research initiative of the Green Chemistry Program, the Alternative Synthetic Pathways research solicitation. Foundational work in chemistry and engineering at the National Science Foundation's program on Environmentally Benign Syntheses and Processes was launched in 1992 and formed a partnership with the EPA through a Memorandum of Understanding the same year. In 1993, the EPA program officially adopted the name "U.S. Green Chemistry Program." Since its inception, the U.S. Green Chemistry Program has served as a focal point for major activities within the United States, such as the Presidential Green Chemistry Challenge Awards and the annual Green Chemistry and Engineering Conference. The first award was presented to Barry M. Trost in 1998 for introducing the concept of atom economy that looks at utilized and wasted atoms in a reaction. In Italy, a multiuniversity consortium (INCA) featured research on green chemistry as one of its central themes. During the last half of the decade, Japan organized the Green and Sustainable Chemistry Network, with an emphasis on promoting research and development on green and sustainable chemistry. The inaugural edition of the journal *Green Chemistry*, sponsored by the Royal Society of Chemistry, appeared in 1999.

CURRENT RESEARCH

The activities in green chemistry research, education, industrial implementation, awards, and outreach are all based on the fundamental definition of green chemistry. Green chemistry can be utilized anywhere in the life cycle, from feedstock origins to beyond end of useful life (Anastas and Kirchoff, 2002). The following section addresses some illustrations where success has been achieved to fulfill the principles of green chemistry.

Bio-based feedstocks: Bio-based feedstocks present several advantages over petroleum-based sources. Biomass-derived carbohydrates are more highly functionalized than hydrocarbon sources, minimizing the need for oxidative chemistry that often requires the use of toxic heavy metals. New crops can provide a continuous supply of raw materials (Kirchoff, 2003). Cargill Dow manufactures Nature Works polylactic acid (PLA) from renewable resources, such as corn and sugar beets. The manufacturing process uses no organic solvents, and the final product is biodegradable, compostable, and recyclable. Products made from PLA can be hydrolyzed with water to lactic acid, yielding prime polymer following purification and polymerization. PLA fibers perform as well as petroleum-based polymers in consumer goods, including beverage containers, food packaging, and clothing.

Reagents: Alcohols are commonly oxidized to carbonyl compounds using heavy metals such as chromium (VI). The Collins group has designed a series of catalysts that activate hydrogen peroxide to bleach wood pulp (Collins et al., 2002). Termed tetra-amido macrocyclic ligand activators, these catalysts are selective, are effective over a wide pH range, consume less energy, and eliminate the problem of chlorinated by-products. This technology has applications in the pulp and paper industry, which is gradually shifting toward better reagents for pulp delignification and bleaching.

Green solvents: Solvents are auxiliary materials used in chemical synthesis. They are not an integral part of the compounds undergoing reaction, yet they play an important role in chemical production and synthesis. By far, the largest amount of "auxiliary waste" in most chemical productions is associated with solvent usage. In a classical chemical process, solvents are used extensively for dissolving reactants, extracting and washing products, separating mixtures, cleaning reaction apparatuses, and dispersing products for practical applications (Li and Trost, 2008). The development of green chemistry redefines the role of a solvent: *an ideal solvent facilitates the mass transfer but does not dissolve*! In addition, a desirable green solvent should be natural, nontoxic, cheap, and readily available. More desirably, it should have additional benefits of aiding the reaction, separation, or catalyst recycling (Li and Trost, 2008).

Water: The only natural solvent on earth is water. Life requires the construction of chemical bonds in an aqueous environment. It is obvious that water is the most inexpensive and environmentally benign solvent. Since it was reported that Diels–Alder reactions could be greatly accelerated by using water as a solvent instead of organic solvents, there has been considerable attention dedicated to the development of organic reactions in water (Li and Trost, 2008). In many cases, because of hydrophobic effects, using water as a solvent not only accelerates reaction rates but also enhances reaction selectivity, even when the reactants are sparingly soluble or insoluble in the medium. Furthermore, the low solubility of oxygen gas in water, an important property in the early development of life in an anaerobic environment, can facilitate air-sensitive transition metal catalysis in open air (Li, 2002). The use of water as a solvent also implies the elimination of tedious protection–deprotection processes for certain acidic hydrogen–containing functional groups, which contributes to the overall synthetic efficiency.

Supercritical CO_2: Breakthroughs in the use of supercritical fluids such as carbon dioxide have met with success in the research laboratory as well as commercially. Supercritical fluids offer a number of benefits, such as the potential to combine reaction and separation processes and the ability to tune the solvent through variations in temperature and pressure. In the supercritical fluids area, CO_2 has received the most attention because its critical temperature and pressure (T_c 31.1°C, P_c 7.7 MPa) are more accessible than those of other solvents (water, e.g., has T_c 374°C and P_c 221 bar). CO_2 offers numerous advantages as a benign solvent: it is nontoxic, nonflammable, and inexpensive and can be separated from the product by simple depressurization. Applications of supercritical CO_2 are found in the dry cleaning industry, where CO_2 replaces per chloroethylene (PCE) as a solvent (Micell Technologies and Hughes Environmental Systems) in semiconductor manufacturing, where the low surface tension of supercritical CO_2 avoids the damage caused by water in conventional processing (Gleason and Ober, 2001) and in chemical processing.

Ionic liquids: Ionic liquids, a relatively new area of solvent investigation, are attractive because of their negligible vapor pressure and their use in polar systems to generate new chemistries. A plethora of ionic liquids can be produced by varying the cations and anions, permitting the synthesis of ionic liquids tailored for specific applications (Anastas and Kirchoff, 2002). The 2001 Kenneth G. Hancock Memorial Student Award in Green Chemistry was presented to

Richard A. Brown Jr. for his work on reactions carried out in ionic liquid and supercritical fluids (Kidwai and Mohan, 2005). Free-radical polymerization of methyl methacrylate in water-soluble ionic liquids and cationic ring-opening polymerizations of 2-(*m*-difluorophenyl)-2-oxazoline and 2-phenyl-2-oxazoline in water-soluble ionic liquids performed by Sanchez et al. (2007) are examples where success has been achieved.

Other solvents: Ester solvents are actually the largest group of green solvents. Special solvents, such as glycerol carbonate, can be used as nonreactive diluents in epoxy or polyurethane systems. Ethyl lactate has been reported as a photo-resist carrier solvent and a cleanup solvent in microelectronics and semiconductor manufacturing, and 2-ethylhexyl lactate can be used as degreaser and as a green solvent in agrochemical formulations, for example, for the protection of paddy rice crops. Another innovative discovery is the recently developed "switchable solvents" by Jessop, Liotta, Eckert, and others (Liu et al., 2006). Such solvents change their properties with different needs. Besides these solvents, other synthetic solvents such as fluorous and property-changing soluble polymer systems (Bergbreiter, 1998) have been evaluated as potential green alternatives.

Biocatalysis: The area of catalysis is sometimes referred to as a "foundational pillar" of green chemistry. The search for new, efficient, and environmentally benign processes for the textile and pulp and paper industries has increased interest in these essentially "green" catalysts, which work with air and produce water as the only by-product, making them more generally available to the scientific community. Laccases (EC 1.10.3.2, *p*-diphenol: dioxygenoxidoreductase) belong to the so-called blue-copper family of oxidases. They are glycoproteins, which are ubiquitous in nature—they have been reported in higher plants and virtually every fungus that has been examined for them. At present, the main technological applications of laccases are in the textile, dye, or printing industries—in processes related to decolorization of dyes (Claus et al., 2002)—and in the pulp and paper industries—for the delignification of woody fibers, particularly during the bleaching process (Bajpai, 1999).

Design of safer chemicals: Design for reduced hazard is a green chemistry principle that is being achieved in classes of chemicals ranging from pesticides to surfactants, from polymers to dyes. Surfactants (Bayer) and polymers (Donlar Corporations, 1996; Cargill, 2001) have been developed to degrade in the environment at the end of their useful lifetime. Dyes without heavy metals (Freeman and Edwards, 2000) are finding applications in the textile industry. BSs and bioemulsifiers are a novel group of molecules and among the most powerful and versatile bioproducts that the modern microbial biotechnology can offer. Hence, in the following section, we discuss some roles and applications of these microbial compounds in oil-related sciences, food industries, bioremediation technologies, and other related fields presenting the processes that exploit commercially available BS technologies and highlighting those in which they may be potentially applied and have a greater impact in the near future.

BIOSURFACTANTS

Microbial surface-active compounds are a group of structurally diverse molecules produced by different microorganisms and are mainly classified by their chemical structure and their microbial origin. They are made up of a hydrophilic moiety, comprising an acid, peptide cations, or anions, mono-, di-, or polysaccharides, and a hydrophobic moiety of unsaturated or saturated hydrocarbon chains or fatty acids. These structures confer a wide range of properties, including the ability to lower surface and interfacial tension of liquids and to form micelles and microemulsions between two different phases. These compounds can be roughly divided into two main classes (Neu, 1996): low-molecular-weight compounds called biosurfactants, such as lipopeptides, glycolipids, and proteins, and high-molecular-weight polymers of polysaccharides, lipopolysaccharide proteins, or lipoproteins that are collectively called bioemulsans (Rosenberg and Ron, 1997) or bioemulsifiers (Smyth et al., 2010). The best studied microbial surfactants are glycolipids. Among these, the best-known compounds are rhamnolipids, trehalolipids, sophorolipids, and mannosylerythritol lipids (MELs). Rhamnolipid production by *Pseudomonas* species has been extensively studied, and potential applications have been proposed (Maier and Soberón–Chávez, 2000). Rhamnolipids from *Pseudomonas aeruginosa* are currently commercialized by Jeneil Biosurfactant, USA, mainly as a fungicide for agricultural purposes or an additive to enhance bioremediation activities (Banat et al., 2010).

PRODUCTION AND OPTIMIZATION

BSs are amphiphilic compounds produced on living surfaces, mostly microbial cell surfaces, or excreted extracellularly. Microorganisms utilize a variety of organic compounds as the source of carbon and energy for their growth. When the carbon source is an insoluble substrate like a hydrocarbon (CxHy), microorganisms facilitate their diffusion into the cell by producing a variety of substances, the BSs. Some bacteria and yeasts excrete ionic surfactants that emulsify the CxHy substrate in the growth medium. Some examples of this group of BS are rhamnolipids, which are produced by different *Pseudomonas* sp., or the sophorolipids that are produced by several *Torulopsis* sp. Some other microorganisms are capable of changing the structure of their cell wall, which they achieve by synthesizing lipopolysaccharides or nonionic surfactants in their cell wall. These are lipopolysaccharides, such as emulsan, synthesized by *Acinetobacter* sp. and lipoproteins or lipopeptides, such as surfactin and subtilisin, produced by *Bacillus subtilis*. Microorganisms also produce surfactants that are in some cases a combination of many chemical types: referred to as the polymeric microbial surfactants.

Several developments in the optimization of culture conditions and downstream processing have been published recently. The use of agroindustrial by-products has been reported for both yeasts and bacteria (Makkar and Cameotra, 2002). Sobrinho et al. (2008) used groundnut oil refinery residues and corn steep liquor as substrates for anionic glycolipid production by *Candida sphaerica*, while the biosynthesis of glycolipids by *P. aeruginosa* was obtained using cashew apple juice as substrate

(Rocha et al., 2007) and vegetable oil refinery wastes (Raza et al., 2007). A very high potential for large-scale industrial application was achieved using the already commercialized Pharmamedia medium for surfactin production by *B. subtilis* MZ-7 (Al-Ajiani et al., 2007). The use of substrates such as soap stick, frying oil, and motor oil have all been explored and has had limited success due to the need for more costly or demanding downstream processing. Novel strains able to produce BSs on renewable and low-cost substrates have also been reported during the past few years. Ruggeri et al. (2009) isolated *Rhodococcus* sp. BS32 that are able to grow on rapeseed oil for the production of extracellular BSs. Glycerol, however, has emerged as an important potential feedstock available in large quantities as a by-product of the biodiesel process (Zheng et al., 2008).

With regard to the development of overproducer strains, genetic manipulation of selected strains remains limited. Although recombinant strains of *Bacillus* sp. and *Acinetobacter* sp. have been described, most genetic manipulation efforts have been directed toward *P. aeruginosa* due in part to its commercial potential and the more detailed knowledge of its genome. The ability to produce a hyperproducer strain of *P. aeruginosa*, however, is quite a difficult task due to the complexity of the transcriptional regulatory network of genes involved in rhamnolipid production. This is further complicated by the fact that rhamnolipids are produced as a mixture of congeners. Attempts have been made to limit the products to mono-rhamnolipids only through cloning the *P. aeruginosa* rhlABRI operon into host organisms such as *Escherichia coli* or nonpathogenic *P. putida* (Cabrera-Valladares et al., 2006; Cha et al., 2008). Wang et al. (2007) also reported the use of genetic engineering to obtain an *E. coli* and *P. aeruginosa* that were able to produce rhamnolipids after transposon-mediated chromosome integration of the rhamnosyltransferase 1 complex. Further yield increase could probably be obtained once the regulation mechanism of BS production is fully elucidated (Hsueh et al., 2007).

FACTORS AFFECTING BIOSURFACTANT PRODUCTION

BS production can be induced by hydrocarbons or other water-insoluble substrates. Another striking phenomenon is the catabolic repression of BS synthesis by glucose and other primary metabolites. The type, quality, and quantity of BS produced are influenced by the nature of the carbon substrate (Lang et al., 1984), the concentration of N, P, Mg, Fe, and Mn ions in the medium, and the culture conditions, such as pH, temperature, agitation, and dilution rate in continuous culture. BS production from *Pseudomonas* strains microbially enhanced oil recovery (MEOR) 171 and MEOR 172 are not affected by temperature, pH, and Ca, Mg concentrations in the ranges found in many oil reservoirs. Their production, on the other hand, in many cases improves with increased salinity. Thus, they are the BSs of choice for the Venezuelan oil industry and in the cosmetics, food, and pharmaceutical markets. The nitrogen source can be an important key to the regulation of BS synthesis. Urea also results in increased BS production. A change in the growth rate of the concerned microorganisms is often sufficient to result in the overproduction of BS. In some cases, addition of multivalent cations to the culture medium can have a positive effect on BS production.

Besides the regulation of BS by chemicals indicated earlier, some compounds like ethambutol, penicillin, chloramphenicol, and EDTA influence the formation of interfacially active compounds. The regulation of BS production by these compounds is either through their effect on solubilization of nonpolar hydrocarbon substrates or by increased production of water-soluble (polar) substrates. In some cases, BS synthesis is regulated by pH and temperature.

POTENTIAL APPLICATIONS

Microbial surface active agents (BSs) are important biotechnological products with a wide range of applications in many industries. Their properties of interest are as follows:

1. In changing surface active phenomena, such as lowering of surface and interfacial tensions
2. Wetting and penetrating actions
3. Spreading
4. Hydrophilicity and hydrophobicity actions
5. Microbial growth enhancement
6. Metal sequestration
7. Antimicrobial action

There are many advantages of BSs when compared to their chemically synthesized counterparts. Some of these are as follows:

- Biodegradability.
- Generally low toxicity.
- *Biocompatibility and digestibility*—which allows their application in cosmetics and pharmaceuticals and as functional food additives.
- *Availability of raw materials*—BSs can be produced from inexpensive raw materials that are available in large quantities; the carbon source may come from hydrocarbons, carbohydrates, and/or lipids, which may be used separately or in combination with each other.
- *Acceptable production economics*—depending upon the application, BSs can also be produced from industrial wastes and by-products, and this is of particular interest for bulk production (e.g., for use in petroleum-related technologies).
- *Use in environmental control*—BSs can be efficiently used in handling industrial emulsions, control of oil spills, biodegradation, and detoxification of industrial effluents and in bioremediation of contaminated soil.
- *Specificity*—BSs, being complex organic molecules with specific functional groups, are often specific in their action (this would be of particular interest in detoxification of specific pollutants): de-emulsification of industrial emulsions, specific cosmetic, pharmaceutical, and food applications.
- *Effectiveness*—at extreme temperatures, pH, and salinity (Kosaric, 2001).

The increasing environmental concern about chemical surfactants triggers attention to microbial-derived surface-active compounds essentially due to their low toxicity and biodegradable nature. At present, BSs are predominantly used in remediation of pollutants; however, they show potential applications in many sectors of food industry. Potential applications of microbial surfactants in food industry and other areas are discussed.

BIOSURFACTANT-ENHANCED BIOREMEDIATION

Oil Bioremediation

Petroleum bioremediation is carried out by microorganisms capable of utilizing hydrocarbons as a source of energy and carbon (Reisfeld et al., 1972). MEOR, exploiting microbial activities and metabolites, is at present gaining increased attention due to some advantages such as the following:

- Natural products are generally harmless and less detrimental to the environment.
- Microbial processes do not require high temperatures and thus there are low energy requirements.
- Costs of microbial products are not affected by the crude oil price and can be produced using inexpensive raw substrates or even waste materials.
- Microbial products/activities can be stimulated in situ within the reservoir, potentially allowing both tailor-made and cost-effective treatments.

Several metabolites are of interest for applications in MEOR including gas (e.g., carbon dioxide, methane, and hydrogen), acids (e.g., acetate and butyrate), solvents (e.g., acetone, n-butanol, and ethanol), biomass for selective plugging, and BSs/biopolymers. BSs in particular have several benefits enhancing oil displacement and movement through oil-bearing rocks by means of three main mechanisms:

1. Reduction of interfacial tension between oil rocks and oil brine
2. Modification of the wettability of porous media
3. Emulsification of crude oil

In addition, BS production contributes to the metabolism of viscous oils by microorganisms that release lighter hydrocarbon fractions, thus making the oil even more fluid. The strategies investigated so far for MEOR involving BSs include the following:

- Injection of ex situ produced BSs into the reservoirs
- Injection of laboratory-selected BS-producing microorganisms into the reservoirs
- Stimulation of indigenous microbial population to produce BSs in situ through supplying suitable nutrients

Only recently, a small fieldscale MEOR experiment has provided for the first-time data of in situ metabolism and activities. Molecular techniques combined with traditional methods showed that *Bacillus* strains injected into oil wells maintained activity, consuming the glucose and nutrients supplied and releasing CO_2 and fermentation products including a lipopeptide BS leading to an increased production estimated as one barrel of oil/day over 7 weeks after the treatment (Perfumo et al., 2010).

Bioremediation involves the acceleration of natural biodegradative processes in contaminated environments by improving the availability of materials (e.g., nutrients and oxygen), conditions (e.g., pH and moisture content), and prevailing microorganisms. Thus, bioremediation usually consists of the application of nitrogenous and phosphorous fertilizers, adjusting the pH and water content, if necessary, supplying air, and often adding bacteria. The addition of emulsifiers is advantageous when bacterial growth is slow (e.g., at cold temperatures or in the presence of high concentrations of pollutants) or when the pollutants consist of compounds that are difficult to degrade, such as polycyclic aromatic hydrocarbons. Bioemulsifiers can be applied as an additive to stimulate the bioremediation process; however, with advanced genetic technologies, it is expected that the increase in bioemulsifier concentration during bioremediation would be achieved by the addition of bacteria that overproduce bioemulsifiers.

ENHANCED BIOREMEDIATION OF ORGANIC CONTAMINATION

The enhanced bioremediation of organic contaminants by BSs is currently being studied. The ability of the rhamnolipid to remove styrene from contaminated soil was evaluated by Guo and Mulligan (2006). It was shown that it was feasible to use rhamnolipid as a washing agent to remove styrene. The results show that more than 70% of removal could be achieved for 32,750 mg/kg of styrene after 1 day and 88.7% removal after 5 days, while a 90% removal was obtained for 16,340 mg/kg of styrene after 1 day. After removal from the soil by rhamnolipid, more than 70% of the styrene could be biodegraded by an anaerobic biomass, in a combined soil flushing, and leachate treatment process. As marine oil spills can pose great threats and cause extensive damage to the marine and coastal ecosystems, the aim of the study by Vasefy and Mulligan (2008) was to evaluate the effectiveness of two commercial biological products, ASAP™ and Degreaser™, and a rhamnolipid BS (JBR 425™) on the biodegradation of weathered light crude oil, heavy crude oil, and diesel fuel spilled on saline water following the USEPA's biological effectiveness test method. The two evaluated products contain bacterial consortia and nutrients and were used as supplementary additives to enhance the biodegradation rate and the extent of hydrocarbon compounds. Chemical analysis was conducted by using gas chromatography/flame ionization detection to determine the fate of weathered oils in saline water over a 28-day experiment. Microbiological analysis was also performed to determine and monitor the viability of the microbial growth by addition of different rhamnolipid and commercial product concentrations to the oil samples. Degradation rates of oil types for most of the treatments were in the order of diesel fuel > light crude oil > heavy crude oil, and generally removal percentage and microbial densities had a direct relationship. Clifford et al. (2007) evaluated the removal of PCE by a rhamnolipid BS. The PCE–BS solution interfacial tension was 10 mN/m, which is quite high and not indicative of mobilization. This is beneficial as it minimizes vertical mobilization.

However, partitioning of PCE did occur and was shown to have a weight solubilization ratio of 1.2 g of PCE/g of rhamnolipid and is thus a good candidate for surfactant-enhanced recovery of PCE. Continuously stirred reactors were used to evaluate the degradation of PCE by a BS (UH) and sodium dodecyl sulfate (SDS) (Harenda and Vipulanandan, 2008). More PCE was solubilized per gram by the BS up to 500 mg/L. Ni/Fe particles were used. The PCE degradation in the presence of the particles occurred in 3 h and was first order. No other residual by-products than chlorine ions were determined. The active groups in the BSs were found to be C–O double bonds.

Metal Removal

Various studies have also been performed to evaluate the feasibility of metal removal by BSs due to their anionic characteristics. Juwarkar et al. (2007) investigated the removal of cadmium and lead by a BS produced by *P. aeruginosa* BS2. Column experiments were performed to determine the removal of Cd and Pb by rhamnolipid. More Cd was removed than Pb. More than 92% of Cd and 88% of Pb was removed by the rhamnolipid (0.1%) within 36 h. The rhamnolipid was also able to decrease toxicity and allow microbial activity (*Azotobacter* and *Rhizobium*) to take place and do not degrade soil quality. Asci et al. (2007) evaluated the potential for removal of Cd(II) from kaolinite. Various sorption models were evaluated for Cd(II). The Kolbe–Corrigan model fitted best. The desorption effects of pH, rhamnolipid concentration, and sorbed Cd(II) concentration were determined. The optimal conditions were pH 6.8, an initial concentration of 0.87 mM, and a rhamnolipid concentration of 80 mM. A removal of 71.9% of Cd(II) was achieved. Kim and Vipulanandan (2006) evaluated the removal of lead from water and contaminated soil (kaolinite). A linear isotherm represented lead desorption from kaolinite. The BS was produced from vegetable oil. Over 75% of the lead could be removed from 100 mg/L contaminated water at 10 times the CMC. The BS to lead ratio for optimal removal was 100:1. Fourier transform infrared spectroscopy indicated that the carboxyl group of the BS was implicated in the removal. Micelle partitioning could also be represented by Langmuir and Freundlich models. The BS micelle partitioning was more favorable than synthetic surfactants (SDS and Triton X-100). A new approach for metal stabilization by BSs was recently discovered. Hexavalent chromium Cr(VI) is an environmental pollutant that is treated by its reduction to the trivalent form Cr(III). The latter can be reoxidized to the toxic form, Cr(VI), under specific conditions. A study was conducted on the removal of Cr(III) to eliminate the hazard imposed by its presence in soil as there has been some evidence that organic compounds can decrease its sorption. The effect of addition of negatively charged BSs (rhamnolipids) on chromium-contaminated kaolinite was studied. Results showed that the rhamnolipids have the capability of extracting a 25% portion of the stable form of chromium, Cr(III), from the kaolinite, under optimal conditions. The removal of hexavalent chromium was also enhanced compared to water by a factor of 2 using a solution of rhamnolipids. Results from the sequential extraction procedure showed that rhamnolipids remove Cr(III) mainly from the carbonate and oxide/hydroxide portions of the kaolinite. The rhamnolipids had also the capability of reducing close to 100% of the extracted Cr(VI) to Cr(III) over a period of 24 days. This study indicated that rhamnolipids could be beneficial for the removal or long-term conversion of Cr(VI) to Cr(III). Much of this information is presented in the later chapters.

CRUDE OIL TRANSPORTATION IN PIPELINES

Crude oil often needs to be transported over long distances from the extraction fields to the refineries. One of the major factors affecting pipelining is oil viscosity that slows the flow. Heavy oils in particular are characterized by viscosities ranging from 1000 cP to more than 100,000 cP at 25°C and cannot be transported through conventional pipelining systems that optimally require viscosities of <200 cP. Heating or diluting with solvents was the traditional method applied to reduce oil viscosity. However, a promising technology consisting of producing a stable oil in water emulsion that facilitates oil motility has been recently developed and introduced new routes to the application of the bioemulsifier type of BSs that have been found particularly suitable for this application. Due to the high number of reactive groups in the molecule, bioemulsifiers bind tightly to oil droplets and form an effective barrier that prevents drop coalescence. Among the bioemulsifiers, emulsan and its analogs synthesized by *A. venetianus* RAG1 are certainly the most powerful, yet others such as alasan and biodipersan produced by different *Acinetobacter* strains have been extensively studied (Gutnick et al., 1987). It was estimated that under optimal conditions, the emulsion could have been transported for 26,000 miles (Hayes et al., 1990). Once transported to the refinery, hydrocarbosols can be either de-emulsified or utilized directly without dewatering or treated with specific enzymes called emulsanes to depolymerize the bioemulsifier, thus breaking the emulsion before use (Hayes et al., 1987).

In the case of waxy crude oils, their transportation is generally affected by the problem of paraffin precipitation that can cause numerous negative consequences from reduction and eventually block of the internal diameter of pipes to changes in the oil composition. Traditional techniques for treating wax included thermal, mechanical, and chemical methods, but they all failed to be fully successful as energy consuming, detrimental to the pipes, and highly toxic, respectively. Thus, over the past decade, microbial treatments became an increasing valuable alternative (Etoumi, 2007). Many bacteria are known to be able to grow on paraffinic hydrocarbons while producing BSs that act as dispersing and solubilizing agents and make the paraffinic fractions more available for the uptake by cells. In this way, not only wax deposits can be dissolved and prevented but also heavy crude oil fractions can be degraded by bacteria to lighter fractions.

CLEANUP OF OIL CONTAINERS/STORAGE TANKS

Large amounts of crude oil are daily moved and distributed to refineries with oil tankers, barges, tank cars, and trucks, thus increasing the problem of the cleanup and maintenance of the containers. In 1991, Banat et al. (1991) described the application of microbial BSs for the cleanup of oil storage tanks. Sludge and oil deposits normally accumulate at the bottom and on the walls of storage tanks, thus requiring periodical cleaning operations. BSs can effectively drive the cleaning activity as demonstrated in a field trial conducted at the Kuwait Oil Company. Two tons of rhamnolipid BS containing culture broth were produced, sterilized, and added to an oil sludge tank along with fresh crude oil and water and circulated continuously for 5 days at an ambient temperature of 40°C–50°C. The oil sludge was effectively lifted and mobilized from the bottom of the tank and solubilized within the emulsion formed. The treatment recovered 91% of hydrocarbons in the sludge. The value of the recovered crude covered the cost of the cleaning operation.

POTENTIAL FOOD APPLICATIONS

Although BSs can be explored for several food processing applications, in this section, we emphasize their potential as food formulation ingredients and antiadhesive agents.

Food formulation ingredients: Apart from their obvious role as agents that decrease surface and interfacial tension, thus promoting the formation and stabilization of emulsions, surfactants can have several other functions in food, for example, to control the agglomeration of fat globules, stabilize aerated systems, improve texture and shelf life of starch-containing products, modify rheological properties of wheat dough, and improve consistency and texture of fat-based products (Kachholz and Schlingmann, 1987). In bakery and ice cream formulations, BSs act controlling consistency, retarding staling, and solubilizing flavor oils; they are also utilized as fat stabilizer and antispattering agent during cooking of oil and fats (Kosaric, 2001). An improvement of dough stability, texture, volume, and conservation of bakery products was obtained by the addition of rhamnolipid surfactants (Van Haesendonck and Vanzeveren, 2004). The authors also suggested the use of rhamnolipids to improve properties of butter cream, croissants, and frozen confectionery products. Recently, a bioemulsifier isolated from a marine strain of *Enterobactercloacae* was described as a potential viscosity enhancement agent of interest in food industry especially due to the good viscosity observed at acidic pH allowing its use in products containing citric or ascorbic acid (Iyer et al., 2006).

Antiadhesive agents: Bacterial biofilms present in food industry surfaces are potential sources of contamination, which may lead to food spoilage and disease transmission (Hood and Zottola, 1995). Due to the fact that food processors have a zero tolerance levels for pathogens like *Salmonella* and also (in most countries) for *Listeria monocytogenes*, a single adherent cell may be as significant as a well-developed biofilm; thus, controlling the adherence of microorganisms to food contact surfaces is an essential step in providing safe and quality products to consumers (Hood and Zottola, 1995).

The involvement of BSs in microbial adhesion and detachment from surfaces has been investigated. A surfactant released by *Streptococcus thermophilus* has been used for fouling control of heat-exchanger plates in pasteurizers as it retards the colonization of other thermophilic strains of *Streptococcus* responsible for fouling

(Busscher et al., 1996). The bioconditioning of surfaces through the use of microbial surfactants has been suggested as a new strategy to reduce adhesion. Pretreatment of silicone rubber with *S. thermophilus* surfactant inhibited the adhesion of *Candida albicans* by 85% (Busscher et al., 1997), whereas surfactants from *Lactobacillus fermentum* and *Lactobacillus acidophilus* adsorbed on glass reduced the number of adhering uropathogenic cells of *Enterococcus faecalis* by 77% (Velraeds et al., 1996). Lately, the BS from *L. fermentum* was reported to inhibit *Staphylococcus aureus* infection and adherence to surgical implants (Gan et al., 2002). The use of BSs released by *Lactobacilli* strains is very promising since these microorganisms are naturally present in human flora and have also a probiotic effect (Singh and Cameotra, 2004).

The use of BSs, which disrupts biofilms and reduces adhesion, in combination with antibiotics could represent a novel antimicrobial strategy. Since antibiotics are in general less effective against biofilms than planktonic cells, the disruption of biofilm by BS can facilitate the antibiotic access to the cells (Irie et al., 2005). The promising results of this work, with a medical focus, suggest a potential application of BSs for surface conditioning in food industry, since both the surface materials and microorganisms involved are of common interest.

An interesting work regarding the use of BSs to inhibit the adhesion of the pathogen *L. monocytogenes* in two types of surfaces classically used in food industry has been conducted by the group of Meylheuc et al. (2001). The preconditioning of stainless steel and PTFE surfaces with a BS obtained from *Pseudomonas fluorescens* inhibits the adhesion of *L. monocytogenes* L028 strain. A significant reduction (>90%) was attained in microbial adhesion levels in stainless steel whereas no significant effect was observed in PTFE. Further work demonstrated that the prior adsorption of *P. fluorescens* surfactant in stainless steel also favored the bactericidal effect of disinfectants (Meylheuc et al., 2006). The ability of adsorbed BSs obtained from gram-negative (*P. fluorescens*) and gram-positive (*Lactobacillus helveticus*) bacteria isolated from foodstuffs in inhibiting the adhesion of *L. monocytogenes* to stainless steel was recently investigated. Adhesion tests showed that both BSs were effective by strongly decreasing the level of contamination of the surface. The antiadhesive biological coating reduced either the total adhering flora or the viable/cultivable adherent *L. monocytogenes* on stainless steel surfaces (Meylheuc et al., 2006). Preliminary studies regarding the corrosion effect of *P. fluorescens* surfactant in stainless steel suggested that it has also a good potential as corrosion inhibitor (Dagbert et al., 2006).

Food processors, however, do not yet use BSs on a large scale as many regulations regarding the approval of new food ingredients are required by governmental agencies, and this process could be quite long. Nevertheless, an increasing number of patents have been issued on BSs (bioemulsifiers) claiming their use as additives for food, cosmetics, and pharmaceutical products (Shete et al., 2006), demonstrating the crescent interest in using these microbial-derived products.

BIOMEDICAL APPLICATIONS

A valuable function of BSs for medical application is their ability to disrupt membranes leading to cell lysis through increased membrane permeability leading to metabolite leakage. This occurs due to changes in physical membrane structure or through disrupting protein conformations that alter important membrane functions such as transport and energy generation (Van Hamme et al., 2006).

Antimicrobial activity of biosurfactants: The high demand for new antimicrobial agents following increased resistance shown by pathogenic microorganisms against existing antimicrobial drugs has drawn attention to BSs as antibacterial agents (Běhal, 2006). Some BSs have been reported to be suitable alternatives to synthetic medicines and antimicrobial agents and may therefore be used as effective and safe therapeutic agents (Cameotra and Makkar, 2004; Singh and Cameotra, 2004).

Lipopeptides form the most widely reported class of BSs with antimicrobial activity. Surfactin, produced by *B. subtilis*, is the best-known lipopeptide. Other antimicrobial lipopeptides include fengycin, iturin, bacillomycins, and mycosubtilins produced by *B. subtilis* (Vater et al., 2002). Lichenysin, pumilacidin, and polymyxin B (Naruse et al., 1990) are other antimicrobial lipopeptides produced by *B. licheniformis, B. pumilus,* and *B. polymyxa,* respectively. The production of antimicrobial lipopeptides by *Bacillus* probiotic products is one of the main mechanisms by which they inhibit the growth of pathogenic microorganisms in the gastrointestinal tract (Hong et al., 2005). Other reported BSs having antimicrobial activity are daptomycin, a cyclic lipopeptide from *Streptomyces roseosporus* (Baltz et al., 2005), viscosin, a cyclic lipopeptide from *Pseudomonas* (Neu et al., 1990; Saini et al., 2008), rhamnolipids produced by *P. aeruginosa* (Abalos et al., 2001; Benincasa et al., 2004), and sophorolipids produced by *C. bombicola* (Kim et al., 2002; Van Bogaert et al., 2007). MEL-A and MEL-B produced by *C. antarctica* strains have also been reported to exhibit antimicrobial action against gram-positive bacteria (Kitamoto et al., 1993).

Recently, a lipopeptide BS produced by a marine organism, *B. circulans,* was found to be active against *Proteus vulgaris, Alcaligenes faecalis,* methicillin-resistant *Staphylococcus aureus* (MRSA), and other multidrug-resistant pathogenic strains (Das et al., 2008a), while not having any hemolytic activity. A rhamnolipid surfactant produced from soybean oil waste had antimicrobial activity against several bacteria and fungi, namely, *B. cereus, S. aureus, Micrococcus luteus, Mucor miehei,* and *Neurospora crassa* (Nitschke et al., 2009b). Flocculosin, a cellobiose lipid produced by the yeast-like fungus *Pseudozyma flocculosa,* was tested against clinical bacterial isolates and the pathogenic yeast *C. albicans* (Mimee et al., 2009). The glycolipid was particularly effective against *Staphylococcus* species, including MRSA, and its antibacterial activity was not influenced by the presence of common resistance mechanisms (e.g., against methicillin and vancomycin) in tested strains. In addition, flocculosin was able to kill *C. albicans* cells in a very short period of time. Huang et al. (2007) observed that a lipopeptide antimicrobial substance produced by the strain *B. subtilis* fmbj, which is mainly composed of surfactin and fengycin, was able to inactivate endospores of *B. cereus.* Observation by TEM indicated that the lipopeptide could damage the surface structure of the spores. The antifungal activities of BSs have long been known, although their action against human pathogenic fungi has been rarely described (Tanaka et al., 1997; Chung et al., 2000; Abalos et al., 2001). Recently, a glycolipid isolated from the yeast-like fungus *P. flocculosa,* named flocculosin, was shown to display in vitro antifungal activity against several pathogenic yeasts, associated with human mycoses (Mimee et al., 2005). This product positively inhibited all pathogenic strains tested under acidic conditions and showed synergistic activity with amphotericin B, increasing its efficacy while decreasing any toxicity and other side effects.

The antifungal activity against phytopathogenic fungi has been demonstrated for glycolipids, such as cellobiose lipids (Teichmann et al., 2007; Kulakovskaya et al., 2009, 2010) and rhamnolipids (Debode et al., 2007; Varnier et al., 2009), and cyclic lipopeptides (Tran et al., 2007), including surfactin, iturin, and fengycin (Arguelles–Arias et al., 2009; Chen et al., 2009; Mohammadipour et al., 2009; Snook et al., 2009; Velmurugan et al., 2009).

Other biomedical and therapeutic applications: BSs have been shown to have many other roles in biomedical application. Surfactin is one of the most powerful BSs and is known to have anti-inflammatory, antibiotic, and antitumor functions (Seydlová and Svobodová, 2008). Cao et al. (2010) demonstrated that surfactin induces apoptosis in human breast cancer MCF-7 cells through a ROS/JNK-mediated mitochondrial/caspase pathway, whereas Byeon et al. (2008) observed that surfactin was able to downregulate LPS-induced NO production in RAW264.7 cells and primary macrophages by inhibiting NF-κB activation. Park and Kim (2009) studied the role of surfactin in the inhibition of the immunostimulatory function of macrophages through blocking the NK-κB, MAPK, and Akt pathways. This provided a new insight into the immunopharmacological role of surfactin in autoimmune disease and transplantation. Their work indicated that surfactin has potent immunosuppressive capabilities, which suggested important therapeutic implications for transplantation and autoimmune diseases, including allergy, arthritis, and diabetes.

Selvam et al. (2009) studied the effect of *B. subtilis* PB6, a natural probiotic, on plasma cytokine levels in inflammatory bowel disease and colon mucosal inflammation. The strain was found to secrete surfactin, which is known to inhibit phospholipase A2, involved in the pathophysiology of inflammatory bowel disease. In animal experiments carried out in rat models for trinitrobenzene sulfonic acid–induced colitis, oral administration of PB6 as a probiotic suppressed colitis as measured by mortality rate and changes in colon morphology and weight gain. Plasma levels of pro-inflammatory cytokines were also significantly lowered and the anti-inflammatory cytokine significantly increased after the oral administration of PB6, supporting the concept that PB6 inhibits PLA2 by secreting surfactin. Han et al. (2008) observed that high-surfactin micelle concentration affected the aggregation of amyloid β-peptide (Aβ [1-40]) into fibrils, a key pathological process associated with Alzheimer's disease. Another interesting property of surfactin and its synthetic analogs is the ability to alter the nanoscale organization of supported bilayers and to induce nanoripple structures with intriguing perspectives for biomedical and biotechnological applications (Bouffioux et al., 2007; Brasseur et al., 2007; Francius et al., 2008). Fengycin, another lipopeptide BS, is also able to cause membrane perturbations (Deleu et al., 2008). Recent results by Eeman et al. (2009) emphasized the ability of fengycin to interact with the lipid constituents of the stratum corneum extracellular matrix and with cholesterol.

In another work, a liposome vector containing betasitosterol beta-D-glucoside BS-complexed DNA was successfully used for herpes simplex virus thymidine kinase gene therapy (Maitani et al., 2006). More recently, nanovectors containing a BS have been successfully used to increase the efficacy for gene transfection in vitro and in vivo (Nakanishi et al., 2009). On the other hand, Morita et al. (2009), using a three-dimensional cultured human skin model, observed that the viability of the SDS damaged cells was markedly improved by the addition of MEL-A in a dose-dependent manner. This demonstrated that MEL-A had a ceramide-like moisturizing activity toward the skin cells. Another interesting application for natural surfactant is the possibility to synthesize metal-bound nanoparticles using an environmentally friendly and benign technology (Palanisamy and Raichur, 2009). The use of gold nanoparticles, in particular, is currently undergoing a dramatic expansion in the field of drug and gene delivery, targeted therapy, and imaging (Boisselier and Astruc, 2009). Recently, Reddy et al. (2009) synthesized, for the first time, surfactin-mediated gold nanoparticles, opening the way to a new and fascinating application of BSs in the biomedical field.

FLAVOLIPIDS: NOVEL BIOSURFACTANTS

Bodour et al. (2004) reported that the genus *Flavobacterium* produces a BS. *Flavobacterium* is an aerobic, nonfermenting, gram-negative, rod-shaped microorganism that exhibits gliding motility. It belongs to the Flavobacteriaceae family in the *Cytophaga–Flavobacterium–Bacteroides* phylum. The small-subunit rRNA of *Flavobacterium* suggests that it is closely related to the sulfur bacteria (Woese et al., 1990; Gherna and Woese, 1992). This organism is ubiquitous in the environment. *Flavobacterium* is known to produce pigments ranging in color from yellow to orange, pink, red, and brown. Some species of *Flavobacterium* degrade organic contaminants, such as pentachlorophenol, nylon oligomers, polyaromatics, and pesticides.

Flavolipid as an emulsifier: The flavolipid is a strong and stable emulsifier. Flavolipid concentrations as low as 19 mg/L exhibited an emulsification index of 100%, indicating complete emulsification of the oil layer. Emulsions were stable even after 1 week.

Remediation applications: BSs have been intensively studied for application in the remediation of organic chemical- and metal-contaminated sites. Therefore, the flavolipid was subjected to a series of tests to begin evaluation of its ability to enhance solubilization and biodegradation of hydrocarbons and to determine whether it has the ability to complex metals. Success has already been achieved in the following cases:

1. Solubilization and biodegradation of hexadecane
2. Effect of BS on complex formation with heavy metals

The flavolipids described herein represent a new class of BSs with strong surface activity and emulsifying ability. The polar moiety of flavolipid features citric acid and two cadaverine molecules, which is quite different from the polar moieties found in any of the currently reported classes of BSs, which are glycolipids, lipoproteins, phospholipids, fatty acid salts, and polymeric BSs. This new class of BSs will be of interest for potential use in a wide variety of industrial and biotechnology applications (Bodour et al., 2003).

NANOTECHNOLOGY

New developments have been made in the area of nanotechnology and BSs. Nanorods of NiO were produced by a water-in-oil microemulsion (Palanisamy, 2008). A BS was added to heptane. A microemulsion was then formed by adding a nickel chloride solution mixed into the BS/heptane, and for the other microemulsion, ammonium hydroxide was added to the same hydrocarbon mixture. The microemulsions were then mixed and centrifuged, and the precipitates were then washed with ethanol to remove the BS and heptane. The nanorods were 22 nm in diameter and 150–250 µ min length (pH 9.6). The particle shape depended on the pH. This could be due to the effect of pH on the BS morphology. The use of the BS is a more ecofriendly approach.

Although not strictly an environmental application, rhamnolipid was evaluated for its effect on the electrokinetic and rheological behaviors on nanozirconia particles (Biswas and Raichur, 2008). The BSs adsorbs onto the zirconia increasingly with the concentration. The BS was able to disperse the zirconia particles at pH 7 and above. Evidence of this was through zeta potential measurements, sedimentation, and viscosity tests. It can serve as an ecofriendly product for flocculation and dispersion of high solid contents of microparticles.

FUTURE CHALLENGES

A totally unexplored area for potential applications of BSs is the formulation of petrochemical products. Biotechnological alternatives to the existing bulk petroleum–derived products have generally failed for various reasons and mostly for not satisfying economic criteria. Diesel fuel blended with water has been known since the early 1900s and is currently applied especially in Europe for public transport fleets, marine engines, and locomotives but also heat facilities in industrial and institutional complexes. The advantages of diesel emulsions are

- Reduction of emission of hazardous pollutants such as nitrogen oxides ($\leq25\%$), carbon oxide ($\leq5\%$), black smoke ($\leq80\%$), and particulate matter ($\leq60\%$)
- Reduction of diesel consumption

An additional aspect is that such fuels are easily applicable without the need of engine modification.

Emulsified fuels are technically water in diesel emulsions with a typical content of water of 10%–20% (v/v). They are prepared using specific surfactant packages along with a variety of additives (e.g., detergents, lubricity enhancers, antifoaming agents, ignition improvers, antirust agents, and metal deactivators). Surfactants are expected to stabilize the emulsion and ensure that the finely dispersed water droplets remain in suspension within the diesel fuel. Nonionic surfactants such as alcohol ethoxylates, fatty acid ethoxylates, and sugar esters of fatty acids are currently the most used (Lif and Holmberg, 2006; Clark et al., 2007).

Despite their environmentally favorable characteristics of higher biodegradability, lower toxicity, and better foaming properties compared to their synthetic chemical counterparts while also showing better stability at extreme pH, salinity, and temperature, the commercialization of microbial surfactants has not been fully achieved largely due to production costs. At present, the production costs for most BSs do not compete with those of chemical surfactants. Different strategies have been proposed to make the process more cost-effective including

1. Development of more efficient bioprocesses, including optimization of fermentative conditions and downstream recovery processes
2. Use of cheap and waste substrates (Thavasi et al., 2007; Raza et al., 2009)
3. Development of overproducing strains

The prospect of new types of surface-active compounds from microorganisms can contribute for the detection of different molecules in terms of structure and properties, but the toxicological aspects of new and current BSs should be emphasized in order to certify the safety of these compounds for various applications.

GREEN CHEMISTRY AND SURFACTANT RESEARCH

The development of green chemistry as a new science and BS research go hand in hand. In fact the roots of green chemistry go back to the 1950s, when Henkel, a chemical company, started monitoring surfactant concentrations in the Rhine River and developed the closed bottle test in order to study the biodegradability of surfactants and since then has strategically steered research and development according to environmental principles, resulting in, for example, the development of zeolite A as an alternative to phosphates in detergents (in order to avoid eutrophication of sweet water lakes in Germany due to overfertilization).

Development of *surfactants for carbon dioxide*, enabling CO_2 to be used as a solvent (e.g., in dry cleaning), won the Presidential Green chemistry Challenge Awards. The discovery of a new surfactant with high surface activity in supercritical carbon dioxide opened a way to new processes in textile and metal industries and for dry cleaning of clothes.

The production and use of BSs also imply to the principles of Green Chemistry. Some examples are cited in the following:

- Olive oil mill effluent, a major pollutant of the agricultural industry in Mediterranean countries, has been used as the raw material for rhamnolipid BS production by *Pseudomonas* sp. JAMM. The use of agroindustrial by-products has been reported both for yeasts and bacteria (Makkar and Cameotra, 2002). Sobrinho et al. (2008) used groundnut oil refinery residues and corn steep liquor as substrates for anionic glycolipid production by *Candida sphaerica*, while the biosynthesis of glycolipids by *P. aeruginosa* was obtained using cashew apple juice as substrate (Rocha et al., 2007) and vegetable oil refinery wastes (Raza et al., 2007), thus complying with the first principle of green chemistry, that is, to Prevent Waste.

• A BS from *P. aeruginosa* was compared with a synthetic surfactant (Marlon A-350) widely used in industry in terms of toxicity and mutagenic properties. Both assays indicated the higher toxicity and mutagenic effect of the chemical-derived surfactant, whereas BS was considered slightly to nontoxic and nonmutagenic (Flasz et al., 1998). The comparison of acute and chronic toxicity of three synthetic surfactants (Corexit, 9500, Triton X-100, and PSE-61) and three microbial-derived surfactants (rhamnolipid, emulsan, and biological cleanser PES-51) commonly used in oil spill remediation revealed that PES-61 (synthetic surfactant) and emulsan (BS) were the least toxic whereas Triton X-100 (synthetic) was the most toxic (Edwards et al., 2003), complying with principle 4 of green chemistry, that is, Design for Safer Chemicals.

• Unlike synthetic surfactants, microbial-produced compounds are easily degraded (Mohan et al., 2006) and particularly suited for environmental applications such as bioremediation (Deleu and Paquot, 2004; Mulligan, 2005). The increasing environmental concern among consumers and the regulatory rules imposed by governments forced industry to search for alternative products such as BSs; thus, they are biodegradable fulfilling Principle 11 of the 12 principles of green chemistry given in the earlier pages of this chapter.

CONCLUSIONS

Green chemistry is not a new branch of science. It is a new philosophical approach that through application and extension of the principles of green chemistry can contribute to sustainable development. Great efforts are still undertaken to design an ideal process that starts from nonpolluting initial materials, leads to no secondary products, and requires no solvents to carry out the chemical conversion or to isolate and purify the product. BS research is a promising approach toward this goal. The BSs seem to enhance the solubilization and emulsification of the contaminants. Due to their biodegradability and low toxicity, BSs such as rhamnolipids are very promising for use in remediation technologies. In addition, there is the potential for in situ production, a distinct advantage over synthetic surfactants. This needs to be studied further. Further research regarding prediction of their behavior in the fate and transport of contaminants will be required. More investigation into the solubilization mechanism of both hydrocarbons and heavy metals by BSs is required to enable model predictions for transport and remediation. New applications for the BSs regarding nanoparticles are developing. Future research should focus on the stabilization of the nanoparticles by BSs before addition during remediation procedures.

The commercial success of microbial surfactants is currently limited by the high cost of production. Optimized growth/production conditions using cheaper renewable substrates and novel and efficient multistep downstream processing methods could make BS production more profitable and economically feasible. Furthermore, recombinant and mutant hyperproducer microbial strains, able to grow on a wide range of cheap substrates, could produce BSs in high yield and, potentially, bring the required breakthrough for their economic production.

Finally, the old adage "An ounce of prevention is worth a pound of cure" applies here. Green chemistry is pollution prevention at the molecular level. The 12 principles of green chemistry provide a structured framework for scientists and engineers to engage in when designing new materials, products, processes, and systems that are benign to human health and the environment (Anastas and Zimmerman, 2003). Engineers use these principles as guidelines to help ensure that designs for products, processes, or systems have the fundamental components, conditions, and circumstances necessary to be more sustainable. Furthermore, the breadth of the principles' applicability is important (Anastas and Zimmerman, 2003). Otherwise, these would not be principles, but a mere compilation of list of useful techniques that have been successfully demonstrated under specific conditions. Indeed, sustainability will be one of the main drivers for innovation in order to allow the technical industries to care for the well-being of consumers in a safe and healthy environment. The hope or long-term vision is that a strong, just, and wealthy society can be consistent with a clean environment, healthy ecosystems, and a beautiful planet. For this education, fundamental research and knowledge transfer are the necessary tools to achieve.

REFERENCES

Abalos, A., Pinazo, A., Infante, M.R., Casals, M., García, F., and Manresa, A. 2001. Physicochemical and antimicrobial properties of new rhamnolipids produced by *Pseudomonas aeruginosa* AT10 from soybean oil refinery wastes. *Langmuir*, 17:1367–1371.

Al-Ajlani, M.M., Sheikh, M.A., Ahmad, Z., and Hasnain, S. 2007. Production of surfactin from *Bacillus subtilis* MZ-7 grown on pharmamedia commercial medium. *Microbial Cell Factory*, 6:17–20.

Anastas, P.T. and Kirchoff, M.M. 2002. Origins, current status, and future challenges of green chemistry. *Accounts of Chemical Research*, 35:686–694.

Anastas, P.T and Warner, J.C. 1998. *Green Chemistry: Theory and Practice*. Oxford, U.K.: Oxford University Press, p. 2.

Anastas, P.T. and Zimmerman, J.B. 2003. Design through the 12 principles of green engineering. *Environmental Science and Technology*, 37:94–101.

Arguelles-Arias, A., Ongena, M., Halimi, B., Lara, Y., Brans, A., Joris, B., and Fickers, P. 2009. *Bacillus amyloliquefaciens* GA1 as a source of potent antibiotics and other secondary metabolites for biocontrol of plant pathogens. *Microbial Cell Factory*, 8:63.

Asci, Y., Nurbas, M., and Acikel, Y.S. 2007. Sorption of Cd(II) onto kaolin as a soil component and desorption of Cd(II) from kaolin using rhamnolipid biosurfactant. *Journal of Hazardous Materials*, B139:50–56.

Bajpai, P. 1999. Application of enzymes in the pulp and paper industry. *Biotechnology Progress*, 15:147–157.

Banat, I.M., Franzetti, A., Gandolfi, I., Bestetti, G., Martinotti, M.G., Fracchia, L., Smyth, T.J., and Marchant, R. 2010. Microbial biosurfactants production, applications and future potential. *Applied Microbiology and Biotechnology*, 87:427–445.

Banat, I.M. et al. 1991. Biosurfactant production and use in oil tank clean-up. *World Journal of Microbiology and Biotechnology*, 7:80–84.

Bayer Corporation, Bayer, A.G. Preparation and use of iminodisuccinic acid salts. US Patent 6107518.

Běhal, V. 2006. Mode of action of microbial bioactive metabolites. *Folia Microbiologia*, 51:359–369.

Benincasa, M., Abalos, A., Oliveira, I., and Manresa, A. 2004. Chemical structure, surface properties and biological activities of the biosurfactant produced by *Pseudomonas aeruginosa* LBI from soapstock. *Anton Leeuw International Journal G*, 85:1–8.

Bergbreiter, D.E. 1998. The use of soluble polymers to effect homogeneous catalyst separation and reuse. *Catalysis Today*, 42:389–397.

Biswas, M. and Raichur, A.M. 2008. Electrokinetic and rheological properties of nano zirconia in the presence of rhamnolipid biosurfactant. *Journal of American Ceramics Society*, 91:3197–3201.

Bodour, A.A., Drees, K.P., and Maier, R.M. 2003. Distribution of biosurfactant–producing microorganisms in undisturbed and contaminated arid southwestern soils. *Applied and Environmental Microbiology*, 69:3280–3287.

Bodour, A.A. et al. 2004. Structure and characterization of flavolipids, a novel class of biosurfactants produced by *Flavobacterium* sp. strain MTN11. *Applied and Environmental Microbiology*, 70:114–120.

Boisselier, E. and Astruc, D. 2009. Gold nanoparticles in nanomedicine: Preparations, imaging, diagnostics, therapies and toxicity. *Chemical Society Reviews*, 38:1759–1782.

Bouffioux, O., Berquand, A., Eeman, M., Paquot, M., Dufrêne, Y.F., Brasseur, R., and Deleu, M. 2007. Molecular organization of surfactin–phospholipid monolayers: Effect of phospholipid chain length and polar head. *Biochimica and Biophysica Acta Biomembrane*, 768:1758–1768.

Brasseur, R., Braun, N., El Kirat, K., Deleu, M., Mingeot–Leclercq, M.P., and Dufrêne, Y.F. 2007. The biologically important surfactin lipopeptide induces nanoripples in supported lipid bilayers. *Langmuir*, 23:9769–9772.

Busscher, H.J., van der Kuij-Booij, M., and van der Mei, H.C. 1996. Biosurfactants from thermophilic dairy streptococci and their potential role in the fouling control of heat exchanger plates. *Journal of Industrial Microbiology and Biotechnology*, 16(1):15–21.

Busscher, H.J., van Hoogmoed, C.G., Geertsema-Doornbusch, G.I., van der Kuij-Booij, M., and van der Mei, H.C. 1997. *Streptococcus thermophilus* and its biosurfactants inhibit adhesion by *Candida* spp. on silicone rubber. *Applied and Environmental Microbiology*, 63:3810–3817.

Byeon, S.E., Lee, Y.G., Kim, B.H., Shen, T., Lee, S.Y., Park, H.J., Park, S.C., Rhee, M.H., and Cho, J.Y. 2008. Surfactin blocks NO production in lipopolysaccharide-activated macrophages by inhibiting NF-κB activation. *Journal of Microbiology and Biotechnology*, 18:1984–1989.

Cabrera-Valladares, N., Richardson, A.P., Olvera, C., Trevino, L.G., Deziel, E., Lepine, F., and Soberon-Chavez, G. 2006. Monorhamnolipids and 3-(3-hydroxyalkanoyloxy) alkanoic acids (HAAs) production using *Escherichia coli* as a heterologous host. *Applied Microbiology and Biotechnology*, 73:187–194.

Cameotra, S.S. and Makkar, R.S. 2004. Recent applications of biosurfactants as biological and immunological molecules. *Current Opinion in Microbiology*, 7:262–266.

Cao, X.H., Wang, A.H., Wang, C.L., Mao, D.Z., Lu, M.F., Cui, Y.Q., and Jiao, R.Z. 2010. Surfactin induces apoptosis in human breast cancer MCF-7 cells through a ROS/JNK–mediated mitochondrial/caspase pathway. *Chemical Biology Interactions*, 183:357–362.

Cargill, D. 2001. Polymers, LLC process to produce biodegradable polylactic acid polymers. In *The Presidential Green Chemistry Challenge Awards Program: Summary of 2000 Award Entries and Recipients.* Washington, DC: EPA744-R-00-001, U.S. Environmental Protection Agency, Office of Pollution Prevention and Toxics, p. 51.

Cha, M., Lee, N., Kim, M., and Lee, S. 2008. Heterologous production of *Pseudomonas aeruginosa* EMS1 biosurfactant in *Pseudomonas putida. Bioresource Technology*, 99:2192–2199.

Chen, X.H., Koumoutsi, A., Scholz, R., Schneider, K., Vater, J., Süssmuth, R., Piel, J., and Borriss, R. 2009. Genome analysis of *Bacillus amyloliquefaciens* FZB42 reveals its potential for biocontrol of plant pathogens. *Journal of Biotechnology*, 140:27–37.

Chung, Y.R., Kim, C.H., Hwang, I., and Chun, J. 2000. *Paenibacillus koreensis* sp. nov. A new species that produces an iturin-like antifungal compound. *International Journal of System Evolution and Microbiology*, 50:1495–1500.

Clark, J.H. 1999. Green chemistry: Challenges and opportunities. *Journal of Green Chemistry*, 1:1–8.

Clark, R.H., Morley, C., and Stevenson, P.A. 2007. Diesel fuel compositions. US Patent 7229481.

Claus, H. et al. 2002. Redox-mediated decolorization of synthetic dyes by fungal laccase. *Applied Microbiology and Biotechnology*, 59:672–678.

Clifford, J.S., Ionnidis, M.A., and Legge, R.L. 2007. Enhanced aqueous solubilization of tetrachloroethylene by a rhamnolipid biosurfactant. *Journal of Colloid and Interface Science*, 305:361–364.

Collins, T.J. et al. 2002. In *Advancing Sustainability through Green Chemistry and Engineering.* Lankey, R.L. and Anastas, P.T. (eds.) Washington, DC: American Chemical Society, Chapter 4, Tetraamido Macrocyclic Ligand Catalytic Oxidant Activators in the Pulp and Paper Industry.

Coupland, C. 2003. *Corporate Identities on the Web: An Exercise in the Construction and Deployment of "Morality".* Nottingham, England: International Centre for Corporate Social Responsibility Research Paper Series No. 02-2003.

Dagbert, C., Meylheuc, T., and Bellon-Fontaine, M.N. 2006. Corrosion behavior of AISI 304 stainless steel in presence of a biosurfactant produced by *Pseudomonas fluorescens. Electrochimica Acta*, 51:5221–5227.

Das, P., Mukherjee, S., and Sen, R. 2008a. Antimicrobial potential of a lipopeptide biosurfactant derived from a marine *Bacillus circulans. Journal of Applied Microbiology*, 104:1675–1684.

Debode, J., De Maeyer, K., Perneel, M., Pannecoucque, J., De Backer, G., and Höfte, M. 2007. Biosurfactants are involved in the biological control of *Verticillium microsclerotia* by *Pseudomonas* spp. *Journal of Applied Microbiology*, 103:1184–1196.

Deleu, M. and Paquot, M. 2004. From renewable vegetables resources to microorganisms: New trends in surfactants. *Comptes Rendus Chimie*, 7:e641–e646.

Deleu, M., Paquot, M., and Nylander, T. 2008. Effect of fengycin, a lipopeptide produced by *Bacillus subtilis*, on model biomembranes. *Biophysics Journal*, 94:2667–2679.

Donlar Corporation. 1996. Production and use of thermal polyaspartic acid. In *The Presidential Green Chemistry Challenge Awards Program: Summary of 1996 Award Entries and Recipients.* Washington, DC: EPA744-K-96-001, U.S. Environmental Protection Agency, Office of Pollution Prevention and Toxics, p. 5.

Edwards, K.R., Lepo, J.E., and Lewis, M.A. 2003. Toxicity comparison of biosurfactants and synthetic surfactants used in oil spill remediation to two estuarine species. *Marine Pollution Bulletin*, 46(10):1309–1316.

Eeman, M., Francius, G., Dufrêne, Y.F., Nott, K., Paquo, T.M., and Deleu, M. 2009. Effect of cholesterol and fatty acids on the molecular interactions of fengycin with stratum corneum mimicking lipid monolayers. *Langmuir*, 25:3029–3039.

Etoumi, A. 2007. Microbial treatment of waxy crude oils for mitigation of wax precipitation. *Journal of Petroleum Science and Engineering*, 55:111–121.

Flasz, A., Rocha, C.A., Mosquera, B., and Sajo, C. 1998. A comparative study of the toxicity of a synthetic surfactant and one produced by *Pseudomonas aeruginosa* ATCC 55925. *Medical Science Research*, 26(3):181–185.

Francius, G., Dufour, S., Deleu, M., Paquot, M., Mingeot-Leclercq, M.P., and Dufrêne, Y.F. 2008. Nanoscale membrane activity of surfactins: Influence of geometry, charge and hydrophobicity. *Biochimica and Biophysica Acta*, 1778:2058–2068.

Freeman, H.S., and Edwards, L.C. 2000. Iron-complexed dyes: Colorants in green chemistry. In *Green Chemical Syntheses and Processes*. Anastas, P.T., Heine, L.G., and Williamson, T.C. (eds.) Washington, DC: American Chemical Society, Chapter 3.

Gan, B.S., Kim, J., Reid, G., Cadieux, P., and Howard, J.C. 2002. *Lactobacillus fermentum* RC-14 inhibits *Staphylococcus aureus* infection of surgical implants in rats. *Journal of Infectious Diseases*, 185(9):1369–1372.

Gherna, R. and Woese, C.R. 1992. A partial phylogenetic analysis of the "flavobacter–bacteroides" phylum: Basis for taxonomic restructuring. *Systematic and Applied Microbiology*, 15:513–521.

Gleason, K.K. and Ober, C.K. 2001. Environmentally benign lithography for semiconductor manufacturing. In *The Presidential Green Chemistry Challenge Awards Program: Summary of 2000 Award Entries and Recipients*. Washington, DC: EPA744-R-00-001, U.S. Environmental Protection Agency, Office of Pollution Prevention and Toxics, pp. 11–12.

Guo, Y. and Mulligan, C.N. 2006. Combined treatment of styrene–contaminated soil by rhamnolipid washing followed by anaerobic treatment. In *Hazardous Materials in Soil and Atmosphere*. Hudson, R.C. (ed.) New York: Nova Science Publishers, pp. 1–38, Chapter 1.

Gutnick, D.L. and Shabtai, Y. 1987. Exopolysaccharide bioemulsifiers. In *Biosurfactants and Biotechnology*. Kosaric, N., Cairns, W.L., and Gray, N.C.C. (eds.) New York: Marcel Dekker Inc., pp. 211–246.

Han, Y., Huang, X., Cao, M., and Wang, Y. 2008. Micellization of surfactin and its effect on the aggregate conformation of amyloid β(1–40). *Journal of Physical Chemistry B*, 112:15195–15201.

Harenda, S. and Vipulanandan, C. 2008. Degradation of high concentrations of PCE solubilized in SDS and biosurfactant with Fe/Ni metallic particles. *Colloid and Surfaces A Physico Engineering Aspects*, 322:6–13.

Hayes, M.E. et al. 1987. Combustion of viscous hydrocarbons. US Patent 4684372.

Hayes, M.E. et al. 1990. Bioemulsifier-stabilized hydrocarbosols. US Patent 4943390.

Hong, H.A., Duc, L.H., and Cutting, S.M. 2005. The use of bacterial spore formers as probiotics. *FEMS Microbiology Reviews*, 29:813–835.

Hood, S.K. and Zottola, E.A. 1995. Biofilms in food processing. *Food Control*, 6(1):9–18.

Hsueh, Y.H., Somers, E.B., Lereclus, D., Ghelardi, E., and Wong, A.C.L. 2007. Biosurfactant production and surface translocation are regulated by PlcR in *Bacillus cereus* ATCC 14579 under low-nutrient conditions. *Applied and Environmental Microbiology*, 73:7225–7231.

Huang, X., Lu, Z., Bie, X., Lü, F., Zhao, H., and Yang, S. 2007. Optimization of inactivation of endospores of Bacillus cereus by antimicrobial lipopeptides from *Bacillus subtilis* fmbj strains using a response surface method. *Applied Microbiology and Biotechnology*, 74:454–461.

Irie, Y., O'Toole, G.A., and Yuk, M.H. 2005. *Pseudomonas aeruginosa* rhamnolipids disperse *Bordetella bronchiseptica* biofilms. *FEMS Microbiology Letters*, 250:237–243.

Iyer, A., Mody, K., and Jha, B. 2006. Emulsifying properties of a marine bacterial exopolysaccharide. *Enzyme and Microbial Technology*, s38:220–222.

Juwarkar, A.A., Nair, A., Dubey, K.V., Singh, S.K., and Devotta, S. 2007. Biosurfactant technology for remediation of cadmium and lead contaminated soils. *Chemosphere*, 68:1996–2002.

Kachholz, T. and Schlingmann, M. 1987. Possible food and agricultural applications of microbial surfactants: An assessment. In *Biosurfactants and Biotechnology*. Kosaric, N., Carns, W.L., and Gray, N.C.C. (eds.) New York: Marcel Dekker Inc., pp. 183–210.

Kidwai, M. and Mohan, R. 2005. Green chemistry: An innovative technology. *Foundations of Chemistry*, 7:269–287.

Kim, J. and Vipulanandan, C. 2006. Removal of lead from contaminated water and clay soil using a biosurfactant. *Journal of Environmental Engineering*, 132(7):777–786.

Kim, K., Yoo, D., Kim, Y., Lee, B., Shin, D., and Kim, E.K. 2002. Characteristics sophorolipid as an antimicrobial agent. *Journal of Microbiology and Biotechnology*, 12:235–241.

Kirchhoff, M.M. 2003. Promoting green engineering through green chemistry. *Environmental Science and Technology*, 37:5349–5353.

Kitamoto, D., Yanagishita, H., Shinbo, T., Nakane, T., Kamisawa, C., and Nakahara, T. 1993. Surface active properties and antimicrobial activities of mannosylerythritol lipids as biosurfactants produced by *Candida antarctica*. *Journal of Biotechnology*, 29:91–96.

Kosaric, N. 2001. Biosurfactants and their application for soil bioremediation. *Food Technology and Biotechnology*, 39(4):295–304.

Kulakovskaya, T., Shashkov, A., Kulakovskaya, E., Golubev, W., Zinin, A., Tsvetkov, Y., Grachev, A., and Nifantiev, N. 2009. Extracellular cellobiose lipid from yeast and their analogues: Structures and fungicidal activities. *Journal of Oleofin Science*, 58:133–140.

Kulakovskaya, T.V., Golubev, W.I., Tomashevskaya, M.A., Kulakovskaya, E.V., Shashkov, A.S., Grachev, A.A., Chizhov, A.S., and Nifantiev, N.E. 2010. Production of antifungal cellobiose lipids by *Trichosporon porosum*. *Mycopathologia*, 169:117–123.

Lang, S., Gilbon, A., Syldatk, C., and Wagner, F. 1984. In *Surfactants in Solution*. Mittal, K.L. and Lindman, B. (eds.), Comparison of Interfacial Active Properties of Glycolipids from Microorganisms, New York: Plenum Press, p. 1365.

Li, C.J. 2002. Quasi-nature catalysis: Developing C–C bond formations catalyzed by late-transition metals in air and water. *Accounts of Chemical Research*, 35:533–538.

Li, C.J. and Trost, B.M. 2008. Green chemistry for chemical synthesis. *PNAS*, 105:13197–13202.

Lif, A. and Holmberg, K. 2006. Water-in-diesel emulsions and related systems. *Advances in Colloid Interface and Science*, 123–126:231–239.

Liu, Y., Jessop, P.G., Cunningham, M., Eckert, C.A., and Liotta, C.L. 2006. Switchable surfactants. *Science*, 313:958–960.

Maier, R.M. and Soberón-Chávez, G. 2000. *Pseudomonas aeruginosa* rhamnolipids: Biosynthesis and potential applications. *Applied Microbiology and Biotechnology*, 54:625–633.

Maitani, Y., Yano, S., Hattori, Y., Furuhata, M., and Hayashi, K. 2006. Liposome vector containing biosurfactant-complexed DNA as herpes simplex virus thymidine kinase gene delivery system. *Journal of Liposome Research*, 16:359–372.

Makkar, R.S. and Cameotra, S.S. 2002. An update on the use of unconventional substrates for biosurfactant production and their new applications. *Applied Microbiology and Biotechnology*, 58:428–434.

Meylheuc, T., Methivier, C., Renault, M., Herry, J.M., Pradier, C.M., and Bellon-Fontaine, M.N. 2006. Adsorption on stainless steel surfaces of biosurfactants produced by gram-negative and gram-positive bacteria: Consequence on the bioadhesive behaviour of *Listeria monocytogenes*. *Colloids and Surfaces B: Biointerfaces*, 52:128–137.

Meylheuc, T., Renault, M., and Bellon-Fontaine, M.N. 2006. Adsorption of a biosurfactant on surfaces to enhance the disinfection of surfaces contaminated with *Listeria monocytogenes*. *International Journal of Food Microbiology*, 109:71–78.

Meylheuc, T., van Oss, C.J., and Bellon-Fontaine, M.N. 2001. Adsorption of biosurfactant on solid surfaces and consequences regarding the bioadhesion of *Listeria monocytogenes* LO 28. *Journal of Applied Microbiology*, 91:822–832.

Mimee, B., Labbé, C., Pelletier, R., and Bélanger, R.R. 2005. Antifungal activity of flocculosin, a novel glycolipid isolated from *Pseudozyma flocculosa*. *Antimicrobial Agents and Chemotherapy*, 49:1597–1599.

Mimee, B., Pelletier, R., and Bélanger, R.R. 2009. In vitro antibacterial activity and antifungal mode of action of flocculosin, a membrane-active cellobiose lipid. *Journal of Applied Microbiology*, 107:989–996.

Mohammadipour, M., Mousivand, M., Salehi, J.G., and Abbasalizadeh, S. 2009. Molecular and biochemical characterization of Iranian surfactin-producing *Bacillus subtilis* isolates and evaluation of their biocontrol potential against *Aspergillus flavus* and *Colletotrichum* gloeosporioides. *Canadian Journal of Microbiology*, 55:395–404.

Mohan, P.K., Nakhla, G., and Yanful, E.K. 2006. Biokinetics of biodegradability of surfactants under aerobic, anoxic and anaerobic conditions. *Water Research*, 40:533–540.

Morita, T., Kitagawa, M., Suzuki, M., Yamamoto, S., Sogabe, A., Yanagidani, S., Imura, T., Fukuoka, T., and Kitamoto, D. 2009. A yeast glycolipid biosurfactant, mannosylerythritol lipid, shows potential moisturizing activity toward cultured human skin cells: The recovery effect of MEL-A on the SDS-damaged human skin cells. *Journal of Oleofin Science*, 58:639–642.

Mulligan, C.N. 2005. Environmental applications for biosurfactants. *Environmental Pollution*, 133:183–198.

Nakanishi, M., Inoh, Y., Kitamoto, D., and Furuno, T. 2009. Nano vectors with a biosurfactant for gene transfection and drug delivery. *Journal of Drug Delivery Science and Technology*, 19:165–169.

Naruse, N., Tenmyo, O., Kobaru, S., Kamei, H., Miyaki, T., Konishi, M., and Oki, T. 1990. Pumilacidin, a complex of new antiviral antibiotics: Production, isolation, chemical properties, structure and biological activity. *Journal of Antibiotics (Tokyo)*, 43:267–280.

Neu, T., Hartner, T., and Poralla, K. 1990. Surface active properties of viscosin: A peptidolipid antibiotic. *Applied Microbiology and Biotechnology*, 32:518–520.

Neu, T.R. 1996. Significance of bacterial surface-active compounds in interaction of bacteria with interfaces. *Microbiology Reviews*, 60:151–166.

Nitschke, M., Costa, S.G., and Contiero, J. 2009b. Structure and applications of a rhamnolipid surfactant produced in soybean oil waste. *Applied Biochemistry and Biotechnology*, 160:2066–2074.

Palanisamy, P. 2008. Biosurfactant mediated synthesis of NiO nanorods. *Material Letters*, 62:743–746.

Palanisamy, P. and Raichur, A.M. 2009. Synthesis of spherical NiO nanoparticles through a novel biosurfactant mediated emulsion technique. *Material Science and Engineering C: Biomimicry and Supramolecular Systems*, 29:199–204.

Park, S.Y. and Kim, Y. 2009. Surfactin inhibits immunostimulatory function of macrophages through blocking NK-κB, MAPK and Akt pathway. *International Immunopharmacology*, 9:886–893.

Perfumo, A., Rancich, I., and Banat, I.M. 2010. Possibilities and challenges for biosurfactants uses in petroleum industry. *Advances in Experimental and Medical Biology*, 672:135–145.

Raza, Z.A., Khalid, Z.M., and Banat, I.M. 2009. Characterization of rhamnolipids produced by a *Pseudomonas aeruginosa* mutant strain grown on waste oils. *Journal of Environmental Science and Health Part A—Toxic/Hazard Substances and Environmental Engineering*, 44:1367–1373.

Raza, Z.A., Rehman, A., Khan, M.S., and Khalid, Z.M. 2007. Improved production of biosurfactant by *Pseudomonas aeruginosa* mutant using vegetable oil refinery wastes. *Biodegradation*, 18:115–121.

Reddy, A.S., Chen, C.Y., Chen, C.C., Jean, J.S., Fan, C.W., Chen, H.R., Wang, J.C., and Nimje, V.R. 2009. Synthesis of gold nanoparticles via an environmentally benign route using a biosurfactant. *Journal of Nanoscience and Nanotechnology*, 9:6693–6699.

Reisfeld, A., Gutnick, D., and Rosenberg, E. 1972. Microbial degradation of crude oil: Factors affecting the dispersion in sea water by mixed and pure cultures. *Applied Microbiology*, 24:363–368.

Rocha, M.V., Souza, M.C.M., Benedicto, S.C., Bezerra, M.S., Macedo, G.R., Pinto, G.A.S., and Goncalves, L.R.B. 2007. Production of biosurfactant by *Pseudomonas aeruginosa* grown on cashew apple juice. *Applied Biochemistry and Biotechnology*, 137–140:185–194.

Rosenberg, E. and Ron, E.Z. 1997. Bioemulsans: Microbial polymeric emulsifiers. *Current Opinion in Biotechnology*, 8:313–316.

Ruggeri, C., Franzetti, A., Bestetti, G., Caredda, P., La Colla, P., Pintus, M., Sergi, S., and Tamburini, E. 2009. Isolation and characterization of surface active compound-producing bacteria from hydrocarbon contaminated environments by a high-throughput screening procedure. *International Biodeterioration and Biodegradation*, 63:936–942.

Saini, H.S., Barragán-Huerta, B.E., Lebrón-Paler, A., Pemberton, J.E., Vázquez, R.R., Burns, A.M., Marron, M.T., Seliga, C.J., Gunatilaka, A.A., and Maier, R.M. 2008. Efficient purification of the biosurfactant viscosin from *Pseudomonas libanensis* strain M9-3 and its physicochemical and biological properties. *Journal of Natural Products*, 71:1011–1015.

Sanchez, C.G., Lobert, M., Hoogenboom, R., and Schubert, U.S. 2007. Microwave–assisted homogeneous polymerizations in water-soluble ionic liquids: An alternative and green approach for polymer synthesis. *Macromolecular Rapid Communications*, 28(4):456–464.

Selvam, R., Maheswari, P., Kavitha, P., Ravichandran, M., Sas, B., and Ramchand, C.N. 2009. Effect of *Bacillus subtilis* PB6, a natural probiotic on colon mucosal inflammation and plasma cytokines levels in inflammatory bowel disease. *Indian Journal of Biochemistry and Biophysics*, 46:79–85.

Seydlová, G. and Svobodová, J. 2008. Review of surfactin chemical properties and the potential biomedical applications. *Century European Journal of Medicine*, 3:123–133.

Shete, A.M., Wadhawa, G., Banat, I.M., and Chopade, B.A. 2006. Mapping of patents on bioemulsifier and biosurfactant: A review. *Journal of Scientific and Industrial Research*, 65(2):91–115.

Singh, P. and Cameotra, S.S. 2004. Potential applications of microbial surfactants in biomedical sciences. *Trends in Biotechnology*, 22(3):142–146.

Smyth, T.J.P., Perfumo, A., McClean, S., Marchant, R., and Banat, I.M. 2010. Isolation and analysis of lipopeptides and high molecular weight biosurfactants. In *Handbook of Hydrocarbon and Lipid Microbiology*. Timmis, K.N. (ed.) Berlin, Germany: Springer, pp. 3689–3704.

Snook, M.E., Mitchell, T., Hinton, D.M., and Bacon, C.W. 2009. Isolation and characterization of Leu7-surfactin from the endophytic bacterium *Bacillus mojavensis* RRC 101, a biocontrol agent for Fusarium verticillioides. *Journal of Agricultural and Food Chemistry*, 57:4287–4292.

Sobrinho, H.B.S., Rufino, R.D., Luna, J.M., Salgueiro, A.A., Campos-Takaki, G.M., Leite, L.F.C., and Sarubbo, L.A. 2008. Utilization of two agroindustrial by-products for the production of a surfactant by *Candida sphaerica* UCP0995. *Process Biochemistry*, 43:912–917.

Tanaka, Y., Tojo, T., Uchida, K., Uno, J., Uchida, Y., and Shida, O. 1997. Method of producing iturin A and antifungal agent for profound mycosis. *Biotechnology Advances*, 15:234–235.

Teichmann, B., Linne, U., Hewald, S., Marahiel, M.A., and Bölker, M. 2007. A biosynthetic gene cluster for a secreted cellobiose lipid with antifungal activity from *Ustilago maydis*. *Molecular Microbiology*, 66:525–533.

Thavasi, R., Jayalakshmi, S., Balasubramanian, T., and Banat, I.M. 2007. Biosurfactant production by *Corynebacterium kutscheri* from waste motor lubricant oil and peanut oil cake. *Letters in Applied Microbiology*, 45:686–691.

Tran, H., Ficke, A., Asiimwe, T., Höfte, M., and Raaijmakers, J.M. 2007. Role of cyclic lipopeptide massetolide A in biological control of *Phytophthora infestans* and in colonization of tomato plants by *Pseudomonas fluorescens. New Phytology*, 175:731–742.

Van Bogaert, I.N.A., Saerens, K., De Muynck, C., Develter, D., Wim, S., and Vandamme, E.J. 2007. Microbial production and application of sophorolipids. *Applied Microbiology and Biotechnology*, 76:23–34.

Van Haesendonck, I.P.H. and Vanzeveren, E.C.A. 2004. Rhamnolipids in bakery products. W.O. 2004/040984, International application patent (PCT).

Van Hamme, J.D., Singh, A., and Ward, O.P. 2006. Physiological aspects. Part 1 in a series of papers devoted to surfactants in microbiology and biotechnology. *Biotechnology Advances*, 24:604–620.

Varnier, A.L. et al. 2009. Bacterial rhamnolipids are novel MAMPs conferring resistance to *Botrytis cinerea* in grapevine. *Plant Cell Environment*, 32:178–193.

Vater, J., Kablitz, B., Wilde, C., Frank, P., Mehta, N., and Cameotra, S.S. 2002. Matrix-assisted laser desorption ionization time of flight mass spectrometry of lipopeptide biosurin whole cells and culture filtrates of *Bacillus subtilis* C-1 isolated from petroleum sludge. *Applied and Environmental Microbiology*, 68:6210–6219.

Velmurugan, N., Choi, M.S., Han, S.S., and Lee, Y.S. 2009. Evaluation of antagonistic activities of *Bacillus subtilis* and *Bacillus licheniformis* against wood-staining fungi: In vitro and in vivo experiments. *Journal of Microbiology*, 47:385–392.

Velraeds, M.M.C., van der Mei, H.C., Reid, G., and Busscher, H.J. 1996. Inhibition of initial adhesion of uropathogenic *Enterococcus faecalis* by biosurfactants from *Lactobacillus* isolates. *Applied and Environmental Microbiology*, 62(6):1958–1963.

Wang, Q., Fang, X., Bai, B., Liang, X., Shuler, P.J., Goddard, W.A. III, and Tang, Y. 2007. Engineering bacteria for production of rhamnolipid as an agent for enhanced oil recovery. *Biotechnology and Bioengineering*, 98:842–853.

Woese, C.R., Mandelco, L., Yang, D., Gherna, R., and Madigan, M.T. 1990. The case for relationship of the flavobacteria and their relatives to the green sulfur bacteria. *Systematic and Applied Microbiology*, 13:258–262.

Zheng, Y.G., Chen, X.L., and Shen, Y.C. 2008. Commodity chemicals derived from glycerol, an important biorefinery feedstock. *Chemical Reviews*, 108:5253–5277.

2 Amphiphilic Molecules of Microbial Origin

Classification, Characteristics, Genetic Regulations, and Pathways for Biosynthesis

Gunaseelan Dhanarajan and Ramkrishna Sen

CONTENTS

INTRODUCTION

Amphiphilic molecules with hydrophobic moiety at one end and hydrophilic moiety at the other end possess the ability to reduce the interfacial/surface tension (ST) at a liquid–liquid interface or a gas–liquid interface. In general, synthetic surfactants produced from petroleum feedstock are used in food, cosmetic, and pharmaceutical industries as emulsifiers, lubricants, stabilizers, wetting agents, etc. (Banat et al., 2000; Mukherjee et al., 2006). Since they are very harmful to the environment, use of more efficient and ecofriendly surfactants is necessary. Various microorganisms produce such risk-free surface active metabolites known as biosurfactants. When compared to synthetic surfactants, biosurfactants are less toxic, highly biodegradable, and stable at extremes of pH, temperature, and salinity (Mukherjee et al., 2006). Furthermore, microbial surfactants can be produced by using low-cost agro-based raw materials as potential carbon sources. Hence, biosurfactants are considered to have strategic advantages over their chemically synthesized counterparts.

Microorganisms synthesize amphiphilic molecules or biosurfactants extracellularly or as part of the cell membrane. These molecules facilitate the growth of their producers by increasing the substrate availability, transporting nutrients, and by acting as biocide agents (Rodrigues et al., 2006). Biosurfactants are classified into many groups, since they are produced from wide variety of microorganisms and have very different chemical structures (Ron and Rosenberg, 2001). These structurally diverse molecules have different properties and inherent functions. In addition to their surface active and emulsification properties, biosurfactants possess antiadhesive, antimicrobial, and anticarcinogenic properties, which make them a versatile class of biomolecules. Owing to their diverse physicochemical characteristics, biosurfactants find myriad applications in pharmaceutical, food, and cosmetic industries as well as in enhanced oil recovery and environmental bioremediation (Sen, 2008; Sen et al., 2011). This chapter deals with the microbial synthesis, classification, and properties of the biosurfactants.

MAJOR CLASSIFICATION OF MICROBIAL AMPHIPHILES

Biosurfactants are classified according to their chemical composition, molecular weight, and microbial source of origin. Consequently, the major classes of biosurfactants include glycolipids, lipopeptides and lipoproteins, fatty acids, phospholipids and neutral lipids, polymeric surfactants, and particulate surfactants (Desai and Banat, 1997) as reported in Table 2.1. The lipophilic moiety of the biosurfactant is usually the hydrocarbon chain of a fatty acid or fatty acid derivatives, but it can also be a protein or a peptide with a high proportion of hydrophobic side chains. The hydrophilic moiety can be a peptide, a carbohydrate, or an ester group (Nitschke and Costa, 2007). Bacteria account for the greatest production of biosurfactants followed by yeasts.

GLYCOLIPIDS

The most commonly described microbial surfactants are the glycolipids. Glycolipids are carbohydrates (hydrophilic moiety) linked with long-chain aliphatic acids or hydroxy aliphatic acids (hydrophobic moiety). The linkage is by means of either

TABLE 2.1
Various Types of Microbial Amphiphiles and Their Producer Microbes

Biosurfactants	Microorganisms	Reference
Glycolipids		
Rhamnolipids	*Pseudomonas aeruginosa*	Hisatsuka et al. (1971)
	P. chlororaphis	Gunther et al. (2005)
	Acinetobacter calcoaceticus	Rooney et al. (2009)
Trehalose lipids	*R. erythropolis*	Rapp et al. (1979)
	Arthrobacter sp.	Li et al. (1984)
	Mycobacterium sp.	Cooper et al. (1989)
Sophorolipids	*C. bombicola*	Cavalero and Cooper (2003)
	C. borgoriensis	Cutler and Light (1979)
	T. apicola	Hommel et al. (1987)
	Ustilago maydis	Hewald et al. (2005)
Mannosylerythritol lipids	*C. antarctica*	Kitamoto et al. (1992)
Cellobiolipids	*T. petrophilum*	Cooper and Paddock (1983)
	U. zeae	Boothroyd et al. (1956)
Lipopeptides		
Surfactin/iturin/fengycin	*B. subtilis*	Arima et al. (1968)
Viscosin	*P. fluorescens*	Neu et al. (1990)
Lichenysin	*B. licheniformis*	Yakimov et al. (1995)
Serrawettin	*S. marcescens*	Matsuyama et al. (1992)
Subtilisin	*B. subtilis*	Bernheimer and Avigad (1970)
Arthrofactin	*Arthrobacter* sp. strain MIS38	Morikawa et al. (1993)
Plipastatin	*B. subtilis*	Volpon et al. (2000)
Amphisin	*Pseudomonas* sp.	Sørensen et al. (2001)
Putisolvin	*P. putida*	Kuiper et al. (2004)
Surface active antibiotics		
Gramicidin	*B. brevis*	Marahiel et al. (1977)
Polymyxin	*B. polymyxa*	Suzuki et al. (1965)
Fatty acids	*C. lepus*	Cooper et al. (1978)
Neutral lipids	*N. erythropolis*	MacDonald et al. (1981)
Flavolipids	*Flavobacterium* sp. strain MTN11	Bodour et al. (2004)
Phospholipids	*T. thiooxidans*	Beebe and Umbreit (1971)
Polymeric surfactants		
Emulsan	*A. calcoaceticus*	Zosim et al. (1982)
	P. fluorescens	Persson et al. (1988)
Alasan	*A. radioresistens*	Navon-Venezia et al. (1995)
Liposan	*C. lipolytica*	Cirigliano and Carman (1985)
Lipomanan	*C. tropicalis*	Kappeli et al. (1984)
Biodispersan	*A. calcoaceticus*	Rosenberg et al. (1988)
Yansan	*Yarrowia lipolytica*	Amaral et al.(2006)
Particulate biosurfactants		
Vesicles	*A. calcoaceticus*	Kappeli and Finnerty (1979)
Emulcyan	*Phormidium J-1*	Fattom and Shilo (1985)

ether or an ester group. The best studied glycolipids are rhamnolipids, sophorolipids, and trehalolipids (Desai and Banat, 1997).

Rhamnolipids

Rhamnolipids are extracellular biosurfactants mainly produced by *Pseudomonas aeruginosa*. It can produce up to 100 g/L of rhamnolipids, and hence, its cost becomes competitive against the production cost of synthetic surfactants (Maier and Soberón-Chávez, 2000). Major forms of rhamnolipid contain one (monorhamnolipid) or two (dirhamnolipid) rhamnose sugars linked to one or two 3-hydroxydecanoic acid moieties. At first, Jarvis and Johnson reported the production of rhamnose-containing glycolipids by *P. aeruginosa* in peptone–glycerol broth. The glycolipid was found to contain two units each of L-rhamnose and β-hydroxydecanoic acid (Jarvis and Johnson, 1949). Mostly, *P. aeruginosa* strains produce rhamnolipid mixtures, which are influenced by the carbon source used, and they vary in the length and the saturation of the fatty acid moiety (Herman and Maier, 2002). Benincasa et al. reported that up to 28 different structural homologues of rhamnolipids were produced by *P. aeruginosa* strains (Benincasa et al., 2004). Difference in the form and the composition of the rhamnolipids produced causes variation in physicochemical properties. Rhamnolipids are low–molecular weight biosurfactants, which are capable of reducing ST and interfacial tension (IFT). When excreted into the medium, they can emulsify hydrocarbon or other hydrophobic substrates and thereby make them available for cell metabolism. Rhamnolipids are also reported to have antimicrobial activity against several bacterial, yeast, and fungal strains (Benincasa et al., 2004).

Sophorolipids

Sophorolipids are largely produced by yeasts, mainly *Candida* sp. The strain *Candida bombicola* ATCC 22214 can produce over 400 g/L sophorolipids and now is used for commercial production and applications (Pekin et al., 2005). Sophorolipids are composed of a disaccharide sophorose, β-glycosidically linked to a long chain hydroxy fatty acid. Sophorolipid can occur in the acidic form, in which the fatty acid tail is free, or in the lactonic form, where the carboxylic end of the fatty acid is connected to the sophorose head by internal esterification. Sophorolipid mixtures also vary in fatty acid chain length, and the variation depends on the type of hydrophobic organic substrate utilized by the microorganism for growth. Properties of sophorolipid mixtures depend on their form and composition. For instance, lactonic sophorolipids have better ST lowering and antimicrobial activity, whereas the acidic ones display a better foam production and solubility (Van Bogaert et al., 2011).

Trehalolipids

Different types of trehalolipids are known to be produced by several microorganisms, including the genera of *Mycobacterium*, *Rhodococcus*, *Arthrobacter*, *Nocardia*, and *Gordonia*. Among the trehalolipids, the trehalose esters produced by *Rhodococcus erythropolis* have been studied most extensively. Trehalolipids consist of a disaccharide trehalose linked to fatty acid groups by ester bond. The ester of fatty acids with trehalose was first noticed by Anderson and Newman in 1933, but not as a surface active agent (Anderson and Newman, 1933). Trehalolipids caught the attention as a general surfactant after an emulsion layer containing trehalose dimycolates was discovered in *Arthrobacter paraffineus* culture broths when the cells were grown on hydrocarbon substrates (Shao, 2011). Trehalolipids are most widely used in bioremediation technologies as such compounds are known to enhance the bioavailability of hydrocarbons (Franzetti et al., 2010).

There are quite a few more microbially produced glycolipids other than the aforementioned glycolipid biosurfactants. Powalla et al. reported the production of a novel pentasaccharide lipid together with two trehalose corynomycolates by *Nocardia corynebacteroides* SM1. The pentasaccharide lipid showed significant surface and interfacial active properties (Powalla et al., 1989). Vollbrecht et al. reported the production of di- and oligosaccharide lipids by *Tsukamurella* sp. with significant surface active and antimicrobial properties (Vollbrecht et al., 1998). Li et al. discovered the production of eight different glycolipids by *Arthrobacter* sp. when incubated with different sugars. Among different glycolipids, more hydrophilic substances such as cellobiose and maltose monocorynomycolates were found to have surface and interfacial active properties (Li et al., 1984). Hewald et al. described the production of mannosylerythritol lipids (MELs) and cellobiose lipids by *Ustilago maydis* under nitrogen-limited conditions (Hewald et al., 2005).

LIPOPEPTIDES

Lipopeptides consist of short linear chains or cyclic structures of amino acids, linked to a fatty acid via ester or amide bonds or both. Lipopeptides contain the rare and modified amino acids, which are not used for ribosomal protein synthesis (Sen, 2010). They vary in terms of the types of amino acids present in the peptide ring, as well as in the chain length and structure of the fatty acid component.

Surfactin

Surfactin, the most studied and potent lipopeptide biosurfactant, is produced by various strains of *Bacillus subtilis*. The cyclic lipopeptide surfactin consists of 3-hydroxy-13-methyl-tetradecanoic acid amidated to the N-terminal amine of a heptapeptide moiety with the carboxy terminal end of the peptide being further esterified to the hydroxyl group of the fatty acid (Sen, 2010). Surfactin was discovered by Arima et al. from the culture broth of *B. subtilis* as a potent inhibitor of blood clotting, but was also found to be a powerful surface active agent (Arima et al., 1968). Besides surfactant property, surfactin possesses a number of biological activities, which makes it a versatile biomolecule with tremendous commercial application potentials.

Lichenysin

Lichenysins are most potent lipopeptide biosurfactants produced by *B. licheniformis* and are named lichenysin A, B, C, D, G, and surfactant BL86 with regard to producing strains. The production of lichenysin was first described by Jenneman et al. in a halotolerant *Bacillus* sp., which was then identified as *B. licheniformis* JF-2 (Jenneman et al., 1983). Lichenysin A was reported to exhibit critical micelle concentration (CMC) of 12 mg/L, which is around one-half of the CMC of surfactin (25 mg/L). Rationale behind this fact was the less polar peptide moiety and the presence of a longer β-hydroxy fatty acid giving lichenysin A delicate hydrophile–lipophile balance (Yakimov et al., 1995).

Bacillus subtilis produces two further families of lipopeptide biosurfactants, namely, iturins and fengycins, which possess excellent surface and biological activities (Besson et al., 1987; Sivapathatsekaran et al., 2009). Other effective surface active lipopeptides include serrawettin from *Serratia marcescens* (Matsuyama et al., 1992), viscosin from *P. fluorescens* (Neu et al., 1990) and *P. viscose* (Neu et al., 1990), and arthrofactin from *Arthrobacter* sp. strain MIS38 (Morikawa et al., 1993). Antibiotics like gramicidins from *B. brevis* (Marahiel et al., 1979) and polymyxins from *B. polymyxa* (Suzuki et al., 1965) were also reported to have pronounced surface active properties.

FATTY ACIDS AND PHOSPHOLIPIDS

Various bacteria and yeasts were able to secrete surface active fatty acids and phospholipids when grown on hydrocarbons. Lipids can be minor components of a cell or excreted into the medium facilitating efficient product recovery. Cooper and Goldenberg described the production of an extracellular neutral lipid biosurfactant by *Bacillus* sp. strain IAF 343 grown on a medium containing only water-soluble substrates. The neutral lipid did not show appreciable surfactant activity but exhibited significant emulsifying activity (Cooper and Goldenberg, 1987). An extracellular lipid produced by *Thiobacillus thiooxidans* was found to be a heterogeneous mixture of phospholipid and neutral lipid. Wetting ability of the mixture facilitated oxidation of elemental sulfur that aided the growth of *T. thiooxidans* on sulfur particles (Beebe and Umbreit, 1971). Extracellular membrane vesicles produced by hexadecane-grown *Acinetobacter* sp. HO1-N was rich in phospholipids, mainly phosphatidylethanolamine and phosphatidylglycerol, and also had trace amount of neutral lipids. The accumulated vesicles represented emulsifying properties by forming microemulsions of hexadecane in water (Kappeli and Finnerty, 1979). *Candida ingens* was reported to produce a fatty acid biosurfactant. Culture medium composition had a strong influence on the production and total fatty acid content of the biosurfactant and thereby its surface activity (Amézcua-Vega et al., 2007). Neutral lipids from *Nocardia erythropolis* (MacDonald et al., 1981) and phosphatidylethanolamines from *R. erythropolis* (Kretschmer et al., 1982) were also shown to have significant surface and interfacial activities.

POLYMERIC BIOSURFACTANTS

Various microbial genera, including *Acinetobacter, Arthrobacter, Pseudomonas, Halomonas, Bacillus*, and *Candida*, have been reported to produce polymeric biosurfactants. Polymeric biosurfactants do not necessarily reduce ST, but they effectively reduce the IFT between immiscible liquids and form stable emulsions. The most studied polymeric biosurfactants are emulsan and liposan. Emulsan from *Acinetobacter calcoaceticus* RAG-1 is a 1000 kDa lipopolysaccharide composed of a heteropolysaccharide linked to a fatty acid via ester and amide bonds (Rosenberg et al., 1979). Previously emulsan was known to be a single polymer, but Mercaldi et al. revealed emulsan as a complex of approximately 80% (w/w) lipopolysaccharide and 20% (w/w) high–molecular weight exopolysaccharide (Mercaldi et al., 2008). Production of extracellular bioemulsifiers is a widespread phenomenon in *Acinetobacter* species. *Acinetobacter radioresistens* KA53 produced a bioemulsifier called alasan, composed of a heteropolysaccharide containing covalently bound alanine and proteins. Liposan is another effective emulsifier produced by *Candida lipolytica* capable of forming stable oil-in-water emulsions with a variety of commercial vegetable oils. It is composed of approximately 83% carbohydrate and 17% protein, and the carbohydrate content is almost similar to that of the emulsan (Cirigliano and Carman, 1985). Generally, the production of emulsifying agents from yeast sources occurs in the presence of water-immiscible substrates, but Sarubbo et al. reported the production of an emulsifier by *C. lipolytica* using glucose as the carbon source (Sarubbo et al., 2001). They described that the biosynthesis of a bioemulsifier was not simply a prerequisite for the degradation of extracellular hydrocarbon. Amaral et al. isolated and characterized an emulsifier from the glucose-based culture medium of *Yarrowia lipolytica* IMUFRJ50682, which exhibited emulsification activities superior to liposan and named it as yansan. It was able to form stable emulsions of both aromatic and aliphatic hydrocarbons (Amaral et al., 2006). *Streptomyces* sp. S1 produced a bioemulsifier composed of 82% protein, 17% polysaccharide and 1% reducing sugar. The bioemulsifier had significant emulsifying properties and also reduced ST of the medium (Kokare et al., 2007). Gutiérrez et al. reported the production of high–molecular weight glycoprotein emulsifiers from two marine *Halomonas* species. They showed the highest reported emulsifying activities derived from a *Halomonas* species (Gutiérrez et al., 2007). Polymeric biosurfactants from *P. aeruginosa* (Koronelli et al., 1983) and *P. nautica* (Husain et al., 1997) have also been shown to have significant emulsification properties.

Particulate Biosurfactants

Extracellular vesicles that help in hydrocarbon uptake by cells and microbial cells with surface active properties are referred to as particulate biosurfactants. *Serratia marcescens* NS 38 produced extracellular vesicles, and their main lipid component had strong wetting activity. In addition to the wetting agent, a red pigment (prodigiosin) was also present in the vesicle (Matsuyama et al., 1986). *Acinetobacter* sp. HO1-N produced extracellular membrane vesicles with 20–50 nm diameter and a buoyant density of 1.158 g/cm^3. They were rich in phospholipid and lipopolysaccharide, exhibiting good emulsification activity (Kappeli and Finnerty, 1979). Surfactant activity in most hydrocarbon-degrading and pathogenic bacteria is attributed to several cell surface components.

SALIENT CHARACTERISTICS OF MICROBIAL AMPHIPHILES

Surface and Interface Activity

Surfactin and lichenysin are the most powerful biosurfactants in reducing ST at low CMC values. Lichenysin A, a lipopeptide from *B. licheniformis*, can reduce the ST of water to 28 mN/m at a CMC as low as 12 mg/L (Yakimov et al., 1995).

Surfactin from *B. subtilis* was reported to lower the ST of water to 27 mN/m, with CMC of 25 mg/L, and IFT of water/hexadecane to 1 mN/m (Cooper et al., 1981). Arthrofactin from *Arthrobacter* sp. strain MIS38 lowered the ST of water to 24 mN/m at 1×10^{-5} M CMC and was reported to be five times more effective than surfactin (Morikawa et al., 1993). Rhamnolipid, a glycolipid produced by *P. aeruginosa*, was reported to decrease the ST of water to 26 mN/m and IFT of water/hexadecane to less than 1 mN/m (Hisatsuka et al., 1971). Recently, a biosurfactant from *P. aeruginosa* isolated from the formation water of petroleum reservoir was able to lower the ST of water from 71.2 to 22.56 mN/m (Xia et al., 2011). Sophorolipid from *Candida bombicola* reduced the ST of water to 34 mN/m at a CMC of 27.17 mg/L (Daverey and Pakshirajan, 2010). A trehalolipid from *R. erythropolis* decreased the IFT of water/hexadecane to less than 1 mN/m and ST of water to 26 mN/m with CMC of 15 mg/L (Lang and Philp, 1998).

EMULSIFICATION ACTIVITY

Biosurfactants are capable of stabilizing and destabilizing the emulsion, hence widely used in dairy, food, and cosmetic industries. While low–molecular weight biosurfactants behave as efficient surface and interfacial active agents, high–molecular weight biosurfactants are effective in forming stable emulsions. Emulsan from *Acinetobacter calcoaceticus* was reported to be an effective emulsifying agent for hydrocarbons in water at 0.001%–0.01% (w/v) (Zosim et al., 1982). Alasan from *Acinetobacter radioresistens* showed emulsifying activity against n-alkanes with different chain lengths and variety of oils. Interestingly, it was found to be 2.5–3 times more active after being heated at 100°C under neutral or alkaline conditions (Navon-Venezia et al., 1995). Liposan from *C. lipolytica* effected emulsification activity of 0.75 U with hexadecane at a hexadecane-to-liposan ratio of 50:1 (w/w) (Cirigliano and Carman, 1985). On the other hand, yansan produced by *Yarrowia lipolytica* displayed superior emulsification activity to those for liposan at a hydrocarbon-to-yansan ratio of 25:1 (w/w). It was capable to form oil-in-water emulsions of various aliphatic and aromatic hydrocarbons (Amaral et al., 2006). A marine bacterium *Planococcus maitriensis* produced an exopolymer that emulsified jatropha, silicone, and paraffin oil with an emulsification index, $E_{1,080} = 100\%$ (Kumar et al., 2007).

TEMPERATURE, pH, AND IONIC STRENGTH TOLERANCE

The stability of biosurfactants at extremes of pH, temperature, and salinity is one of their unique properties that lead them to find countless industrial and environmental applications. Surfactin exhibited good surface activity at a temperature of 100°C, over the pH range of 5–12, and up to 20% NaCl and 0.5% $CaCl_2$ (Gong et al., 2009). *B. licheniformis* 86 produced a lipopeptide biosurfactant, which was active from pH 4 to 13, temperature 25°C–120°C, and 30% NaCl (Horowitz et al., 1990). Sophorolipids from *Candida bombicola* showed unhindered surface activity after 2 h incubation in boiling water, over a broad pH range of 2–10 and up to 20% NaCl (Daverey and Pakshirajan, 2010). Recently, a biosurfactant produced by

P. aeruginosa was stable over the pH range of 5–12, temperature up to 120°C, and up to 20 g/L NaCl concentrations (Xia et al., 2011). Few other bioemulsifiers and biosurfactants have also been reported to tolerate wide range of pH, temperature, and salinity (Amaral et al., 2006; Chen et al., 2012; Kim et al., 1997).

LOW TOXICITY

It is extremely essential to investigate the pharmacological properties and toxicity of the antibiotic biosurfactants to exploit them as safe drugs. Hwang et al. examined the subacute toxicity of surfactin from *B. subtilis* in adult Sprague–Dawley rats for 28 days. Rats survived even with a high dose of surfactin (2 g/kg), and the no-observed-adverse-effect-level of surfactin was found to be 500 mg/kg (Hwang et al., 2009). The cytotoxicity of the MEL from *C. antarctica* was examined using a neutral red assay in mouse fibroblast L929 cells. It was much less toxic than the synthetic surfactants and was reported to be not harmful to human skin and eyes, indicating its potential applicability to industries such as cosmetics or personal care (Kim et al., 2002). A glycolipid biosurfactant from *Rhodococcus ruber* was examined for its acute toxicity against outbred male albino mice. No effect on central nervous system or weight loss was found during the 14-day observation (Kuyukina et al., 2007). Few other reports have also compared the toxicity of biosurfactants and chemical counterparts, indicating the low toxicity of biosurfactants (Dehghan-Noudeh et al., 2005; Edwards et al., 2003; Hirata et al., 2009; Poremba et al., 1991).

BIODEGRADABILITY

As the concern about the environment is increasing, easily biodegradable microbial surfactants are preferred over the synthetic surfactants for environmental applications (Nitschke and Costa, 2007). Pei et al. examined the biodegradability of rhamnolipids by incubating it in black loamy soil and red sandy soil for 1 week. The degradation rate was slow initially, but 92% of rhamnolipid was found to be mineralized in both kinds of soil at the end of the week (Pei et al., 2009). Kim et al. compared the biodegradability of MEL and two chemical surfactants by modified biochemical oxygen demand method. MEL was readily degraded by activated sludge microorganisms in just 5 days, whereas the chemical surfactants showed a lower degradation rate even after 7 days (Kim et al., 2002). Recently, the biodegradability of biosurfactants produced by five different bacterial strains was compared with sodium dodecyl sulfate (SDS). The biosurfactants were highly biodegradable when compared to SDS (Lima et al., 2011).

BIOSYNTHESIS AND GENETICS OF MAJOR MICROBIAL AMPHIPHILES

Hydrophilic and hydrophobic domains of the biosurfactants are synthesized by separate pathways and are then amalgamated to form an amphipathic structure (Hommel and Ratledge, 1993). These two domains may be synthesized de novo by two independent pathways or derived from carbon substrates present in the environment or one domain by de novo synthesis while the other is substrate dependent (Desai and Banat, 1997). Good knowledge about the biosynthetic pathway and substrate dependency would improve the structure and, hence, characteristics of the biosurfactants through metabolic and genetic engineering approach. Thus, the aspects of biosynthesis and genetic regulation of few well-characterized biosurfactants are discussed here.

SURFACTIN

Surfactin is synthesized nonribosomally by a large multienzyme peptide synthetase complex known as surfactin synthetase. This enzyme complex consists of four enzymatic subunits, namely, SrfA (402 kDa), SrfB (401 kDa), SrfC (144 kDa), and SrfD (40 kDa) (Shaligram and Singhal, 2010). Surfactin synthetase catalyses the incorporation of seven amino acids into the peptide moiety of surfactin by a process referred to as the thiotemplate mechanism. The mechanism involves the activation of amino acids by ATP and assembly of amino acids into a peptide chain. The lipopeptide is then formed by linking the peptide to a hydroxy fatty acid using an acyltransferase enzyme (Sen, 2010). The role of SrfA is to thioesterify L-Glu and two leucine residues, while SrfB incorporates L-Asp, L-Val, and L-Leu as thioesters. SrfC adds the C-terminal L-leucine residue and SrfD functions as the thioesterase/acyltransferase enzyme in the initiation process (Steller et al., 2004).

Nakano et al. identified two genetic loci (*sfp* and *srf*) that are involved in the biosynthesis of surfactin (Nakano et al., 1988). *sfp* codes for phosphopantetheinyl transferase Sfp, which activates the peptidyl carrier protein domains of surfactin synthetase. *srf* operon encodes four subunits of surfactin synthetase, SrfA–D. The regulation of the expression of the *srf* operon entails ComX and a two-component regulatory system composed of ComP and ComA. Surfactin synthesis is regulated by a cell density–responsive mechanism, and when the cell density is high, the signal peptide ComX gets accumulated in the growth medium (Das et al., 2008a). The histidine protein kinase ComP, which is present in the cytoplasmic membrane, recognizes ComX and donates a phosphate to the response regulator ComA and activates it. ComA, in this state, stimulates the transcription and thus the expression of the *srf* operon (Cosby et al., 1998). Enzyme RapC acts on ComA–ComP system in order to dephosphorylate ComA and inhibits the production of surfactin by inactivating ComA. CSF (competence and sporulation stimulatory factor), a chemical signal that participates in the regulation of sporulation, prevents inactivation of ComA by inhibiting RapC activity. CSF is imported into the cell through the oligopeptide permease enzyme Spo0K encoded by the *spo0K* gene (Solomon et al., 1996).

RHAMNOLIPIDS

Rhamnolipids are produced in two forms, rhamnosyl-β-hydroxydecanoyl-β-hydroxydecanoate (monorhamnolipid) and rhamnosyl-rhamnosyl-β-hydroxydecanoyl-β-hydroxydecanoate (dirhamnolipid). The central metabolic pathways that are crucial to rhamnolipid biosynthesis are fatty acid synthesis and deoxythymidine diphosphate (dTDP)-activated sugar synthesis. Biosynthesis of rhamnolipids occurs sequentially in three steps (Burger et al., 1963). First step is the synthesis of the fatty acid dimer moiety of rhamnolipids and free 3-(3-hydroxyalkanoyloxy) alkanoic acid (HAA), catalyzed by RhlA enzyme. The fatty acid moiety required for the synthesis deviates from the general fatty acid biosynthetic pathway at the level of the ketoacyl reduction with the help of RhlG enzyme (Campos-Garcia et al., 1998). The next reaction is the formation of monorhamnolipids from dTDP-L-rhamnose and an HAA, catalyzed by RhlB (rhamnosyltransferase1). Finally, dirhamnolipids are formed by the addition of dTDP-L-rhamnose to the monorhamnolipids, catalyzed by RhlC (rhamnosyltransferase2). dTDP-L-rhamnose is formed from dTDP-D-glucose, and the conversion takes place upon the action of enzymes encoded by *rmlA*, *rmlB*, *rmlC*, and *rmlD*, which form the *rmlBCAD* operon (Rahim et al., 2000).

The production of rhamnolipids in *P. aeruginosa* is regulated by *rhl* quorum sensing system at the transcriptional level (Soberón-Chávez et al., 2005). Different genes that encode enzymes involved in the rhamnolipid biosynthesis are *rhlA*, *rhlB*, *rhlR*, and *rhlI*. The genes *rhlA and rhlB* are arranged as an operon and encode RhlB rhamnosyltransferase, while *rhlR* and *rhlI* act as regulators of the *rhlAB* expression. The *rhlC* gene that codes for RhlC rhamnosyltransferase is not linked to other *rhl* genes and forms an operon with a gene whose function is not known (Rahim et al., 2001). Another system of genes called *las* system regulates the *rhl* system and in turn regulates rhamnolipid synthesis. The quorum sensing response depends on certain autoinducers, which are responsible for the activation of *rhlAB* transcription, synthesized by *rhlI and lasI* (Pesci et al., 1997). The quorum sensing response is primarily expressed under nutrient-limited conditions (Maier and Soberón-Chávez, 2000).

SOPHOROLIPIDS

Sophorolipid production is most optimal when both a hydrophobic and a hydrophilic carbon source are supplied in the production medium (Van Bogaert et al., 2011). Synthesis of sophorolipids requires glucose and a common fatty acid, both of which can be supplied as such in the medium. The fatty acid constituents can also be synthesized de novo from acetate or by stepwise oxidation of alkanes in the growth medium (Van Bogaert et al., 2008). The first step in the biosynthesis is hydroxylation of the fatty acid to a terminal (ω) or subterminal (ω-1) hydroxy fatty acid. Next, nucleotide-activated glucose (UDP-glucose) is linked to the hydroxyl group of the fatty acid upon action of glycosyltransferase I. Subsequently, another UDP-glucose is coupled to the first glucose moiety by glycosyltransferase II to get acidic nonacetylated sophorolipid (Saerens et al., 2011). It could either be retained in the native sophorolipid mixture or further modified by both internal esterification (lactonization) and acetylation of the carbohydrate head. Lactonization occurs by an esterification reaction between the carboxyl group of fatty acid and hydroxyl group of sugar moiety while acetylation is carried out by acetyl-CoA-dependent acetyltransferase enzyme (Sen et al., 2011). Sophorolipids are often formed as mixtures, which differ in the degree of acetylation of the sugar moiety and fatty acid saturation and lactonization. The gene regulation of sophorolipid synthesis is being studied extensively.

UDP-glucosyltransferase genes, *UGTA1* and *UGTB1*, have been identified and confirmed to play an important role in glucosylation steps. However, other aspects of gene regulation have to be explored (Saerens et al., 2011).

FUTURE PERSPECTIVES

- The commercialization of microbial surfactants is limited because of high production cost and low yields. Since the raw material cost constitutes 30% of the total production cost, use of low-cost substrates can make the biosurfactant production economically viable. Also optimal growth and production conditions with cheap and efficient downstream processing methods are essential (Mukherjee et al., 2006).
- Use of recombinant and mutant hyperproducing microbial strains can improve the biosurfactant yield. The genes responsible for the biosurfactant synthesis have to be extensively studied, so that the production technologies can be controlled and the product yields can be improved (Das et al., 2008b).
- Use of various statistical and mathematical optimization tools can improve the yield and reduce the cost of biosurfactant production (Sen, 2007; Sen and Swaminathan, 1997; Sivapathasekaran et al., 2010).
- Even though a variety of microorganisms have been found to produce different types of biosurfactants, *Bacillus, Pseudomonas*, and *Candida* sp. are primarily focused. Since many other genera are known to produce biosurfactants, they have to be closely examined (Mukherjee et al., 2006).
- Biosurfactants with new structures and new properties have to be developed through the application of molecular biotechnology. The structural modification of biosurfactants would probably widen the potential use of these compounds.
- Biosurfactants from extremophiles and hyperextremophiles should be given more attention, which are more suitable to use in oil fields (Perfumo et al., 2010).

REFERENCES

Amaral, P.F.F., da Silva, J.M., Lehocky, M. et al. 2006. Production and characterization of a bioemulsifier from *Yarrowia lipolytica*. *Process Biochemistry*, 41:1894–1898.
Amézcua-Vega, C., Poggi-Varaldo, H.M., Esparza-García, F., Ríos-Leal, E., and Rodríguez-Vázquez, R. 2007. Effect of culture conditions on fatty acids composition of a biosurfactant produced by *Candida ingens* and changes of surface tension of culture media. *Bioresource Technology*, 98(1):237–40.
Anderson, R.J. and Newman, M.S. 1933. The chemistry of the lipids of tubercle *Bacilli*: xxxiii. Isolation of trehalose from the acetone-soluble fat of the human tubercle *Bacillus*. *The Journal of Biological Chemistry*, 101:499–504.
Arima, K., Kakinuma, A., and Tamura, G. 1968. Surfactin, a crystalline peptide lipid surfactant produced by *Bacillus subtilis*: Isolation, characterization and its inhibition of fibrin clot formation. *Biochemical and Biophysical Research Communications*, 31(3):488–494.
Banat, I.M., Makkar, R.S., and Cameotra, S.S. 2000. Potential commercial applications of microbial surfactants. *Applied Microbiology and Biotechnology*, 53:495–508.
Beebe, J.L. and Umbreit, W.W. 1971. Extracellular lipid of *Thiobacillus thiooxidans*. *Journal of Bacteriology*, 108(1):612–614.

Benincasa, M., Abalos, A., Oliveira, I., and Manresa, A. 2004. Chemical structure, surface properties and biological activities of the biosurfactant produced by *Pseudomonas aeruginosa* LBI from soapstock. *Antonie van Leeuwenhoek*, 85:1–8.

Bernheimer, A.W. and Avigad, L.S. 1970. Nature and properties of a cytolytic agent produced by *Bacillus subtilis*. *Journal of General Microbiology*, 61:361–369.

Besson, F., Chevanet, C., and Michel, G. 1987. Influence of the culture medium on the production of iturin A by *Bacillus subtilis*. *Journal of General Microbiology*, 133(3):767–772.

Bodour, A.A., Guerrero-Barajas, C., Jiorle, B.V. et al. 2004. Structure and characterization of flavolipids, a novel class of biosurfactants produced by *Flavobacterium* sp. strain MTN11. *Applied and Environmental Microbiology*, 70:114–120.

Boothroyd, B., Thorn, J.A., and Haskins, R.H. 1956. Biochemistry of the ustilaginales. XII. Characterization of extracellular glycolipids produced by *Ustilago* sp. *Canadian Journal of Biochemistry and Physiology*, 34:10–14.

Burger, M.M., Glaser, L., and Burton, R.M. 1963. The enzymatic synthesis of a rhamnose-containing glycolipid by extracts of *Pseudomonas aeruginosa*. *The Journal of Biological Chemistry*, 238(8):2595–2602.

Campos-García, J., Caro, A.D., Nájera, R., Miller-Maier, R.M., Al-Tahhan, R.A., and Soberón-Chávez, G. 1998. The *Pseudomonas aeruginosa* rhlG gene encodes an NADPH-dependent beta-ketoacyl reductase which is specifically involved in rhamno-lipid synthesis. *Journal of Bacteriology*, 180(17):4442–51.

Cavalero, D.A. and Cooper, D.G. 2003. The effect of medium composition on the structure and physical state of sophorolipids produced by *Candida bombicola* ATCC 22214. *Journal of Biotechnology*, 103(1):31–41.

Chen, J., Huang, P.T., Zhang, K.Y., and Ding, F.R. 2012. Isolation of biosurfactant producers, optimization and properties of biosurfactant produced by *Acinetobacter* sp. from petroleum contaminated soil. *Journal of Applied Microbiology*, 112:660–671.

Cirigliano, M.C. and Carman, G.M. 1985. Purification and characterization of liposan, a bioemulsifier from *Candida lipolytica*. *Applied and Environmental Microbiology*, 50(4):846–850.

Cooper, D.G. and Goldenberg, B.G. 1987. Surface-active agents from two *Bacillus* species. *Applied and Environmental Microbiology*, 53(2):224–229.

Cooper, D.G., Liss, S.N., Longay, R., and Zajic, J.E. 1989. Surface activities of *Mycobacterium* and *Pseudomonas*. *Journal of Fermentation Technology*, 59:97–101.

Cooper, D.G., Macdonald, C.R., Duff, S.J.B., and Kosaric, N. 1981. Enhanced production of surfactin from *Bacillus subtilis* by continuous product removal and metal cation additions. *Applied and Environmental Microbiology*, 42:408–412.

Cooper, D.G. and Paddock, D.A. 1983. *Torulopsis petrophilum* and surface activity. *Applied and Environmental Microbiology*, 46:1426–1429.

Cooper, D.G., Zajic, J.E., and Gerson, D.F. 1978. Production of surface active lipids by *Corynebacterium lepus*. *Applied and Environmental Microbiology*, 37:4–10.

Cosby, W.M., Vollenbroich, D., Lee, O.H., and Zuber, P. 1998. Altered srf expression in *Bacillus subtilis* resulting from changes in culture PH is dependent on the SpoOK oligopeptide permease and the ComQX system of extracellular control. *Journal of Bacteriology*, 180(6):1438–1445.

Cutler, A.J. and Light, R.J. 1979. Regulation of hydroxydocosanoic and sophoroside production in *Candida bogoriensis* by the level of glucose and yeast extract in the growth medium. *The Journal of Biological Chemistry*, 254:1944–1950.

Das, P., Mukherjee, S., and Sen, R. 2008a. Genetic regulations of the biosynthesis of microbial surfactants: An overview. *Biotechnology and Genetic Engineering Reviews*, 25:165–186.

Das, P., Mukherjee, S., and Sen, R. 2008b. Antimicrobial potential of a lipopeptide biosurfactant derived from a marine *Bacillus circulans*. *Journal of Applied Microbiology*, 104(6):1675–1684.

Daverey, A. and Pakshirajan, K. 2010. Sophorolipids from *Candida bombicola* using mixed hydrophilic substrates: Production, purification and characterization. *Colloids and Surfaces B: Biointerfaces*, 79:246–253.

Dehghan-Noudeh, G., Housaindokht, M., and Bazzaz, B.S.F. 2005. Isolation, characterization, and investigation of surface and hemolytic activities of a lipopeptide biosurfactant produced by *Bacillus subtilis* ATCC 6633. *The Journal of Microbiology*, 43(3):272–276.

Desai, J.D. and Banat, I.M. 1997. Microbial production of surfactants and their commercial potential. *Microbiology and Molecular Biology Reviews*, 61(1):47–64.

Edwards, K.R., Lepo, J.E., and Lewis, M.A. 2003. Toxicity comparison of biosurfactants and synthetic surfactants used in oil spill remediation to two estuarine species. *Marine Pollution Bulletin*, 46:1309–1316.

Fattom, A. and Shilo, M. 1985. Production of emulcyan by *Phormidium* J-1: Its activity and function. *FEMS Microbiology Ecology*, 31:3–9.

Franzetti, A., Gandolfi, I., Bestetti, G., Smyth, T.J.P., and Banat, I.M. 2010. Production and applications of trehalose lipid biosurfactants. *European Journal of Lipid Science and Technology*, 112(6):617–627.

Gong, G., Zheng, Z., Chen, H. et al. 2009. Enhanced production of surfactin by *Bacillus subtilis* E8 mutant obtained by ion beam implantation. *Food Technology and Biotechnology*, 47(1):27–31.

Gunther, N.W., Nuñez, A., Fett, W., and Solaiman, D.K.Y. 2005. Production of rhamnolipids by *Pseudomonas chlororaphis*, a nonpathogenic bacterium. *Applied and Environmental Microbiology*, 71(5):2288–2293.

Gutiérrez, T., Mulloy, B., Black, K., and Green, D.H. 2007. Glycoprotein emulsifiers from two marine *Halomonas* species: Chemical and physical characterization. *Journal of Applied Microbiology*, 103(5):1716–1727.

Herman, D.C. and Maier, R.M. 2002. Biosynthesis and applications of glycolipid and lipopeptide biosurfactants. In *Lipid Biotechnology*, eds. Gardner, H.W. and Kuo, T.M. pp. 629–654. New York: Marcel Dekker.

Hewald, S., Josephs, K., and Bölker, M. 2005. Genetic analysis of biosurfactant production in *Ustilago maydis*. *Applied and Environmental Microbiology*, 71(6):3033–3040.

Hirata, Y., Ryu, M., Oda, Y. et al. 2009. Novel characteristics of sophorolipids, yeast glycolipid biosurfactants, as biodegradable low-foaming surfactants. *Journal of Bioscience and Bioengineering*, 108(2):142–146.

Hisatsuka, K., Nakahara, T., Sano, N., and Yamada. K. 1971. Formation of rhamnolipid by *Pseudomonas aeruginosa* and its function in hydrocarbon fermentation. *Agricultural Biology and Chemistry*, 35:686–692.

Hommel, R.K. and Ratledge, C. 1993. Biosynthetic mechanisms of low molecular weight surfactants and their precursor molecules. In *Biosurfactants: Production, Properties, Applications*, ed. Kosaric, N. pp. 3–63. New York: Marcel Dekker.

Hommel, R., Stiiwer, O., Stuber, W., Haferburg, D. and Kleber H.P. 1987. Production of water-soluble surface-active exolipids by *Torulopsis apicola*. *Applied Microbiology and Biotechnology*, 26(3):199–205.

Horowitz, S., Gilbert, J.N., and Griffin, W.M. 1990. Isolation and characterization of a surfactant produced by *Bacillus licheniformis* 86. *Journal of Industrial Microbiology*, 6:243–248.

Husain, D.R., Goutx, M., Acquaviva, M., Gilewicz, M., and Bertrand. J.C. 1997. The effect of temperature on eicosane substrate uptake modes by a marine bacterium *Pseudomonas nautica* strain 617: Relationship with the biochemical content of cells and supernatants. *World Journal of Microbiology and Biotechnology*, 13:587–590.

Hwang, Y., Kim, M., Song, I. et al. 2009. Subacute (28 day) toxicity of surfactin C, a lipopeptide produced by *Bacillus subtilis*, in rats. *Journal of Health Science*, 55(3):351–355.

Jarvis, F.G. and Johnson, M.J. 1949. A glycolipid produced by *Pseudomonas aeruginosa*. *Journal of the American Chemical Society*, 71:4124–4126.

Jenneman, G.E., McInerney, M.J., Knapp, R.M. et al. 1983. A halotolerant, biosurfactant-producing *Bacillus* species potentially useful for enhanced oil recovery. *Developments in Industrial Microbiology*, 24:485–492.

Kappeli, O. and Finnerty, W.R. 1979. Partition of alkane by an extracellular vesicle derived from hexadecane-grown *Acinetobacter*. *Journal of Bacteriology*, 140(2):707–712.

Kappeli, O., Walther, P., Mueller, M., and Fiechter, A. 1984. Structure of the cell surface of the yeast *Candida tropicalis* and its relation to hydrocarbon transport. *Archives of Microbiology*, 138:279–282.

Kim, H.S., Jeon, J.W., Kim, S.B., Oh, H.M., Kwon, T.J., and Yoon, B.D. 2002. Surface and physico-chemical properties of a glycolipid biosurfactant, mannosylerythritol lipid, from *Candida antarctica*. *Biotechnology Letters*, 24:1637–1641.

Kim, P., Oh, D., Kim, S., and Kim, J. 1997. Relationship between emulsifying activity and carbohydrate backbone structure of emulsan from *Acinetobacter calcoaceticus* RAG-1. *Biotechnology Letters*, 19(5):457–459.

Kitamoto, D., Fuzishiro, T., Yanagishita, H., Nakane, T., and Nakahara, T. 1992. Production of mannosylerythritol lipids as biosurfactants by resting cells of *Candida antarctica*. *Biotechnology Letters*, 14:305–310.

Kokare, C.R., Kadam, S.S., Mahadik, K.R., and Chopade, B.A. 2007. Studies on bioemulsifier production from marine *Streptomyces* sp. S1. *Indian Journal of Biotechnology*, 6:78–84.

Koronelli, T.V., Komarova, T.I., and Denisov, V. 1983. Chemical composition and role of *Pseudomonas aeruginosa* peptidoglycolipid in hydrocarbon assimilation. *Mikrobiologiia*, 52(5):767–770.

Kretschmer, A., Bock, H., and Wagner, F. 1982. Chemical and physical characterization of interfacial-active lipids from *Rhodococcus erythropolis* grown on n-alkanes. *Applied and Environmental Microbiology*, 44(4):864–870.

Kuiper, I., Lagendijk, E.L., Pickford, R. et al., 2004. Characterization of two *Pseudomonas putida* lipopeptide biosurfactants, putisolvin I and II, which inhibit biofilm formation and break down existing biofilms. *Molecular Microbiology*, 51(1):97–113.

Kumar, A.S., Mody, K., and Jha, B. 2007. Evaluation of biosurfactant/bioemulsifier production by a marine bacterium. *Bulletin of Environmental Contamination and Toxicology*, 79:617–621.

Kuyukina, M.S., Ivshina, I.B., Gein, S.V., Baeva, T.A., and Chereshnev, V.A. 2007. In vitro immunomodulating activity of biosurfactant glycolipid complex from *Rhodococcus ruber*. *Bulletin of Experimental Biology and Medicine*, 144(3):326–330.

Lang, S. and Philp, J.C. 1998. Surface-active lipids in *Rhodococci*. *Antonie van Leeuwenhoek*, 74:59–70.

Li, Z., Lang, S., Wagner, F., Witte, L., and Wray, V. 1984. Formation and identification of interfacial-active glycolipids from resting microbial cells. *Applied and Environmental Microbiology*, 48(3):610–617.

Lima, T.M.S., Procopio, L.C., Brandao, F.D., Carvalho, A.M.X., Totola, M.R., and Borges, A.C. 2011. Biodegradability of bacterial surfactants. *Biodegradation*, 22:585–592.

Macdonald, C.R., Cooper, D.G., and Zajic, J.E. 1981. Surface-active lipids from *Nocardia erythropolis* grown on hydrocarbons. *Applied and Environmental Microbiology*, 41(1):117–123.

Maier, R.M. and Soberón-Chávez, G. 2000. *Pseudomonas aeruginosa* rhamnolipids: Biosynthesis and potential applications. *Applied Microbiology and Biotechnology*, 54:625–633.

Marahiel, M.A., Danders, W., Krause, M., and Kleinkauf, H. 1979. Biological role of gramicidin S in spore functions. Studies on gramicidin-S-negative mutants of *Bacillus brevis* ATCC9999. *European Journal of Biochemistry*, 99(1):49–55.

Matsuyama, T., Kaneda, K., Nakagawa, Y., Isa, K., Hara-Hotta, H., and Yano, I. 1992. A novel extracellular cyclic lipopeptide which promotes flagellum-dependent and -independent spreading growth of *Serratia marcescens*. *Journal of Bacteriology*, 174(6):1769–1776.

Matsuyama, T., Murakami, T., Fujita, M., Fujita, S., and Yano, I. 1986. Extracellular vesicle formation and biosurfactant production by *Serratia marcescens*. *Journal of General Microbiology*, 132:865–875.

Mercaldi, M.P., Dams-Kozlowska, H., Panilaitis, B., Joyce, A.P., and Kaplan, D.L. 2008. Discovery of the dual polysaccharide composition of emulsan and the isolation of the emulsion stabilizing component. *Biomacromolecules*, 9:1988–1996.

Morikawa, M., Daido, H., Takao, T., Murata, S., Shimonishi, Y., and Imanaka, T. 1993. A new lipopeptide biosurfactant produced by *Arthrobacter* sp. Strain MIS38. *Journal of Bacteriology*, 175(20):6459–6466.

Mukherjee, S., Das, P., and Sen, R. 2006. Towards commercial production of microbial surfactants. *Trends in Biotechnology*, 24(11):509–515.

Nakano, M.M., Marahiel, M.A., and Zuber, P. 1988. Identification of a genetic locus required for biosynthesis of the lipopeptide antibiotic surfactin in *Bacillus subtilis*. *Journal of Bacteriology*, 170(2):5662–5668.

Navon-Venezia, S., Zosim, Z., Gottlieb, A., Legmann, R., Carmeli, S., Ron E.Z., and Rosenberg, E. 1995. Alasan, a new bioemulsifier from *Acinetobacter radioresistens*. *Applied and Environmental Microbiology*, 61(9):3240–3244.

Neu, T.R., Härtner, T., and Poralla, K. 1990. Surface active properties of viscosin: A peptidolipid antibiotic. *Applied Microbiology and Biotechnology*, 32:518–520.

Nitschke, M. and Costa, S.G.V.A.O. 2007. Biosurfactants in food industry. *Trends in Food Science & Technology*, 18:252–259.

Pei, X., Zhan, X., and Zhou, L. 2009. Effect of biosurfactant on the sorption of phenanthrene onto original and H_2O_2-treated soils. *Journal of Environmental Sciences*, 21(10):1378–1385.

Pekin, G., Vardar-Sukan, F., and Kosaric, N. 2005. Production of sophorolipids from *Candida bombicola* ATCC 22214 using Turkish corn oil and honey. *Engineering in Life Sciences*, 5(4):357–362.

Perfumo, A., Rancich, I., and Banat, I.M. 2010. Possibilities and challenges for biosurfactants uses in petroleum industry. In *Biosurfactants*, ed. Sen, R. pp. 135–145. Austin, TX: Landes Biosciences.

Persson, A., Oesterberg, E., and Dostalek. M. 1988. Biosurfactant production by *Pseudomonas fluorescens* 378: Growth and product characteristics. *Applied Microbiology and Biotechnology*, 29:1–4.

Pesci, E.C., Pearson, J.P., Seed, P.C., and Iglewski, B.H. 1997. Regulation of las and rhl quorum sensing in *Pseudomonas aeruginosa*. *Journal of Bacteriology*, 179:3127–3132.

Poremba, K., Gunkel, W., Lang S., and Wagner, F. 1991. Toxicity testing of synthetic and biogenic surfactants on marine microorganisms. *Environmental Toxicology and Water Quality*, 6:157–163.

Powalla, M., Lang, S., and Wray, V. 1989. Penta- and disaccharide lipid formation by *Nocardia corynebacteroides* grown on n-alkanes. *Applied Microbiology and Biotechnology*, 31:473–479.

Rahim, R., Burrows, L.L., Monteiro, M.A., Perry, M.B., and Lam, J.S. 2000. Involvement of the rml locus in core oligosaccharide and O polysaccharide assembly in *Pseudomonas aeruginosa*. *Microbiology*, 146:2803–2814.

Rahim, R., Ochsner, U.A., Olvera, C., Graninger, M., Messner, P., Lam, J.S., and Soberón-Chávez, G. 2001. Cloning and functional characterization of the *Pseudomonas aeruginosa* rhlC gene that encodes rhamnosyltransferase 2, an enzyme responsible for di-rhamnolipid biosynthesis. *Molecular Microbiology*, 40(3):708–718.

Rapp, P., Bock, H., Wray, V., and Wagner, F. 1979. Formation, isolation and characterization of trehalose dimycolates from *Rhodococcus erythropolis* grown on n-alkanes. *Journal of General Microbiology*, 115:491–503.

Rodrigues, L., Banat, I.M., Teixeira, J., and Oliveira, R. 2006. Biosurfactants: Potential applications in medicine. *Journal of Antimicrobial Chemotherapy*. 57:609–618.

Ron, E.Z. and Rosenberg, E. 2001. Natural roles of biosurfactants. *Environmental Microbiology*, 3(4):229–236.

Rooney, A.P., Price, N.P.J., Ray, K.J., and Kuo, T.M. 2009. Isolation and characterization of rhamnolipid-producing bacterial strains from a biodiesel facility. *FEMS Microbiology Letters*, 295:82–87.

Rosenberg, E., Rubinovitz, C., Gottlieb, A., Rosenhak, S., and Ron, E.Z. 1988. Production of biodispersan by *Acinetobacter calcoaceticus* A2. *Applied and Environmental Microbiology*, 54:317–322.

Rosenberg, E., Zuckerberg, A., Rubinovitz, C., and Gutnick, D.L. 1979. Emulsifier of *Arthrobacter* RAG-1: Isolation and emulsifying properties. *Applied and Environmental Microbiology*, 37(3):402–408.

Saerens, K.M., Roelants, S.L., Van Bogaert, I.N., and Soetaert, W. 2011. Identification of the UDP-glucosyltransferase gene UGTA1, responsible for the first glucosylation step in the sophorolipid biosynthetic pathway of *Candida bombicola* ATCC 22214. *FEMS Yeast Research*, 11(1):123–132.

Sarubbo, L.A., Marçal, M.C., Neves, M.L., Silva, M.P., Porto, A.L., and Campos-Takaki, G.M. 2001. Bioemulsifier production in batch culture using glucose as carbon source by *Candida lipolytica*. *Applied Biochemistry and Biotechnology*, 95:59–67.

Sen, R. 1997. Response surface optimization of the critical media components for the production of surfactin. *Journal of Chemical Technology and Biotechnology*, 68:263–270.

Sen, R. 2008. Biotechnology in petroleum recovery: The microbial EOR. *Progress in Energy and Combustion Science*, 34:714–724.

Sen, R. 2010. Surfactin: Biosynthesis, genetics and potential applications. In: *Biosurfactants*, ed. Sen, R. pp. 316–323. Austin, TX: Landes Biosciences.

Sen, R., Mudhoo, A., and Gunaseelan, D. 2011. Biosurfactants: Synthesis, properties and applications in environmental bioremediation. In *Bioremediation and Sustainability: Research and Applications*, eds. Mohee, R. and Mudhoo, A. pp. 137–211. Beverly, MA: Scrivener Publishing.

Sen, R. and Swaminathan, T. 1997. Application of response-surface methodology to evaluate the optimum environmental conditions for the enhanced production of surfactin. *Applied Microbiology and Biotechnology*, 47:358–363.

Shaligram, N.S. and Singhal, R.S. 2010. Surfactin—A review on biosynthesis, fermentation, purification and applications. *Food Technology and Biotechnology*, 48(2):119–134.

Shao, Z. 2011. Trehalolipids. In *Biosurfactants*, ed. Soberón-Chávez, G. pp. 121–143. Berlin, Germany: Springer.

Sivapathasekaran, C., Mukherjee, S., Ray, A., Gupta, A., and Sen, R., 2010. Artificial neural network modeling and genetic algorithm based medium optimization for the improved production of marine biosurfactant. *Bioresource Technology*, 101:2884–2887.

Sivapathasekaran, C., Mukherjee, S., Samanta, R., and Sen, R. 2009. High-performance liquid chromatography purification of biosurfactant isoforms produced by a marine bacterium. *Analytical and Bioanalytical Chemistry*, 395:845–854.

Soberón-Chávez, G., Lépine, F., and Déziel, E. 2005. Production of rhamnolipids by *Pseudomonas aeruginosa*. *Applied Microbiology and Biotechnology*, 68:718–725.

Solomon, J.M., Lazazzera, B.A., and Grossman, A.D. 1996. Purification and characterization of an extracellular peptide factor that affects two different developmental pathways in *Bacillus subtilis*. *Genes and Development*, 10:2014–2020.

Sørensen, D., Nielsen, T.H., Christophersen, C., Sørensen, J., and Gajhede, M. 2001. Cyclic lipoundecapeptide amphisin from *Pseudomonas* sp. strain DSS73. *Acta Crystallographica Section C*, 57:1123–1124.

Steller, S., Sokoll, A., Wilde, C., Bernhard, F., Franke, P., and Vater, J. 2004. Initiation of surfactin biosynthesis and the role of the SrfD-thioesterase protein. *Biochemistry*, 43:11331–11343.

Suzuki, T., Hayashi, K., Fujikawa, K., and Tsukamoto, K. 1965. The chemical structure of polymyxin E: The identities of polymyxin E_1 with colistin A and of polymyxin E_2 with colistin B. *The Journal of Biochemistry*, 57(2):226–227.

Van Bogaert, I.N.A., Develter, D., Soetaert, W., and Vandamme, E.J. 2008. Cerulenin inhibits de novo sophorolipid synthesis of *Candida bombicola*. *Biotechnology Letters*, 30(10):1829–1832.

Van Bogaert, I.N.A., Zhang, J., and Soetaert, W. 2011. Microbial synthesis of sophorolipids. *Process Biochemistry*, 46:821–833.

Vollbrecht, E., Heckmann, R., Wray, V., Nimtz, M., and Lang, S. 1998. Production and structure elucidation of di- and oligosaccharide lipids (biosurfactants) from *Tsukamurella* sp. nov. *Applied Microbiology and Biotechnology*, 50:530–537.

Volpon, L., Besson, F., and Lancelin, J. 2000. NMR structure of antibiotics plipastatins A and B from *Bacillus subtilis* inhibitors of phospholipase A_2. *FEBS Letters*, 485(1):76–80.

Xia, W., Dong, H., Yu, L., and Yu, D. 2011. Comparative study of biosurfactant produced by microorganisms isolated from formation water of petroleum reservoir. *Colloids and Surfaces A: Physicochemical and Engineering Aspects*, 392:124–130.

Yakimov, M.M., Timmis, K.N., Wray, V., and Fredrickson, H.L. 1995. Characterization of a new lipopeptide surfactant produced by thermotolerant and halotolerant subsurface *Bacillus licheniformis* BAS50. *Applied and Environmental Microbiology*, 61(5):1706–1713.

Zosim, Z., Gutnick, D.L., and Rosenberg, E. 1982. Properties of hydrocarbon-in-water emulsions stabilized by *Acinetobacter* RAG-1 emulsan. *Biotechnology and Bioengineering*, 24:281–292.

3 Rhamnolipids
Characteristics, Production, Applications, and Analysis

Fereshteh Arab and Catherine N. Mulligan

CONTENTS

INTRODUCTION

Many microorganisms, such as bacteria, fungi, and yeasts, growing on various substrates such as sugars, oil, alkenes, and wastes, synthesize a wide range of surface-active agents, called biosurfactants. These surface-active agents are amphiphilic compounds, with two opposing parts. Their hydrophilic part consists of mono-, di-, or polysaccharides, amino acids, cyclic peptides, phosphate groups, carboxylic acid, or alcohol. Their hydrophobic part is composed of long-chain fatty acids, hydroxy fatty acids or α-alkyl-β-hydroxy fatty acids, or fatty alcohols (Mulligan, 2009). These active agents reduce the interfacial tension, resulting in reducing the free energy of the system by replacing the bulk molecules of higher energy in the interface. Biosurfactants' amphiphilic nature, low toxicity, biodegradability, and excellent surface-active properties, along with their ecological acceptability and efficiency at extreme temperatures, pH, and salinity make them excellent candidates for a variety of applications. Some of these applications are in the fields of environmental engineering, pharmaceutics, cosmetics, detergents, and food industries (Mulligan and Gibbs, 2004).

The environmental applications and the ability of these surface-active agents to increase the solubility of the hydrophobic materials, and the fact that they are produced from renewable resources, are the main reasons that many researchers are interested in these environmentally friendly surface-active agents. They can be used in enhanced oil recovery, remediation of organic compounds from soil, and recovery of heavy metals from contaminated soil and mine tailings (Mulligan, 2005).

Typical desirable properties of biosurfactants include solubility enhancement, surface tension reduction, and low critical micelle concentration (CMC) (Mulligan, 2005). CMCs of biosurfactants vary between 1 and 200 mg/L and their molecular weight ranges from 500 to 1500 Da (Lang and Wagner, 1987; Mulligan, 2005).

The CMC of rhamnolipids depends on the source of the production and the type of bacteria used to produce rhamnolipids. For example, rhamnolipids produced by different strains of *Pseudomonas aeruginosa* display different CMC values. This confirms that the CMC is strain dependent. According to Mulligan (2009), some strains of *P. aeruginosa* were able to produce rhamnolipids that display CMCs varying from 10 to 230 mg/L and were able to decrease the surface tensions of water from 72.9 to 29 mN/m. The other factor affecting the CMC is the ratio of monorhamnolipids to dirhamnolipids in the solutions.

With respect to their chemical compositions, biosurfactants are divided into groups such as lipopeptides, glycolipids, fatty acids, polysaccharide protein complexes, peptides, phospholipids, and neutral lipids (Banat et al., 2010). They could also be classified according to their microbial origins, physicochemical properties, and mode of actions (Banat et al., 2010; Nie et al., 2010).

Altogether, in the practical sense, they are divided into two main classes: low-molecular-weight and high-molecular-weight biosurfactants (Banat et al., 2010; Franzetti et al., 2010). The low-molecular-weight biosurfactants are able to reduce the surface tension at the air/water or oil/water interface, while the high-molecular-weight biosurfactants act as a bioemulsifiers and are able to stabilize oil-in-water emulsions (Banat et al., 2010). Most of the biosurfactants are either anionic or neutral, while only a few are cationic.

Although biosurfactants have shown excellent potential in different fields, their high cost of production has caused them to not have the same commercial success as their synthetic rivals. Therefore, increasing the production of biosurfactants and decreasing the expenses related to production have been the focus of many investigations during the past decade. This objective could be achieved by choosing high-yield strains and inducing mutation in the strains in order to produce desired congeners and homologues. As well, this could be achieved by optimizing the condition of production, using low-cost substrates, developing economical methods for filtration, and finding methods for recycling them.

During the past decade, pioneers have done research and countless experiments on the production of different types of biosurfactants. There also have been numerous investigations on the efficiency of adding biosurfactants in enhancing the bioremediation of hydrocarbon- or heavy metal–contaminated water, soil, and mine tailings. Additionally, there have been some studies on the comparison of the effectiveness of bioremediation by using biosurfactants versus other conventional methods of remediation. Numerous studies revealed that rhamnolipid efficiency has surpassed the other biosurfactants. For that reason, rhamnolipids are known as one of the most promising and most investigated biosurfactants. Table 3.1 summarizes the characteristics of commercial rhamnolipids from Jeneil Company (2006).

Rhamnolipids, glycolipid biosurfactants, are mainly produced by *P. aeruginosa*. The first report about rhamnolipid isolation from *P. aeruginosa* was in 1949 *by* Jarvis and Johnson (Gunther et al., 2005). Bergstrom (1946), Hauser and Karnovsky (1954), Jarvis and Johnson (1949), and Edwards and Hayashi (1965) were known as the pioneers on the research on rhamnolipids (Abdel-Mawgoud et al., 2011). They took the first steps toward discovering the exact chemical nature of these biomolecules and their structural units. Due to rhamnolipids' significant tension-active and

TABLE 3.1
Properties of Rhamnolipids

Property	Description
Surface tension (mN/m)	30
pH of aqueous solution	6.5–7.5
Critical micelle concentration (mg/L)	25
Appearance	Dark reddish brown solution
Solubility	Soluble at neutral pH
Volatility	Not volatile
Specific gravity at 25°C	1.05–1.15
Molecular weight	504 (Rha1), 650 (Rha2)
Formula	$C_{26}H_{45}O_9$ (monorhamnolipids), $C_{32}H_{58}O_{13}$ (dirhamnolipids)

Source: Adapted from Jeneil Biosurfactants Co., 2006.

emulsifying properties, it has shown a vast potential in different fields (Lotfabad et al., 2010). There have been many investigations focused on this biosurfactant in order to decode its biosynthetic pathway (Abdel-Mawgoud et al., 2011). In recent years, with the advancement of accurate methods of analysis, there have been significant advances in understanding these biomaterials.

The research on rhamnolipids and their production resulted in developing a range of techniques for the isolation of few different rhamnolipid-producing bacterial strains, characterization of rhamnolipid homologues and congeners, and unraveling of the genetic details underlying rhamnolipid production in *P. aeruginosa* (Abdel-Mawgoud et al., 2010; Rezanka et al., 2011). *P. aeruginosa* is a common bacterium that is able to survive in a variety of environments. These bacteria are easy to be cultivated, and after relatively short incubation periods, they are able to produce relatively high yields of rhamnolipids (Nguyen and Sabatini, 2011).

RHAMNOLIPID-PRODUCING MICROORGANISMS

In recent years, there have been many studies to find new species of microorganisms that are able to produce rhamnolipid. By reviewing the articles, it has been noticed that most rhamnolipid-producing strains were found and isolated from the oil-contaminated soils around refineries. In those harsh environments, with high salinity and lack of other sources of carbon and nutrients, these microorganisms evolved to survive and to use hydrocarbons as a source of carbon. As well, they adapt to survive in both aerobic and anaerobic environments. One of the survival strategies of these microorganisms consists of producing rhamnolipids. These organic compounds help the bacteria to move easily toward the source of food or away from a place where the level of nutrients is lower and has a high level of toxins, as well as eliminating the competitors. Rhamnolipids solubilize hydrocarbons in order to be absorbed through the cell membrane of the bacteria. Furthermore, Kaczorek et al. (2012) reported that the rhamnolipid changes the cell surface of the bacteria and increases the hydrophobicity of the cells, to make them able to absorb hydrophobic compounds such as petroleum hydrocarbons. This process increases the biodegradation of these toxic compounds. Moreover, Kaczorek et al. (2012) reported that among all their chosen genera (*Bacillus, Pseudomonas, Aeromonas, Achromobacter,* and *Flavimonas*), *the* addition of rhamnolipids has affected the cell membrane of *Pseudomonas* genus than other genera. Their experiments proved that the *Pseudomonas* genus was the most effective genus in the bioremediation of diesel.

The genus *Pseudomonas* and its strains are known to be the first known producers of the rhamnolipids. There are many strains in this genus that are able to produce rhamnolipids: strains such as *P. aeruginosa, P. chlororaphis, P. alcaligenes, P. putida, P. chlororaphis, P. chlororaphis,* and *P. stutzeri* (Abdel-Mawgoud et al., 2011; Celik et al., 2008; Kaczorek et al., 2011; Lotfabad et al., 2010; Mulligan et al., 1989b; Gunther et al., 2005; Rahman et al., 2010a). Among the strains in this genus, *P. aeruginosa* is known as the main producer of the rhamnolipids.

The *Burkholderia* genus, which used to be known as a branch of *Pseudomonas*, has many similarities with *Pseudomonas* genus, as well as the ability to produce rhamnolipid. Abdel-Mawgoud et al. (2011) introduces some less known but nonpathogenic strains of *Burkholderia* such as *B. glumae, B. plantarii,* and *B. thailandensis* as rhamnolipid producers.

There are a number of reports of some other known genera that are able to synthesize rhamnolipids; a few of these genera are *Enterobacter* sp., *Pseudoxanthomonas* sp., *Pantoea* sp., *Nocardioides* sp., *Renibacterium* (strain *R. salmoninarum* 27BN), *Rhizobacteria,* and *Tetragenococcus* (strain *Tetragenococcus koreensis*) (Abdel-Mawgoud et al., 2011; Christova et al., 2004; Lee et al., 2005; Sastoque-Calal et al., 2010). Also there have been some reports on three strains of *thermophilic* bacteria that are able to produce rhamnolipids; these strains are *Thermus aquaticus* CCM 3488, *Thermus* sp. CCM 2842, and *Meiothermus ruber* CCM 4212 (Rezanka et al., 2011).

PSEUDOMONAS STRAINS

The genus *Pseudomonas* and its strains have been extensively studied due to their ability to produce the highest rate of rhamnolipids among the other rhamnolipid-producing genera; they are the first known producers of rhamnolipids, although their pathogenic characteristics motivate researchers to find other alternative rhamnolipid producers.

Pseudomonades are gram-negative bacteria and are rod-shaped with a polar flagellum. Most of the species in this genus have been known as opportunistic pathogens. As a facultative aerobe, these bacteria have their optimal metabolism in an aerobic condition, nonetheless, in an anaerobic condition; they are able to respire by using nitrate or other alternative electron acceptors. More than 202 species of *Pseudomonas* are known to this date (Tindall et al., 2006).

These opportunistic pathogenic bacteria could produce up to 60 homologues and congeners of rhamnolipids depending on the condition and media that they are grown in. They could produce a variety of different homologues due to different carbon sources.

Lotfabad et al. (2010) stated that their experiments revealed that chemical composition and the rate of the production of rhamnolipids are strain specific.

Some of the known *Pseudomonas* strains that have been introduced in the literature as rhamnolipid-producing strains are *P. aeruginosa, P. chlororaphis, P. alcaligenes, P. putida, P. chlororaphis, P. chlororaphis, P. stutzeri* (strain G11), and *P. stutzeri* (Abdel-Mawgoud et al., 2011; Celik et al., 2008; Janiyani et al., 1992; Kaczorek et al., 2011; Lotfabad et al., 2010; Mulligan and Gibbs, 1989; Rahman et al. 2010). Moreover, many researchers have acknowledged that among all the known rhamnolipid-producing strains, *P. aeruginosa* is the main producer of the rhamnolipids.

PSEUDOMONAS AERUGINOSA

P. aeruginosa, a gram-negative gammaproteobacterium, is an opportunistic pathogen widely distributed in soil, marshes, and coastal marine habitats, as well as on plant and animal tissues throughout the world. They are able to live in a wide variety of environments, in temperatures as low as 0°C and as high as 42°C, and in environments with high or low salinity. Moreover, they tolerate high concentrations of dyes, weak antiseptics, heavy metals, and many commonly used antibiotics. They are able to live in aerobic and anaerobic environments. In the absence of oxygen, they survive by using nitrate or other alternative electron acceptors (Hamamoto et al., 1994; Hassan and Fridovich, 1980; Nakahara et al., 1977; Stover et al., 2000; Todar, 2004).

According to Todar (2004), *P. aeruginosa* needs very few nutrients, and there are even reports that these bacteria live in distilled water. The metabolic versatility of *P. aeruginosa* makes it possible for these bacteria to grow by using a range of different organic compounds. There have been reports that they are able to grow up on 75 different organic compounds. These bacteria's optimal growth occurs in humid environments and at a temperature around 37°C (Todar, 2004).

P. aeruginosa are rod-shaped motile bacteria (Figure 3.1) measuring 0.5–0.8 μm by 1.5–3.0 μm. They are considered one of the most vigorous, fast-swimming bacteria

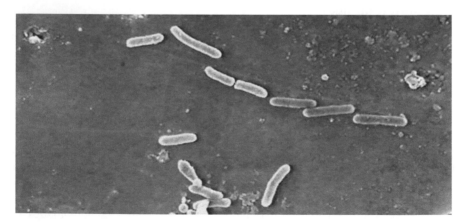

FIGURE 3.1 Scanning electron micrograph (SEM) of *Pseudomonas aeruginosa* bacteria (×5000). (From CDC Public Health Image Library (PHIL), 2011.)

(Todar, 2004). Colonies of different strains of *P. aeruginosa* display different colors that are very helpful in distinguishing the strains from each other. Pimienta et al. (1997) discussed different strains of *P. aeruginosa* and their ability to produce rhamnolipids with different surface activities. They also explained that depending on the strain being cultivated, various colors of colonies—from brownish green to dark green or light green—colonies with smooth borders or uneven borders, and flat colonies or convex ones can be observed.

P. aeruginosa are able to move from a depleted environment toward the environment with optimal conditions, as they are capable of swimming, swarming, and twitching. They swim in aqueous environments with the help of their flagella. The flagella and chemoreceptors help the bacteria to respond rapidly to changes in their environment. *P. aeruginosa* swarming depend on cell elongation, existence of a sufficient number of tendrils, and cell-to-cell contacts (Chrzanowski et al., 2011). Morris et al. (2011) noted that *P. aeruginosa* strains that were producing larger quantities of rhamnolipids were the ones that demonstrated the best swarming ability on solid surfaces.

The twitching movement of the bacteria takes place as a result of stretching and shrinking of their IV pili. Pili are considered to be an important part of the cell structure that not only help in the movement of the cell but also facilitate the attachment of the cell to abiotic surfaces. Over and above, pili have significant influence on the virulence activity of the *P. aeruginosa* cell. Chrzanowski et al. (2011) indicated that production of the pili by bacteria depends on the nitrogen content of the media. High levels of nitrogen in the environment limit the expression of genes that is responsible for the synthesis of pili, therefore limiting the motility of the cell. The nitrogen level in the environment also controls the production of rhamnolipids. Rhamnolipids ease the microbial movement from depleted environments toward nitrogen- and phosphorus-rich environments (Chrzanowski et al., 2011). According to Raya et al. (2010), rhamnolipids are able to decrease the microorganism attachment to the solid surfaces. Their experiments proved that rhamnolipids with concentrations as low as 13 mg/L were able to hinder the microorganism attachment significantly (Raya et al., 2010). Furthermore, rhamnolipids change the microbial cell surface properties that result in an increase in the contact area between the cells and carbon source (Chrzanowski et al., 2011), thus increasing the carbon uptake.

P. aeruginosa strains also produce three types of pigments: pyoverdin (fluorescein), a fluorescent yellow pigment, pyocyanin, a nonfluorescent blue pigment, and pyorubrin, a nonfluorescent red pigment (Propst and Lubin 1979; Todar, 2004). These pigments each have a role in the life and function of the bacteria. For example, pyocyanin works as a bactericidal agent and helps bacteria to compete with other bacteria in their environment; it also facilitates infections caused by *P. aeruginosa* in human and animals. Pyoverdin is generally produced when bacteria live in iron-deficient environments. It enables bacteria to leech iron from its surroundings (Hassan and Fridovich, 1980; Todar, 2004).

P. aeruginosa, depending on its growth phase, could synthesize rhamnolipids and/or polyhydroxyalkanoic acids (PHAs). The main constituents of both rhamnolipids and PHA are β-hydroxyalkanoic acids that are composed of 3-hydroxydecanoic acids linked by ester bonds (Ballot, 2009; Chayabutra et al., 2001a; Pham et al., 2004).

Chayabutra et al. (2001a) performed some research on the production of rhamno-lipids from *P. aeruginosa* (ATCC 10145); their experiments showed that this strain of *P. aeruginosa* was able to produce PHAs during their exponential growth, and then during the stationary phase, the bacterium was able to produce a considerable amount of rhamnolipids (Chayabutra et al., 2001a; Rahman et al., 2002).

The experiments of Chayabutra et al. (2001a) revealed that the rate of the pro-duction of β-hydroxyalkanoic acids in exponential phase is similar to its produc-tion in early stationary phase. In the stationary phase, bacteria stop the process of PHA polymerization and, instead, start to synthesize rhamnolipids. Then again, the experiments of Rahman et al. (2010) that consisted of cultivating *P. aeruginosa* DS10 in a microbioreactor showed that rhamnolipid production took place during the stationary phase of the bacterial growth, although many researchers indicated that the rate of the production of rhamnolipids and its composition are dependent on carbon source and production environment. For example, Lotfabad et al. (2010) reported that a high rate of the production of rhamnolipids was obtained by using lipid sources such as soybean oil as the medium for *P. aeruginosa* MR01. On the other hand, absence or existence of oxygen in the environment could dictate the type of the substrate that can be used. For instance, according to Chayabutra et al. (2001a), *P. aeruginosa* ATCC 10145 is able to break down and use hydrocarbons only under aerobic condition. Their investigations showed that the strain ATCC 10145 was able to produce five different rhamnolipid structures from using hydrocarbons as the sub-strate. HPLC–MS analysis revealed that the produced rhamnolipids were a mixture of two main groups with just β-hydroxydecanoic acids, and the other groups with a β-hydroxydecanoic acid linked to the sugar, as well as a β-hydroxydodecanoic acid or β-hydroxydodecanoic acid (Chayabutra et al., 2001a).

PSEUDOMONAS CHLORORAPHIS

P. chlororaphis are aerobic, mesophilic, rod-shaped, nonspore-forming, gram-negative bacteria belonging to the genus *Pseudomonas,* which are able to move by their polar flagella (Todar, 2004). Although they are aerobic heterotrophs, there are reports of their ability to live anaerobically through denitrification. The natural habi-tat of these bacteria is soil, where these bacteria, by secreting rhamnolipids and anti-biotic phenazine, compete with other strains of bacteria and fungi. *P. chlororaphis* are not pathogens and are not considered compatible with most indigenous animals and plants in the environment, but it can be successfully used against the fungi that attack the roots of crops (EPA, 2001; Rij et al., 2004). These bacteria have the ability to produce and secrete fluorescent pigments when they are in iron-depleted environ-ments (Meyer and Geoffroy, 2002). Tombolini et al. (1999) stated that antifungal properties of *P. chlororaphis* vary and is strain dependent. Therefore, by using the proper strain, a successful in situ fungicide process will take place.

Gunther et al. (2007) conducted experiments for producing rhamnolipids from *P. chlororaphis* strain NRRL B-30761 by using an aqueous medium. Using an aque-ous medium not only gives the bacteria chance of using fresh culture medium, it can also assist in maintaining the medium composition at its optimal composition. Also, the produced rhamnolipids and excess biomass could be removed continuously.

Gunther et al. (2005) have done some experiments on rhamnolipid production by *P. chlororaphis* using glucose as the sole carbon source for growth. Their experiments showed that the rhamnolipid production rate for *P. chlororaphis* was 1 g/L, slightly less than the production rate for *P. aeruginosa* (1.0–1.6 g/L). By addressing the fact that in rhamnolipid production, one of the hurdles was pathogenicity of the producers, it resulted in complicating the production and separation phases. Also it substantially increased the cost of the production. Thus nonpathogenicity of *P. chlororaphis* makes it compatible for many applications and the process of production more simplified. The analysis of the produced rhamnolipids showed that the rhamnolipids produced by *P. chlororaphis* are composed of only monorhamnolipids. This is an indicator of the absence of the rhlC gene, which is responsible for the biosynthesis of dirhamnolipids in these bacteria. The HPLC–MS analysis revealed that the predominant molecular structure in the produced rhamnolipids consisted of a monounsaturated hydroxy fatty acid of 12 carbons and one saturated hydroxy fatty acid of 10 carbons (Gunther et al., 2005).

Optimal growth for *P. chlororaphis* occurs at temperatures between 20°C and 28°C and pH of 6.3–7.5. Gunther et al. (2005) reported that the optimal temperature for the rhamnolipid production by *P. chlororaphis* was room temperature (20°C–23°C), and these bacteria could grow on a static mineral salt media containing 2% glucose. These are considered other advantages over rhamnolipid production by *P. aeruginosa*, which needs rapid mechanical agitation at 37°C for optimal production. The optimal incubation time of 120 h has been reported. Also it was observed that the larger surface area of the medium compared to its volume resulted in the higher production rate. In summary, the ease of culturing, nonpathogenicity, and economical feasibility of producing rhamnolipids make *P. chlororaphis* more desirable for use in many applications and for the commercial production of rhamnolipid (Gunther et al., 2006).

PSEUDOMONAS PUTIDA

P. putida is one of the strains in the *Pseudomonas* genus. These opportunistic, mesophilic, aerobic, gram-negative bacteria can be found in a variety of environments. They are fluorescent rod-shaped, nonspore-forming, as well as chemoheterotrophic. These motile bacteria move with their one or more flagella, swimming near the surface of water, and are considered indigenous to soil and freshwater environments (Davis et al., 2011; Martinez-Garcia and de Lorenzo, 2011).

P. putida are able to break down the aliphatic and aromatic hydrocarbons, as well as many organic toxins. *P. putida* has shown the potential to be used for cleaning the polluted sites from hydrocarbons and toxins (Timmis, 2002). According to Singh and Walker (2006), *P. putida* can break down and metabolize a number of organophosphorus compounds. Although *P. putida* is a pathogenic strain, some scientists believe that because the upper growth temperature of these bacteria is 35°C, it can't survive inside the body of most mammals and human. Also, there is no evidence that *P. putida* cause disease in plants; indeed, there are reports about the ability of *P. putida* to grow inside the root plants and boost the immunity of plants against disease-causing microorganisms (Molina et al., 2000).

There are some reports that a single strain of *P. putida, P. putida* 21BN, is able to produce rhamnolipids (Tuleva et al., 2002). It has been observed that by changing the substrate, these bacteria are able to produce rhamnolipids with different composition and different surface activities. Tuleva et al. (2002) reported that in their experiments, production of rhamnolipid from *P. putida,* growing on the soluble substrates such as glucose or a less soluble substrate such as hexadecane, resulted in yielding 1.2 g/L of rhamnolipids. Further experiments showed that when hexadecane was used as the sole carbon source, the produced rhamnolipids were able to decrease the surface tension of the medium to 29 mN/m. Also, they were able to form stable and compact emulsions with high emulsifying activity of 69%.

Martinez-Toledo and Rodriguez-Vazquez (2011) have performed some experiments to enhance the production rate of rhamnolipids by *P. putida* CB-100. They found that using a high level of glucose and ammonium chloride increased the rate of the production to 18.2 and 1.0 g/L respectively. Their experiments were conducted for 5 days at 37°C and pH 7, while the container was continually shaken at a speed of 150 rpm. It was reported that by altering the medium to achieve the C:N ratio of 5:1 and to a C:P ratio of 10:1, after 46 h, 27 mg/L of rhamnolipids was obtained.

The experiments of Raza et al. (2007) consisted of generating mutations in the strain *P. putida* by inducing gamma rays. The mutated strain, *P. putida* 300-B, was able to produce rhamnolipids. This mutated strain could use a variety of substrates such as hydrocarbons, waste frying oils, vegetable oil refinery wastes, and molasses with different production rates. Experiments conducted by Raza et al. (2007) showed that soybean and corn waste frying oils as the carbon sources had the best outcomes followed by kerosene and paraffin oils. In their experiments, they obtained the highest yield (4.1 g/L) by using substrates made of soybean waste frying oils. The produced rhamnolipids could decrease the surface tension of separated supernatant to 31.2 mN/m and displayed the CMC of 91 mg/L.

Nelson et al. (2002) point out that although *P. putida* and *P. aeruginosa* are so similar and share at least 85% of their genetic coding, *P. putida* misses the main virulence factors on their genes. So it can be safely used for different applications. They stated that *P. putida* is highly capable to act as a safe host for cloning and accepting other genes from other strains.

Ochsner et al. (1995), in their experiments on the recombinant *P. putida* strain, KT2442, showed that this strain was able to reach the rate of 0.6 g/L for rhamnolipid production grown when cultured under nitrogen-limiting condition. Wittgens et al. (2011) also used *P. putida* KT2440 as a host strain for rhamnolipid production. When glucose is chosen as the sole carbon source, these bacteria show a very high growth rate. It has been indicated that *P. putida* is in the same genus and closely related to *P. aeruginosa* and, thus, carries the necessary pathways for the synthesis of rhamnolipid. Therefore, genetically modifying this strain is more profitable than other chosen host strains. On the other hand, the nonpathogenicity of this strain makes it desirable for replacing the traditional rhamnolipid producer, *P. aeruginosa.* Further experiments showed the resilience of *P. putida* in the presence of a high concentration of rhamnolipids (as high as 90 g/L). The results of the Wittgens et al. (2011) experiments showed that they achieved a production rate of 0.15 g rhamnolipids per gram of glucose. They reported that following the production, there was no need for more purification, as they didn't use any oil as the substrate and its carbon metabolism didn't produce any by-products. The produced rhamnolipids were mainly monorhamnolipids (C_{10}:C_{10}). These researchers also performed some metabolic optimization, including separating the biomass synthesis pathway from rhamnolipid synthesis pathway, eliminating polyhydroxyalkanoate (PHA) synthesis pathway, which is competing, and optimizing the pathway of synthesizing rhamnolipids from glucose. These metabolic modifications result in the use of the majority of substrates toward rhamnolipid synthesis.

PSEUDOMONAS FLUORESCENS

P. fluorescens are a group of gram-negative, obligate aerobic, rod-shaped bacteria that live in the soil and water and can be recognized by their green-yellow, water-soluble fluorescent pigments. They have an extremely versatile metabolism, and some of the strains in this group could survive in anaerobic environments by using nitrate as the electron acceptors. Most of the strains of fluorescent pseudomonads grow well on a media composed of nitrate, glucose, and inorganic salts. Their optimal temperature is 25°C–30°C (Rhodes, 1959).

Experiments conducted by Gunther et al. (2005) on *P. fluorescens* strains ATCC 17397, ATCC 17824, ATCC 17577, and ATCC 17816 showed no sign of rhamnolipid production. Bondarenko et al. (2010) conducted some experiments to study the resilience of the indigenous bacteria when they were exposed to rhamnolipids supplemented to a polluted area. Experiments showed that exposure to rhamnolipids, even in low concentration, was able to reduce the bioluminescence of the *P. fluorescens* strains. They reported that the growth rate of *P. fluorescens* was affected by concentrations as low as 0.96 mg/L for 30 min. Researchers such as Ochsner (2005) showed that the recombinant strain *P. fluorescens* ATCC 15453 growing on glycerol as the substrate was able to produce rhamnolipids, but the concentration of the produced rhamnolipids, which was mainly rhamnolipid type 2, was 0.15 g/L, 10% of what *P. aeruginosa* produces.

The experiments of Healy et al. (1996) included growth of the bacterium *P. fluorescens* NCIMB 11712 on virgin olive oil and demonstrated that production of a glycolipid in the form of a rhamnolipid took place. Furthermore, Abouseoud et al. (2008a,b) conducted some research on the production of rhamnolipids by *P. fluorescens* Migula 1895-DSMZ. They studied the effect of a variety of carbon and nitrogen sources and also different C:N ratios on bacterial cultures growing at constant temperature and pH. They reported that optimal productivity was obtained when using olive oil and ammonium nitrate as carbon and nitrogen sources and C:N ratio was 10. The produced rhamnolipids could decrease the surface tension of the supernatant to less than 32 mN/m, and it demonstrated emulsification of 65% in 36–48 h. Recovery of the rhamnolipids was done through the precipitation method by using acetone. Afterward, there were more essays to determine the quality of the recovered product. According to Abouseoud et al. (2008a,b), the recovered rhamnolipids not only showed good foaming and emulsifying and antimicrobial activities, but also were stable throughout its exposure to high temperatures, when it was exposed to 120°C for 15 min. Produced rhamnolipids also showed resilience to high salinity environment (up to 10% NaCl) and exposure to a wide range of pH. Following the investigations of Abouseoud et al. (2008a) on rhamnolipid production by *P. fluorescens* Migula 1895-DSMZ, using olive oil as the sole carbon and for 5 days at ambient temperature and at neutral pH showed that after 40 h, the values of surface tension of supernatant reached 30 mN/m with an emulsification of 67%, but after 72 h, its surface tension was 35 mN/m and emulsification reached to 62%. Furthermore, results from the investigations of Abouseoud et al. (2010) on the production of rhamnolipids by *P. fluorescens* Migula 1895-DSMZ growing on olive oil as the carbon source was aligned with results from the past experiments. In their experiments, the product displayed the CMC of 290 mg/L, and it was stable at higher pH and showed resilience toward high salinity.

PSEUDOMONAS ALCALIGENES

P. alcaligenes is a gram-negative, aerobic, motile bacterium that is found in soils. Oliveira et al. (2009) are the pioneers in research on the production of rhamnolipids by *P. alcaligenes* by using palm oil as the sole carbon source. For their experiments, they isolated the strain of *P. alcaligenes* polyepsilon-caprolactone (PCL) from crude oil–contaminated soil. They conducted some experiments by varying the palm oil concentration (0.5% and 5.5%, v/v), with initial pH of the media (6 and 8), as well as the agitation rate (80 and 220 rpm). Their experiment showed that change in the concentration of palm oil has the highest effect on the rhamnolipid production; on the other hand, agitation could slightly affect the oil biodegradation, and varying pH had no effect on the production. It has been shown that the rhamnolipid mixture produced by the *P. alcaligenes* strain consisted of four types of dirhamnolipids and two types of monorhamnolipids. The product had the CMC value of 30 mg/L and decreased the surface tension of substrate (palm oil) to 31 mN/m. They observed the emulsification of more than 55% for crude oil after 72 h. Their results revealed that the *P. alcaligenes* strain is a rhamnolipid producer. These bacteria can be used in in situ treatment of polluted soils and for many other environmental applications.

BURKHOLDERIA STRAINS

The Burkholderia genus used to be a branch of *Pseudomonas* genus and was classified as the same genus until 1992, before the analysis of their rRNA sequence showed that they belong to two different genera. Still they share many of the physiochemical characteristics such as the ability to produce rhamnolipids (Abdel-Mawgoud et al., 2011). Furthermore, Abdel-Mawgoud et al. (2011) introduced some less known but nonpathogenic strains of *Burkholderia* that are able to produce rhamnolipids, strains such as *B. glumae*, *B. plantarii*, and *B. thailandensis*.

Burkholderia pseudomallei

B. pseudomallei is a gram-negative, aerobic, pathogenic, motile, and nonspore-forming bacteria that can be found in contaminated soil, water, and fruits. This bacterium could live in nutrient-depleted and also highly acidic environment (Pathema, 2006). According to Gunther et al. (2006), this pathogenic strain is known as one of the early rhamnolipid producers. This strain is able to produce rhamnolipids containing fatty acyl chains with 14 carbons in length. Haussler et al. (1998) stated that *B. pseudomallei* is able to produce a heat-stable toxin and is resistant against alkaline and acidic conditions. They also reported about the production of dirhamnolipids (Rha–Rha–C_{14}–C_{14}) with a molecular mass of 762 Da.

Burkholderia glumae

Costa et al. (2011) conducted some experiments on rhamnolipid production by *B. glumae* AU6208, a gram-negative soil bacterium, on canola oil as the carbon source and urea as the nitrogen source. Their experiments verified that *B. glumae*, a pathogen member of *Burkholderia* genus, is able to produce substantial quantities of rhamnolipids. They reported that after 7 days, 1000.7 mg/L of rhamnolipids were produced with a CMC of 25–27 mg/L, and they were able to reduce the interfacial tension of hexadecane from 40 to 18 mN/m. The produced rhamnolipids also demonstrated exceptional emulsifying properties for long-chain alkanes. The rhamnolipids produced by *B. glumae* consisted of a variety of mono-and dirhamnolipid congeners, but the main portion of the mixture is made of dirhamnolipids (Rha–Rha–C_{14}–C_{14}). It has been reported that the side chains in these rhamnolipids varied between C_{12}–C_{12} and C_{16}–C_{16} and are much longer than the one produced by *P. aeruginosa*. Moreover, the produced rhamnolipids showed excellent surface-active properties.

Burkholderia plantarii

Hoermann et al. (2010) studied the rhamnolipid production by *B. plantarii* DSM 9509(T), a gram-negative soil bacterium. Their experiment showed that when they used glucose as the only source of carbon, 45.74 mg/L of rhamnolipids were obtained. Their analysis showed that the structure of produced rhamnolipids by *B. plantarii* DSM 9509(T) was different from the ones produced by *P. aeruginosa*. It was mainly made by rhamnolipid Rha(2)–C_{14}–C_{14}, and its CMC was between 15 and 20 mg/L, and it was able to lower surface tension of distilled water to 29.4 mN/m. Vatsa et al. (2010) stated that *B. plantarii* is able to produce heat-stable dirhamnolipids (Rha–Rha–C_{14}–C_{14}), which are structurally similar to the rhamnolipids produced by *P. aeruginosa* and are able to stimulate mononuclear cells in human to produce tumor necrosis factor.

Burkholderia thailandensis

Dubeau et al. (2009) conducted some experiments on the process of rhamnolipid production by noninfectious, gram-negative bacteria, *B. thailandensis*. It has been indicated that *B. thailandensis* carries the genes (rhlA, rhlB, and rhlC) responsible for the rhamnolipid synthesis (Dubeau et al., 2009; Toribio et al., 2010). Therefore, it is expected that these bacteria produce rhamnolipids using similar pathways as *P. aeruginosa*, and the produced rhamnolipid's structures would be similar, although it has been indicated that *B. thailandensis* had the ability to produce rhamnolipids with long side chains containing 3-hydroxy fatty acid moieties as opposed to the short-chain rhamnolipids produced by *P. aeruginosa*.

The rhamnolipids produced by *B. thailandensis* were mainly composed of dirhamnolipids, and the minor portion of the mixture is made of monorhamnolipids.

The CMC value of these rhamnolipids was 225 mg/L, and they demonstrated good surface-active properties and were able to decrease the surface tension of water to 42 mN/m.

ENTEROBACTER SP.

Jadhav et al. (2011) conducted some research on rhamnolipids produced by *Enterobacter* sp. MS16; their experiment has shown that the *Enterobacter* sp. MS16 carries 16S rRNA gene and is able to produce rhamnolipids with excellent surface-active properties on a variety of substrates such as groundnut oil cake, sunflower oil, and molasses. They stated that using sunflower oil cake as the carbon source resulted in the highest rhamnolipid production rate (1.5 g/L). The produced rhamnolipids were able to reduce the surface tension of substrates by 68%, and their emulsification ability rate depends on the hydrocarbon for emulsification. These rhamnolipids also demonstrated fungicide properties and were able to inhibit the fungal spore germination. Analysis of the structure of these biomolecules showed that they are made of hydroxyl fatty acids of chain lengths of C16 and C18 and sugars.

Darvishi et al. (2011) reported the production of rhamnolipid by *Enterobacter cloacae*. These bacteria are able to continue production even in high salinity and high temperature although the optimal environment for these microorganisms is the temperature of 40°C and the pH equal to 7.0, using a substrate supplemented with 1.0% (w/v) olive oil, 1.0% (w/v) sodium nitrate, and 1.39% (w/v) K_2HPO_4 also with continual agitation of 150 rpm. In that optimal environment, they obtained the highest rate of rhamnolipid production by *E. cloacae* (1.74 g/L). The produced rhamnolipids display extreme oil spreading properties and could spread the olive oil and other vegetable oil into comparatively vast area. They could reduce the surface tension of the medium (olive oil and ammonium sulfate) from 58.3 to 31.7 mN/m and oil/water interfacial tension from 16.9 to 0.65 mN/m, increasing the bioavailability of the oil and, therefore, rapid biodegradation of oil. These properties make these biosurfactants an ideal candidate for environmental applications and clean up the oil spills. Experiments of Darvishi et al. (2011) showed that applying a mixture of biosurfactants from two different strains was much effective in the biodegradation of petroleum hydrocarbons than using biosurfactants from a single strain.

RHAMNOLIPID PRODUCTION

Since 1946, when researchers reported the discovery of rhamnolipid (Jarvis and Johnson, 1949), many studies have been done to explore this valuable material. Technological advances during the past decade enabled researchers to have a closer look at the process of rhamnolipid production and enabled scientists to unravel some genetic details underlying this process and its pathways. These findings allowed scientists to increase the production of rhamnolipids by selecting the optimal environment and finding the economically feasible and renewable substrates to decrease the cost of production. Moreover, by initiating mutation in the cells, they can produce more favorable congeners and also find nonpathogenic microorganisms that could produce rhamnolipids or transfer the genes responsible for rhamnolipid production to other nonpathogenic microorganisms (Abdel-Mawgoud et al., 2011; Mulligan and Gibbs, 2004; Nguyen and Sabatini, 2011). Besides, the focus of other research has been to determine the potential of rhamnolipids for a variety of applications.

Some microorganisms evolved to survive in partially nutrition-depleted environments by the secretion of rhamnolipids, and some microorganisms using the coexistence with these bacteria to survive. There are two known mechanisms that describe how rhamnolipids improve the substrate uptake. It is being done by solubilizing the otherwise insoluble substrate, making them more usable for cells. The second mechanism is by introducing changes into the membrane of cells. Numerous investigations demonstrated that rhamnolipids make changes in microbial cell surface properties by increasing the contact area between the cells and carbon source thus increasing the carbon uptake (Chrzanowski et al., 2011). The mechanism of changes that rhamnolipids initiate on the membrane needs to be investigated further. Rhamnolipids also facilitate the bacterial movement from depleted environments toward nitrogen- and phosphorus-rich ones (Chrzanowski et al., 2011). These findings coincide with the findings of Mulligan and Gibbs (1989a) that the production of rhamnolipids increases in nitrogen- and phosphorus-depleted environments.

Rhamnolipid is a valuable glycolipid biosurfactant that is primarily produced extracellular by *P. aeruginosa* (Mulligan and Gibbs, 1993). There has been some research to find new sources of bacteria for rhamnolipid production, but it should be noted that the producing strains affect the quality of rhamnolipids and their efficiency in specific applications, as well as the effect they have on other microorganisms. For example, rhamnolipids produced by *P. aeruginosa* are able to increase nutrient uptake for specific groups of bacteria but have no effect on other unrelated species (Banat et al., 2010). At the same time, biosurfactants produced by *Rhodococcus erythropolis* are able to stimulate the uptake even in a few unrelated species (Banat et al., 2010).

By using the appropriate reactor, not only the rate of the productions can be increased but also the cost of the production can be decreased. The experiments of Rahman et al. (2010) consisted of cultivating *P. aeruginosa* DS10 in a microbioreactor that was made of polytetra fluoroethylene and compared the result with those from bacteria cultivated in the regular bioreactors. Their experiments, cultivating the bacteria in the microbioreactor, resulted in producing 106 μg/mL of rhamnolipids during the stationary phase of bacterial growth. The produced rhamnolipids were able to reduce the surface tension of distilled water from 72 to 27.9 mN/m and emulsify kerosene by 71.3%.

For cultivating the microorganisms that are aerobic, aeration is an important factor that affects the production rate. Speed of agitation is another factor that affects the production rate. Mulligan and Gibbs (2004) conducted some experiments on comparing the production of rhamnolipid in a continually stirred tank reactor (CSTR) with a sequencing batch reactor (SBR) and by using diesel-contaminated soil. They stated that the production of rhamnolipid occurred only in the SBR reactor, and no production of rhamnolipid was observed in the CSTR reactor. They also reported that the produced rhamnolipid had a CMC value of 70 mg/L and was able to emulsify the diesel. Although this emulsifying property is highly concentration dependent, it was observed that by increasing the concentration of biosurfactant, emulsification of the diesel increased. Mulligan and Gibbs (2004) reported that in a SBR, the diesel removal efficiency reached 96%, and for the CSTR, the removal was just 75%. At the end of the cycle, the biosurfactants were removed through biodegradation.

To achieve the maximum production, the operating conditions in the reactors must be kept optimal (Mulligan and Gibbs, 1993). The pH and temperature are two important factors that have to be monitored continually. Each strain of bacteria has their own preferences, and their optimum temperature and pH are different. They have different tolerances toward changing conditions, so one should consider all the information about the strains they chose for biosurfactant production. For example, for *P. aeruginosa*, the pH of about 6.2 and the temperature range between 33°C and 37°C are considered optimal pH and temperature, and at pH higher than 7.5, they stop producing rhamnolipids (Mulligan and Gibbs, 1993).

Mulligan and Gibbs (2004) reported that nutrient limitation (nitrogen, phosphorous, iron, and oxygen) affects the production rate. Maximum production occurs during the nongrowth period of the cells. When the bacteria reach the stationary phase, the growth rate slows down and the number of bacteria reaches the maximum possible and there will not be any substantial increase in the number of cells. This happens as a result of nutrient depletion and accumulation of by-products. This can be the reason that, in majority of investigations, it was observed that the rhamnolipid production took place during the stationary phase of bacterial growth.

It should be noted that the foaming capability of the rhamnolipids in the process of aerobic production could hinder the process by surrounding the colonies by the produced foams and limiting the access to oxygen for the bacteria (Chayabutra et al., 2001b). Chayabutra et al. (2001b) suggested that to eliminate foaming or oxygen limitation, fermentation can be done under anaerobic conditions and through the denitrification process. However, the rate of the rhamnolipid production will be much less compared to the aerobic production rate. It has been observed that the rate of anaerobic production is one-third of production rate in the aerobic condition. To increase the productivity in anaerobic conditions, Chayabutra et al. (2001b) recommended increasing the number of initial bacteria.

By applying a well-aerated system or continuously separating the foam from cultures, the problem of foaming or respiratory limitation can be prevented. Mulligan and Gibbs (2004) suggested that by using plug flow reactors and also controlling foaming and foam collection, production could be increased. Using aerated bioreactors such as airlift fermenters has been shown to increase the rate of the production.

P. aeruginosa is able to produce rhamnolipids by using a variety of carbon sources such as C_{11} and C_{12} alkanes, *n*-hexadecane, succinate, pyruvate, citrate, fructose, glycerol, olive oil, palmitic acid, stearic acid, oleic acid, linoleic acid, glycerol, vegetable oil, glucose, and mannitol (Chayabutra et al., 2001a; Christova et al., 2011; Lotfabad et al., 2010; Mulligan et al., 2001a). Although all of these carbon sources are able to support the growth of the bacteria and rhamnolipid production, the comparison of the results of the experiments by using different substrates shows a significant difference between the growth rates of bacteria grown on each substrate. And sometimes in order to enable bacteria to grow on some of the substrates, there is a need for some additives to help to break down the molecules of substrate and increase their absorption through the cell membrane. According to Chayabutra et al. (2001b), their experiments showed that for some substrates such as palmitic acid and stearic acid, there was a need to add rhamnolipids prior starting the fermentation to increase the growth rate of bacteria.

The cost of production could be reduced by using less expensive substrates such as carbohydrates, vegetable oils, and food wastes, and using less costly alternative for required minerals. For example, ammonium phosphate could be used instead of other sources of phosphate, which increases the production rate (Mulligan, 2005). Moreover, according to Mulligan and Gibbs (1989), the highest production rate of rhamnolipids was obtained when they used nitrate instead of ammonium as nitrogen source. Their experiments showed that adding a limited amount of ammonia (less than 5 nM) and a small amount of glutamic acid (170 μM) will increase the production of rhamnolipid (Chayabutra et al., 2001; Mulligan and Gibbs, 1989).

The rate of rhamnolipid production has a direct correlation with the composition of the medium, For example, it depends on the carbon source, concentrations of nitrogen, phosphate, and iron (Eswari et al., 2012). A number of reports show that the highest rate of the rhamnolipid production is obtained when the ratio of carbon to nitrogen in the medium is 3:1, where the carbon source was ethanol and the nitrogen source was soy flour and yeast extract. In some other reports, it has been pointed out that when waste frying oil has been used as substrate, using the ratio of 8 to 1 for carbon to nitrogen resulted in the highest rate of production (Mulligan, 2005).

Mulligan and Gibbs (1989) reported about the effect of nutrient limitation on rhamnolipid production. Their experiments showed that the limitation of some nutrient and minerals in the substrate triggered the overproduction of rhamnolipids. For instance, limiting the phosphate level, calcium, iron, potassium, magnesium, sodium, and other trace minerals in the substrate caused an increase in the production of rhamnolipids (Mulligan, 2005). The results from the experiments on limiting different nutrients in the substrate, in the process of culturing the bacteria, have shown that the limitation of phosphate has the greatest effect on the rhamnolipid production rate. It increases the production rate by four- to fivefold (Chayabutra et al., 2001). Chrzanowski et al. (2011) reported that these changes in production rates are the result of activating the rlhAB gene that is triggered with starvation signals. Altogether, the carbon and nitrogen sources, the phosphate, iron, calcium, iron, potassium, magnesium, sodium, and other trace mineral contents are considered critical factors that are able to dictate the rate of rhamnolipid production (Abdel-Mawgoud et al., 2011; Mulligan and Gibbs, 2004).

Mulligan and Gibbs (2004) pointed out that there is a correlation between glutamine production activity and rhamnolipid production. Likewise, adding glutamine during fermentation stops the production of rhamnolipids (Mulligan, 2005). It also has been documented that adding aluminum sulfate and trace metals gradually during the course of fermentation increases the production of rhamnolipids.

Another strategy for lowering the cost of rhamnolipid production is by using aqueous substrates. An aqueous medium not only gives the bacteria chance to access the fresh medium continuously, but can also assist in maintaining the medium composition at its optimal composition. Also, the produced rhamnolipids and excess biomass can be removed continuously.

One of the major costs in the rhamnolipid production is related to its recovery. By finding cost-effective methods of separation and purification, the cost could be decreased significantly. In literatures, many methods have been suggested for the recovery of these biosurfactants and separating it from biomass following its production (Mulligan and Gibbs, 1993; Salleh et al., 2011).

Some of the techniques that have been used to recover the rhamnolipids are centrifugation, diafiltration, adsorption on resins and tangential flow filtration, solvent extraction, and precipitation. The most common isolation techniques for rhamnolipids are precipitation by acidification and extraction using ethyl acetate (Banat et al., 2010; Mulligan et al., 1989a). Banat et al. (2010) indicated that trichloroethanoic acid/acetone, ethanol, and chloroform/methanol can be used for precipitation and therefore extraction and recovery of biosurfactants.

Rhamnolipids are considered acidic, and therefore they participate in the acidic solution and stay aqueous in basic environment. Therefore, rhamnolipids are easily extracted from the solution by acidification and then by using nonpolar solvents such as ethyl acetate (Abdel-Mawgoud et al., 2010).

According to Mulligan (2005), solvents such as chloroform–methanol, butanol, acetic acid, ether, pentane, ethyl acetate, or dichloromethane methanol have been used for extraction of rhamnolipids, and precipitation has been done by using one of the following: ammonium sulfate, ethanol, acidic acid, and acetone. Salleh et al. (2011) reported that in their experiments for recovering rhamnolipid produced by *P. aeruginosa* USM AR-2, the organic solvent extraction by using methanol, 1-butanol, and chloroform resulted in the highest recovery rate (89.7% w/v).

Using these solvents for extraction resulted in obtaining a purified rhamnolipid in the form of brown-colored powder. This product displayed high stability at high temperature (as high as 121°C). As a result, further purification process wouldn't be needed, and the product could be used in different industries, for instance, in bioremediation and the microbial-enhanced oil recovery. Desai and Banat (1997) and Banat et al. (2010) categorized the processes of extraction of rhamnolipids to diafiltration and participation, precipitation with acetone, ultrafiltration, centrifugation, crystallization, and adsorption. For the recovery of rhamnolipid, Salleh et al. (2011) used precipitation by ammonium sulfate; this method resulted in white particles, and the rate of recovery was low. Banat et al. (2010) suggested that for biosurfactants with high molecular weight, one of the suitable approaches is precipitation using ammonium sulfate and purification through dialysis. Christova et al. (2011) described the process of rhamnolipid purification through the silica gel columns.

For the recovery of rhamnolipids, Khoshdast et al. (2012) used acid precipitation by using hydrochloric acid (1N). They reported that the rate of recovery was about 9 g of a honey-colored paste of rhamnolipid per liter of the culture medium. Although rhamnolipid recovery by acid precipitation not only produces a lower percentage of rhamnolipids, it also results in a sticky, viscous, brown-colored paste of rhamnolipid. So for many applications, there is a need for further purification. Mulligan and Gibbs (1990) considered using an ultrafiltration membrane as an advantageous method for the recovery and purification of rhamnolipids. They elucidated that by using an ultrafiltration membrane, glucose, amino acids, phosphates, and other small molecular mass compounds will be separated from the product. The molecular mass cutoff of the membrane that was used for separating rhamnolipids was 50,000 Daltons, and it was able to retain 92% of the rhamnolipids. They also noted that these ultrafiltration membranes can be incorporated with the process of continuous biosurfactant production.

OPTIMIZATION OF THE RHAMNOLIPID BIOSYNTHESIS

Researchers are working intensively to determine the genetic details underlying the rhamnolipid production by *P. aeruginosa*. Despite the recent advances in the technology, researchers are in the first steps of solving the unknowns, and there are many questions to be solved in this field.

According to Abdel-Mawgoud et al. (2011), for rhamnolipids, the synthesis of the lipid part takes place through the pathway of fatty acid synthesis from carbon units; therefore, the type and the structure of the lipid part of the rhamnolipid do not depend on the type of carbon source being used. Hence, there wouldn't be any changes in the lipid part following the use of various carbon sources for the cultivation of bacteria.

The lipid part of the rhamnolipids is synthesized through fatty acid synthesis type (II), which is a series of disassociated synthesis systems, and each part of the reactions is controlled by their specific gene. This means that each step of the synthesis and each single reaction are catalyzed by separate proteins that are encoded by a distinct gene. This type of fatty acid synthesis is observed only in bacteria and plants. As it have been noted before, to decrease the cost and optimize the production of the rhamnolipids and enhancing the composition of the product, there have been many approaches, and in this chapter, a brief summary of some of the main methods will be discussed.

ISOLATION OF RHAMNOLIPID PRODUCERS

By increasing the knowledge about the properties of rhamnolipids, new applications for this valuable material are found. The efficiency of rhamnolipids in many applications in different areas, from environmental sector (for the bioremediation of soil and water or fungicide) to cosmetic and food industries, made this material desirable, and by increasing the demand for it, there is a need to find and isolate new high-yield rhamnolipid-producing microorganism and the ones that are able to use low-cost renewable or waste substrates (Mulligan and Gibbs, 2004). Numerous microorganisms have been found to be able to produce rhamnolipids, and these rhamnolipid-producing organisms can be found in a variety of environments. They can be found in both warm and cold climates, they can survive even the toughest environment such as polluted soils with high salinity, and they evolved to use a variety of carbon sources.

Rashedi et al. (2006) reported the isolation of 152 bacterial strains from the contaminated oils in the warm climate of Khuzestan, Iran. They used a hemolysis method for the initial identification of the biosurfactant-producing microorganisms. They reported that 55 of isolated strains showed hemolytic activity. Among those 55 strains, only 12 strains could produce biosurfactants with superior surface tension and emulsification activities. And between those 12 strains, two of them showed the highest rate of biosurfactant production by growing on paraffin and glycerol as the sole carbon source. In this section, a few common methods for isolating rhamno-lipid-producing bacteria are discussed. The advantage of these methods are that they are easy to apply and don't need complicated instruments. For more precise results, one could use more advanced methods, such as real-time PCR and DNA microarray-based assays.

CTAB Agar Test

Pinzon and Lu-Kwang (2009) described a method for the detection of rhamnolipids by culturing the microorganisms on the agar plates containing methylene blue and cetyl trimethylammonium bromide (CTAB). Also according to Abdel-Mawgoud et al. (2010), CTAB agar can be used for identifying rhamnolipid-producing strains. Rhamnolipid produced by microorganisms is able to react with bromide salt, so the agar around the colonies of rhamnolipid-producing microorganisms becomes dark blue. In this test, the microorganisms are cultured on a CTAB methylene blue agar plate. After 24 h if the strain was able to produce anionic biosurfactants (e.g., rham-nolipids), there will be a purple-blue haze with a sharp defined edge around the culture (Jadhav et al., 2011) (Figure 3.2).

Hemolytic Test

The hemolytic property of the rhamnolipids is another method for identifying and isolating the rhamnolipid-producing strains. McClure and Schiller (1992) indicated that the presence of rhamnolipids is responsible for the heat-stable hemolytic activ-ity of the released fluids of *P. aeruginosa*. In the hemolytic method, blood agar is being used for culturing the strains, and after incubating the colonies, a halo is observed around the colonies due to lysis of the cells potentially by biosurfac-tant production. Further investigations are performed to determine if the clearing halos are created by rhamnolipids and not by other hemolytic components (Abdel-Mawgoud et al., 2010; Anandaraj and Thivakaran, 2010; Mulligan et al., 1989c; Siegmund and Wagner 1991). The other approach is to separate the supernatant and then inject into the holes made in the blood agar Petri dishes that are then left for 24–48 h for the observation and measurement of the hemolysis zones (Rahman et al., 2010) (Figure 3.3).

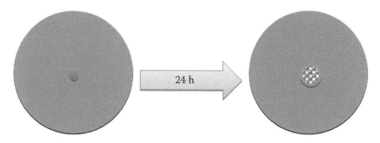

FIGURE 3.2   CTAB assay to identify the rhamnolipid-producing strains.

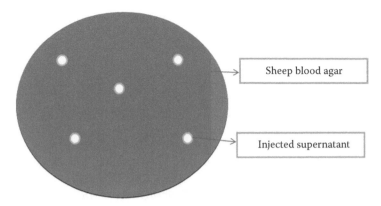

Sheep blood agar

Injected supernatant

FIGURE 3.3 Hemolytic test of rhamnolipids by injecting the supernatants separated from cells.

Although using hemolytic property of the rhamnolipids is a useful and easy method for identifying and isolating the rhamnolipid strains, it should be considered that the presence of other hemolytic substances could interfere with the results.

Isolation of the New Strains

Isolating the high-yielding strains or the strains that produce high-performing products is one of the approaches to optimize the rhamnolipid production. *Pseudomonas* strains have been known as the major producers of rhamnolipids, but the rate of the production varies between different strains of *Pseudomonas*. Also the types of rhamnolipid homologues and congeners that they produce and their ratios are strongly strain dependent. For instance, the absence of an rhlC gene homologue in *P. chlororaphis* makes this strain able to produce only RLs with one rhamnose unit and two hydroxyacyl moieties (Abdel-Mawgoud et al., 2011).

Research has proven that the strain *P. aeruginosa* BN10 is able to produce dir-hamnolipids at higher rates compared to other strains. Moreover, the experiments of Christova et al. (2011) revealed that *P. aeruginosa* is able to keep the high rate of production, growing on all types of selected carbon sources. They also noted that the rhamnolipids produced by *P. aeruginosa* BN10 demonstrated low CMC (40 mg/L) and were able to reduce the surface tension of pure water from 72 to 29 mN/m. Furthermore, Abdel-Mawgoud et al. (2011) noted that the rhamnolipids produced by *P. aeruginosa*, *P. glumae*, *P. plantarii*, *B. pseudomallei*, and *B. thailandensis* have a higher amount of dirhamnolipids in the mixture compared to the amount of monorhamnolipids.

One of the challenges of working with high-yield *P. aeruginosa* is its pathogenic characteristic. Research for finding new sources for rhamnolipids showed that there are many other known genera that are able to synthesize rhamnolipids; few of these genera are *Enterobacter* sp., *Pseudoxanthomonas* sp., *Pantoea* sp., and *Nocardioides* sp. (Abdel-Mawgoud et al., 2011).

According to Banat et al. (2010), the types of strains that produce the biosurfactants have an effect on the efficiency of the biosurfactants. For instance, rhamnolipids produced by most strains of *P. aeruginosa* can stimulate the degradation of n-hexadecane, and the biosurfactants produced by most strains of *Rhodococcus* have no affect on the degradation of n-hexadecane, but the biosurfactants produced by one of *Rhodococcus* strains, *R. erythropolis* strain 3C-9, have a great impact on the degradation rate of n-hexadecane (Banat et al., 2010). The main reason for this difference is that the type and composition of biosurfactants are strain dependent.

It also should be noted that the maximum rate of the rhamnolipid production for different strains occurs at different growth phases, and it could vary among the strains of a genera. For example, most reports indicate that most *P. aeruginosa* strains have the maximum rate of rhamnolipid production during their stationary phase of growth. On the other hand, Cha et al. (2008) reported that the maximum production rate of rhamnolipids by strain *P. aeruginosa* EMS1, which was isolated from activated sludge, occurred during the exponential phase of the bacteria growth.

Isolation of New Nonpathogenic Rhamnolipid Producers

Many researchers have tried to find and isolate other rhamnolipid-producing microorganisms. For example, Rooney (2009) introduced several strains of bacteria that were isolated from a biodiesel polluted area, which were *Acinetobacter calco-aceticus, E. asburiae, E. hormaechei, Pantoea stewartii,* and *P. aeruginosa.* Their experiments showed that all of the isolated strains were able to produce rhamnolip-ids. However, among the isolated strains, *E. hormaechei* and *A. calcoaceticus* were able to produce rhamnolipids at the same rate as *P. aeruginosa.* Their further inves-tigations by mass spectrometry (MS) analyses revealed that the produced rhamno-lipids were composed of both monorhamnolipids (Rha–C_{10}–C_{10}) and dirhamnolipids (Rha–Rha–C_{10}–C_{10}).

According to Costa et al. (2011), *B. glumae* AU6208, a nonpathogenic micro-organism, is able to produce significant amounts of rhamnolipids, a mixture of a variety of mono- and dirhamnolipid congeners. In comparison to the structure of the rhamnolipids produced by *P. aeruginosa* with the rhamnolipids produced by *B. glumae,* the latter biosurfactants have longer side chains and present excellent sur-face-active properties and have shown that they are able to emulsify the long-chain alkanes easily (Costa et al., 2011).

Cha et al. (2008) investigated the rhamnolipid production in a nonpathogenic bac-terium, *P. putida,* a gram-negative rod-shaped saprotrophic soil bacterium. Their observation showed that by introducing rhamnosyltransferase genes into the chro-mosomes of this bacterium, this mutant strain was able to produce rhamnolipids. Furthermore, Wittgens et al. (2011) also introduced the engineered *P. putida* (strain KT2440) as a producer for rhamnolipids. Their experiments showed that by using glucose as a carbon source, this strain has been able to demonstrate the highest rate of the rhamnolipid production in comparison to the other rhamnolipid producers. However, Martinez-Toledo and Rodriguez-Vazquez (2006) investigated the effect of different substrates and environmental conditions on the rhamnolipid production rate in *P. putida.* They concluded that by using a medium made with phenanthrene, hydrated ferrous sulfate, potassium phosphate, glucose, yeast extract, corn oil, and ammonium chloride, the maximum rhamnolipid production rate was obtained.

They also noted that the optimal production conditions were a temperature of 37°C and pH 7.0, with a continuous agitation speed of 150 rpm.

Pantazaki et al. (2010) introduced the strain *Thermus thermophilus* HB8 (ATCC 27634), a gram-negative bacterium, as a promising rhamnolipid producer, which by growing on insoluble substrates (for instance, sunflower seed oil or oleic acid) is able to yield rhamnolipids. In their analysis, it has been shown that produced rhamnolipids are a mixture of both monorhamnolipids and dirhamnolipids.

Christova et al. (2004) stated that the strain *R. salmoninarum* 27BN has the ability to produce rhamnolipid when it has been cultivated on n-hexadecane. *R. salmoninarum* is rod shaped, gram positive, and is identified as fish pathogenic bacterium. They isolated these bacteria from hydrocarbon-polluted wastewater in industrial areas. Their analysis showed that produced rhamnolipids are a mixture of mono- and dirhamnolipids. They also stated that the maximum production rate occurred during the stationary stage of bacterial growth. Therefore, the product had the characteristics of a secondary metabolite. As with other rhamnolipid producers, the limitation of some nutrients stimulates the rhamnolipid production by *R. salmoninarum*.

GENETICALLY ENGINEERED MICROORGANISMS FOR RHAMNOLIPID PRODUCTION

Some researchers attempted to create rhamnolipid-producing microorganisms by introducing the genes responsible for rhamnolipid production to the chromosomes of other microorganisms (Ochsner et al., 1995). Their experiments showed that by using some indigenous bacteria from contaminated areas and transferring the genes responsible for biosynthesis of the rhamnolipids into their chromosomes, in situ rhamnolipid production can be achieved and that will be a great achievement in the process of the bioremediation of contaminated areas. However, it will be difficult to get regulatory approval for use due to potential risk.

In order to find some new hosts to be successfully genetically modified and become a rhamnolipid producer, the potential host should go through screening and the ones that are resilient against rhamnolipids are selected. For this, Wittgens et al. (2011) has conducted some experiments to determine the minimum inhibition concentration (MIC) for some strains of bacteria, to select some highly resistant bacteria in the presence of rhamnolipids to genetically modify and turn them to rhamnolipid producers. In their experiments, the strains *E. coli, B. subtilis, C. glutamicum*, and *P. putida* were exposed to a concentration of 90 g/L of dirhamnolipid with a purity of 95%. They noticed that the inhibition effect of rhamnolipids on gram-positive strains was much greater than that for the gram-negative strains. For example, in the presence of less than 100 mg/L rhamnolipids, the growth rate of *C. glutamicum*, which is a gram-positive strain, was reduced by 60% in contrast to gram-positive *B. subtilis*, which demonstrated a different behavior in the presence of rhamnolipids. By secreting lipases, *B. subtilis* was able to break down the dirhamnolipid and separate its fatty acids from sugar molecules. This caused the prolonged lag phase and unaffected growth rate of *B. subtilis*.

Wittgens et al. (2011) pointed out that gram-negative species are more resilient to the rhamnolipid exposure, and even rhamnolipids at high concentration cannot inhibit the growth of bacteria completely. For example, *E. coli* was not inhibited in the presence of 90 g/L dirhamnolipid. At this concentration, the growth rate of the bacteria was slowed down by 50%.

Ochsner et al. (1995) studied and compared the rhamnolipid production of *P. fluorescens, P. putida, P. oleovorans*, and *E. coli*, which were genetically altered through introducing the genes responsible for rhamnolipid production into their chromosomes. These researchers observed that *P. putida* had the highest yield of rhamnolipids and *P. fluorescens* had the second highest production rate. Furthermore, the studies of other researchers such as Cha et al. (2008) involved transferring the genes responsible for synthesizing rhamnolipids from *P. aeruginosa* to *P. putida*. They reported that this recombinant strain was able to produce rhamnolipids. When soybean oil was used as the substrate, and the rate of production of rhamnolipids was up to 7.3 g/L.

Ochsner et al. (1995) noted that even though the enzyme rhamnosyltransferase was synthesized in the recombinant strains *E. coli* and *P. oleovorans,* these two strains didn't produce rhamnolipids. This latter observation contradicts the results from some other researchers who worked with genetically engineered *E. coli*.

Cabrera-Valladares et al. (2006) observed that by altering the DNA of *E. coli*, it could become a monorhamnolipid producer. Their experiments on the recombinant *E. coli* showed that although this metabolic engineered strain is able to produce rhamnolipids, the rhamnolipid production rate for these *E. coli* was much lower than the rate of the production in *P. aeruginosa*. However, they stated that the rate of the production in recombinant *E. coli* was more than the previous reports on *E. coli*. They noted that by optimization of the substrate and its availability, the rate of the production was increased substantially.

Cabrera-Valladares et al. (2006) reported that recombinant *E. coli* HB101 grown on a hydrophobic substrate (e.g., oleic acid) demonstrated a high production rate of rhamnolipids. They also stated that the recombinant strain of *E. coli* W3110, which expressed the *P. aeruginosa operons* rhlAB *and* rmlBDAC, growing on glucose as the carbon source, was able to produce more than twice the amount of rhamnolipids produced by the recombinant *E. coli HB101*.

Wang et al. (2007) conducted some experiments by transferring the RhlAB gene, the gene responsible for rhamnolipid production, into the chromosomes of two strains, the *P. aeruginosa* PAO1-rhlA and *Escherichia coli* BL21 (DE3), which were not able to produce rhamnolipids before. It was observed that following the integration of the gene into the chromosomes, both strains were able to produce rhamnolipids (Wang et al., 2007). In Table 3.2, some of the known producers of rhamnolipids are introduced.

USE OF LOW-COST SUBSTRATES

As it has been noted before, one of the important approaches to decrease the cost of the rhamnolipid production involves using low-cost substrates such as waste frying oil (Banat et al., 2010; Lotfabad et al., 2010; Mulligan and Gibbs, 2004; Wang et al., 2007; Zhu et al., 2007) and industrial wastes (Nitschke et al., 2005). Pornsunthorntawee et al. (2010) indicated that not only genetic regulatory governs the rhamnolipid production, but it was also affected by the main metabolic pathways such as fatty acid synthesis and activated sugars and enzymes.

TABLE 3.2
Some Common Rhamnolipid Producers

Strain	Producer Category	Author
P. aeruginosa ATCC 9027	Natural	Mulligan et al. (1989a)
P. aeruginosa PG201	Natural	Ochsner et al. (1995)
P. aeruginosa UW-1	Natural	Sim et al. (1997)
P. aeruginosa 47T2 NCIB	Natural	Haba et al. (2000)
P. aeruginosa PAO1	Natural	Jian et al. (2010)
P. aeruginosa LBI	Natural	Benincasa et al. (2002)
P. aeruginosa PAO1	Natural	Wang et al. (2007)
P. aeruginosa PEER02	Recombinant	Wang (2007)
P. aeruginosa zju.u1	Natural	Zhu (2007)
P. aeruginosa zju.u1M	UV mutagenesis	Zhu (2007)
P. aeruginosa PEER02	Recombinant	Wang et al. (2007)
P. aeruginosa EMS1	Natural	Cha et al. (2008)
P. aeruginosa USM AR-2	Natural	Salleh et al. (2011)
P. aeruginosa MSIC02	Natural	de Sousa et al. (2011)
P. aeruginosa DS10-129	Natural	Rahman et al. (2010)
P. aeruginosa PAO1	Natural	Müller et al. (2010)
P. aeruginosa MR01	Natural	Lotfabad et al. (2010)
P. aeruginosa BN10	Natural	Christova et al. (2011)
P. aeruginosa BS-161R	Natural	Kumar and Mamidyala (2011)
P. aeruginosa MIG-N146,	Mutant	Guo et al. (2009)
P. aeruginosa DAUPE 614	Natural	Monteiro et al. (2007)
P. putida KT2442	Recombinant	Ochsner et al. (1995)
P. putida 21BN	Natural	Tuleva et al. (2002)
P. putida CB-100	Recombinant	Martínez-Toledo and Rodriguez-Vazquez (2011)
P. putida KCTC 1067	Recombinant	Cha et al. (2008)
P. putida KT2440	Recombinant	Wittgens et al. (2011)
P. putida KT42C1	Recombinant	Wittgens et al. (2011)
P. fluorescens ATCC 15453	Recombinant	Ochsner et al. (1995)
P. fluorescens ATCC15453	Recombinant	Wittgens et al. (2011)
P. chlororaphis NRRL B-30761	Natural	Gunther et al. (2007)
P. oleovorans	Recombinant	Ochsner et al. (1995)
P. alcaligenes	Natural	Oliveira et al. (2009)
B. plantarii DSM 9509	Natural	Hoermann et al. (2010)
B. glumae AU6208	Natural	Costa et al. (2011)
B. glumae	Natural	Pajarron et al. (1993); Costa et al. (2011)
B. thailandensis	Natural	Dubeau et al. (2009)
Renibacterium salmoninarum 27BN	Natural	Christova et al. (2004)
Thermus thermophilus HB8 (ATCC 27634)	Natural	Pantazaki et al. (2010)
Tetragenococcus koreensis	Natural	Lee et al. (2005)
E. coli HB101 (pRK2013)	Recombinant	Ochsner et al. (1995)
E. coli TNERAB	Recombinant	Wang et al. (2007)
E. coli W3110	Recombinant	Cabrera-Valladares (2006)
E. coli HB101	Recombinant	Cabrera-Valladares (2006)

Rhamnolipid producers could use a variety of substrates for rhamnolipid production. By using low-cost substrates such as carbohydrates, vegetable oils, and industrial wastes, the cost of the production could be reduced. Fats and oil wastes from food industries are considered environmental pollutants; using these materials as a substrate for rhamnolipid production not only decreases the cost of rhamnolipid production, but also solves another environmental problem.

Costa et al. (2008) have done some experiments on rhamnolipid production by using a few different substrates made of soybean soapstock, chicken fat, hydrogenated vegetable fat, and soybean frying oil generated by food industry and restaurants. They reported that *Pseudomonas* LMI 6c and *Pseudomonas* LMI 7a were able to utilize the substrates mentioned earlier and produce a decent amount of rhamnolipids. In their experiments, soybean soapstock showed the best result, and the bacteria by using this substrate produced 9.69 g/L of rhamnolipids that were able to reduce the surface tension of the medium of soybean oil. Along the 18 strains being tested, two of the strains (6c and 7a) were able to lower the surface tension of the medium to 31 mN/m. Furthermore, Costa et al. (2009b) reported that the highest overall production rate for rhamnolipids and PHAs that they achieved was by using the waste frying oil as the carbon source; they reported that with this substrate, the production rate for rhamnolipid was 660 mg/L.

Costa et al. (2010) reported that in their experiments, the rhamnolipids produced by *P. aeruginosa* L2-1, by using a waste cooking oil as the substrate, were a mixture of 16 different rhamnolipid congeners in which monorhamnolipid (Rha–C_{10}–C_{10}) and dirhamnolipid (Rha–Rha–C_{10}–C_{10}) were the largest portion. The L2-1 rhamnolipids demonstrated a similar or, even in some cases, superior surface-active properties than commercial rhamnolipids produced by *P. aeruginosa* (JBR599—Jeneil Biosurfactant Co., Saukville, Wisconsin).

The experiments of Zhu et al. (2007) consisted of the comparison of rhamnolipid production rate by using the sludge from a refinery versus the sludge from a catering as substrate. It was shown that the substrate made with the sludge from catering resulted in the highest rate of the rhamnolipid production. In the experiments of Nitschke et al. (2005), it was observed that *P. aeruginosa* was able to grow and produce rhamnolipids using various waste oils as substrates. In their experiments, they compared the rhamnolipid production rate on the substrates composed of wastes from soybean, cottonseed, babassu, palm, corn oil, and waste obtained from refinery. Their experiments showed that the rate of the production was highest when the bacteria were cultured on the substrate made of the soybean soapstock waste. They reported that the rhamnolipids produced on the substrate made of soybean soapstock waste had the ability to lower the surface tension of the medium to 26.9 mN/m and demonstrated a low CMC of 51.5 mg/L. Their analysis also showed that the produced rhamnolipids are a mixture of 10 different homologues with monorhamnolipid (44%) and dirhamnolipid (29%) as the chief components in the mixture.

Gunther et al. (2005) pointed out that *P. aeruginosa* is able to grow and produce rhamnolipid by using a variety of different carbon sources. However, the highest levels of rhamnolipid production were obtained when they used vegetable-based oils such as soybean oil, olive oil, corn oil, and canola oil as the carbon sources.

On the other hand, Cha et al. (2008) reported that the maximum production rate of rhamnolipids by strain *P. aeruginosa* EMS1 was achieved by using acidified soybean oil as the carbon source.

Similarly, Lotfabad et al. (2010) stated that *P. aeruginosa* MR01 grown on soybean oil as a carbon source demonstrated an excellent performance on the production of rhamnolipids. Moreover, Lotfabad et al. (2010) reported that a high rate of the production of rhamnolipids was obtained by using lipid sources such as soybean oil as the medium for *P. aeruginosa* MR01. They also suggested that the cost of the production could be decreased by using frying oil wastes.

Wittgens et al. (2011) stated that for the production of rhamnolipid by *P. putida*, a variety of carbon sources can be used, but when a hydrophobic substrate is used, the process of recovery and purification would be much more difficult than when glucose is used as the carbon source. Choi et al. (2011) described inducing mutation in *P. aeruginosa* PA14, in order to stop the growth after the biomass reached the optimal concentration for rhamnolipid production. By applying this strategy, the provided substrate is used efficiently and mostly for synthesizing rhamnolipids. Many researchers have shown that the type of substrate could affect the rate of the rhamnolipid production, and this applies to other types of biosurfactants (Mulligan and Gibbs, 2004).

de Sousa et al. (2011) have done some experiments on the production of rhamnolipids by *P. aeruginosa* MSIC02, using a variety of carbon sources. The highest level of the production, 1269.79 mg/L, occurred when they used hydrolyzed glycerin as the carbon source. They noted that by altering the concentration and composition of substrate and environmental conditions, the maximum production rate was achieved. The optimal medium consisted of 18 g/L of glycerol, 4.0 g/L of sodium nitrate, and 62 mM of mono potassium phosphate. The pH and the temperature of the optimal production were 7.0 and 37°C respectively. Further analysis showed that the purified product was a mixture of RL1 (L-rhamnosyl-beta-hydroxydecanoyl-beta-hydroxydecanoate) and RL2 (L-rhamnosyl L-rhamnosyl-beta-hydroxydecanoyl-beta-hydroxydecanoate), and the product demonstrated very good emulsifying properties (65%).

Wittgens et al. (2011) noted that by minimizing the cell growth, a higher rate of rhamnolipid production is achieved, and there wouldn't be any need for cell growth monitoring. The structure of the produced rhamnolipids depends on the substrate being used. Their experiments showed that the best result was achieved by using sucrose or glycerol as the substrate. Wittgens et al. (2011) pointed out that the path for using a substrate depends on the type of substrate. For example, for transporting glucose, the ABC transporter needs a molecule of ATP for each glucose molecule, but on the other hand, glycerol could be easily conveyed by diffusion through an ion channel (Wittgens et al., 2011).

The type of substrate not only affects the rate of the production, but can also determine the structure and the efficiency of the produced rhamnolipids. Costa et al. (2006) showed that by using locally produced substrates, they could reach the desirable results. They conducted some experiments on rhamnolipid production by *P. aeruginosa* LBI, using some Brazilian native oils, such as oils from buriti, cupuacu, passion fruit, andiroba, Brazilian nut, and babassu, as the substrates.

Using Brazilian nut and passion fruit resulted in the highest production rate, 9.9 and 9.2 g/L, respectively. The produced rhamnolipid was a mixture mostly composed of monorhamnolipid (Rha–C_{10}–C_{10}) and a lower amount of dirhamnolipid (Rha–Rha–C_{10}–C_{10}), with a CMC value of 55–163 mg/L, and was able to decrease the surface tension of substrate substantially (29.8–31.5 mN/m). These rhamnolipids also demonstrated 93%–100% emulsifying activity with toluene and 70%–92% when kerosene was used. According to Das and Mukherjee (2005), the maximum yield of biosurfactants was obtained by using NH_4Cl as the nitrogen source and glycerol as the carbon source.

Rashedi et al. (2006) reported that they conducted some experiments on biosurfactant producing-bacteria isolated from contaminated oils in the southwest of Iran. Their experiments consisted of culturing the isolated strains on different carbon sources such as n-hexadecane, paraffin oil, glycerol, and molasses, and by using urea, sodium nitrate, and ammonia as nitrogen sources. Next, the yield of produced glycolipid and the quality of the biosurfactants were measured. These measurements were based on the CMC value, surface tension decrease, and emulsification capability of the produced biosurfactant. The results showed that the best outcome, 690 mg/L rhamnose, was obtained when glycerol and sodium nitrate were used as the carbon and nitrogen sources and C/N ratio was 55:1 (Rashedi et al., 2005). Their experiments showed that the highest yield was obtained when the bacteria were cultivated in glycerol with a concentration of 5% v/v. Although increasing the concentration of glycerol (to 5% v/v) resulted in increasing the yield, concentrations higher than 5% v/v inhibited the growth of the bacteria. They reported that after 7 days, the total rhamnolipid production was 4.2 g/L, yield to biomass was 0.65 g/g, and the product (rhamnolipids) showed a CMC of 19 mg/L.

It also should be noted that using the suitable substrate results in a higher yield and production of desirable homologues and congeners. For example, *P. aeruginosa* demonstrates the highest rate of rhamnolipid production when only vegetable oils are used as the carbon source (Wittgens et al., 2011). Also, the experiments of Mata-Sandova et al. (1999) showed that the rate of rhamnolipid production by the strain *P. aeruginosa* UG2 is controlled by the type of substrate. The strain grown on the hydrophobic substrates could produce a higher amount of rhamnolipids than when the strains were grown on hydrophilic substrates (Mata-Sandova et al., 1999).

For many environmental applications, dirhamnolipids have shown to be a better candidate than its counterpart monorhamnolipid. The mixture with a higher ratio of dirhamnolipids shows lower CMC and surface tension. Lotfabad (2010) reported that by changing the substrate type, this ratio could be changed. Nitschke (2005) also noted that by cultivating bacteria on a hydrophilic substrate, the produced rhamnolipids were chiefly dirhamnolipids, but when they attempted to cultivate bacteria on a hydrophobic substrate, they produced monorhamnolipids predominantly.

Using aqueous substrates is another useful method for decreasing the cost of rhamnolipid production. Gunther et al. (2007) conducted experiments for producing rhamnolipids from *P. chlororaphis* strain NRRL B-30761 by using an aqueous medium. Using an aqueous medium gives the bacteria a chance of using fresh substrate and also assists in controlling the medium composition to make sure that it is kept at its optimal composition and the produced rhamnolipid and excess biomass could be removed continuously.

The cost of recovery is another important factor, and as it has been discussed by choosing a suitable substrate, it can be reduced. Choosing efficient and cost-effective methods of recovery is another approach. As it has been extensively discussed in the previous part, the novel methods such ultrafiltration (Mulligan, 2005) and rhamnolipid purification through the silica gel columns (Christova et al., 2011) have been proven to be more environmentally friendly than previous methods such as acidification and participation.

OPTIMIZATION OF THE GROWTH ENVIRONMENT

One of the approaches for obtaining rhamnolipids in high quality and higher quantity is through finding and applying the optimal growth environment for the producing microorganisms. Altering the parameters such as pH, temperature, aeration, and agitation rate while cultivating the producing microorganisms changes the ratio and composition as well as the production rate of the final product. Therefore, applying optimal condition helps to produce rhamnolipids with desired composition, with better performance and with a higher production rate.

As it has been discussed previously, developing an aeration system (Mulligan et al., 1989c) and also continuous removal of the product prevent the toxic effects of the biosurfactants on the bacteria. Choosing the speed of agitation, optimum temperature, moisture, and the availability of the substrate are important factors that should be noted. Furthermore, the experiments of Nitschke (2005) showed that optimizing the temperature and pH and adjusting the agitation rate and aeration have a great effect on the biosurfactant production rate.

RECOVERY OF BY-PRODUCTS

Rhamnolipid-producing strains such as *P. aeruginosa* are also able to produce a valuable natural polymer, PHAs, that has been described as the biodegradable thermoplastic (Costa et al., 2009b). Elastomeric properties of PHAs make them a suitable candidate for biomedical applications as well as an environmentally friendly packaging material (Nitschke et al., 2011). Hori et al. (2002) pointed out that production of PHAs and rhamnolipids results in economically feasible production of rhamnolipids.

Pantazaki et al. (2011) have done some experiments on the effect of carbon source on rhamnolipid production by *Thermus thermophilus* HB8 and showed that when they used sodium gluconate as the carbon source, the production rate was much higher than when using glucose as the sole carbon source. These bacteria, similar to *P. aeruginosa*, are able to produce both PHAs and rhamnolipids. Their experiments also showed that low initial concentration of the phosphate and long incubation time stimulated rhamnolipid production in the bacteria. On the other hand, high initial concentrations of the phosphate stimulated the bacteria to produce a higher ratio of biomass and also PHAs. Thus by recovering the produced

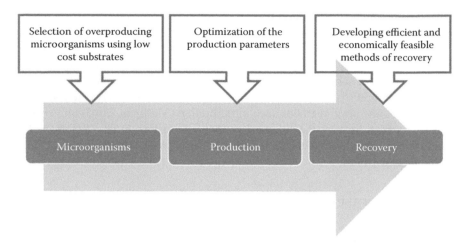

FIGURE 3.4 Different cost reduction strategies for the production of microbial surface-active compounds. (Adapted from Banat, I. et al., *Appl. Microbiol. Biotechnol.*, 87(2), 427, 2010.)

PHA in the process of rhamnolipid production, the feasibility of the process in terms of cost and revenue would be increased.

In summary, the strategies used to decrease the costs associated with rhamnolipid production that make this material able to compete with the synthetic ones are the following: optimizing the production condition, developing overproducing strains and strains that could grow on low-cost substrates, genetically modifying the microorganisms to produce desired congeners, using low-cost substrates, and developing a low-cost extraction and purification procedures (Figure 3.4) (Banat et al., 2010; Lotfabad et al., 2010; Wang and Mulligan, 2008).

RHAMNOLIPID STRUCTURE

Rhamnolipids, the metabolic by-product of *P. aeruginosa*, are one of the most extensively studied biosurfactants. These glycosides are a mixture of a variety of homologues and congeners. According to Abdel-Mawgoud et al. (2011), 60 different rhamnolipids congeners and homologues have been identified. Rhamnolipids produced by *P. aeruginosa* are generally a mixture of two main groups, monorhamnolipids (RL) and dirhamnolipids (RL2), with hydroxyacyl moieties mostly from C_8 up to C_{12}. The main difference between these two groups is the number of rhamnose units that is linked to hydroxyl carboxylic acids (Lotfabad et al., 2010).

On the other hand, rhamnolipids generated by species from the *Burkholderia* genus are mainly dirhamnolipids with two rhamnose units and mostly C_{14} hydroxyacyl chains (Abdel-Mawgoud et al., 2010; Mulligan and Gibbs, 2004). The difference between the structures of these two main groups is the number of rhamnose groups that is linked to one or two 3-hydroxy fatty acids of different chain lengths, which also could contain one double bond (Mulligan, 2005).

According to Abdel-Mawgoud et al. (2011), in rhamnolipids, the glycone part and an aglycone part are linked to each other by the o-glycosidic linkage. The glycone parts are made of one or two rhamnose moieties linked to each other through a-1, 2-glycosidic linkage. The 2-hydroxyl group of rhamnose group usually remains free, though in some homologues, they can be acylated with a long-chain alkenoic acid.

The aglycone part is composed of one, two, or three saturated or even sometimes mono- or polyunsaturated β-hydroxy fatty acid chains. The length of chains varies from 8 to 16 carbons. These chains are linked to each other via an ester bond formed between the β-hydroxyl groups of the distal chain and the carboxyl group of the proximal chain. Although carboxyl group of the distal β-hydroxy fatty acid chain usually remains free, in some homologues, this group is esterified with a short alkyl group and displays the structure of the best known RL congener, α-L-rhamnopyranosyl-α-L-rhamnopyranosyl-β-hydroxydecanoyl-β-hydroxydecanoate, which is typically symbolized as Rha–Rha–C_{10}–C_{10} (Abdel-Mawgoud et al., 2011).

Mulligan (2005) categorized rhamnolipids into four general types: type I (R1), L-rhamnosyl-β-hydroxydecanoyl-β-hydroxydecanoate, with a molecular mass of 504 Da; type II (R2), L-rhamnosyl-β-L-rhamnosyl-β-hydroxydecanoyl-β-hydroxydecanoyl-β hydroxydecanoate, with a molecular mass of 660 Da; type III (R3), one rhamnose attached to β-hydroxydecanoic acid; and type IV (R4), two rhamnoses attached to β-hydroxydecanoic acid (Figure 3.5) (Mulligan, 2009).

As noted previously, many researchers such as Deziel et al. (1999) and Lotfabad et al. (2010) stated that the type of substrate being used for the growth of *P. aeruginosa* has a great impact on the composition of the produced rhamnolipids. For example, the experiments of Deziel et al. (1999) showed that rhamnolipids produced by *P. aeruginosa* strain using mannitol or naphthalene as a carbon source have shown different percentages of congeners and homologues. Using mannitol as the carbon source increases the percentage of the rhamnolipids with two rhamnose groups and two 3-hydroxydecanoic acid groups. On the other hand, using naphthalene increased the production of rhamnolipids with two rhamnose groups and one 3-hydroxydecanoic acid group. Investigations conducted by Deziel et al. (1999) and Mulligan (2009) showed that in the mixture of rhamnolipids produced by *P. aeruginosa*, the presence of 28 different rhamnolipid congeners and up to 7 homologues can be identified. Also according to Lotfabad et al. (2010), *Pseudomonas* sp. grown on different carbon sources has been reported to produce rhamnolipid mixtures of 4–28 different homologues. Furthermore, there have been some reports on other bacterial species that are able to produce different congeners and homologues of rhamnolipids (Abdel-Mawgoud et al., 2010).

Moreover, many researchers reported that *P. aeruginosa* strains are able to produce a variety of mixtures of homologues and congeners of rhamnolipids under different environmental conditions (Abdel-Mawgoud et al. 2010; Dubeau et al., 2009; Nguyen and Sabatini, 2011; Ochsner et al. 1994; Van Gennip et al. 2009). According to Mulligan (2009), parameters such as fermenter design, pH, nutrient composition, substrate, and temperature dictate the composition and the structure of produced rhamnolipids (Mulligan, 2009).

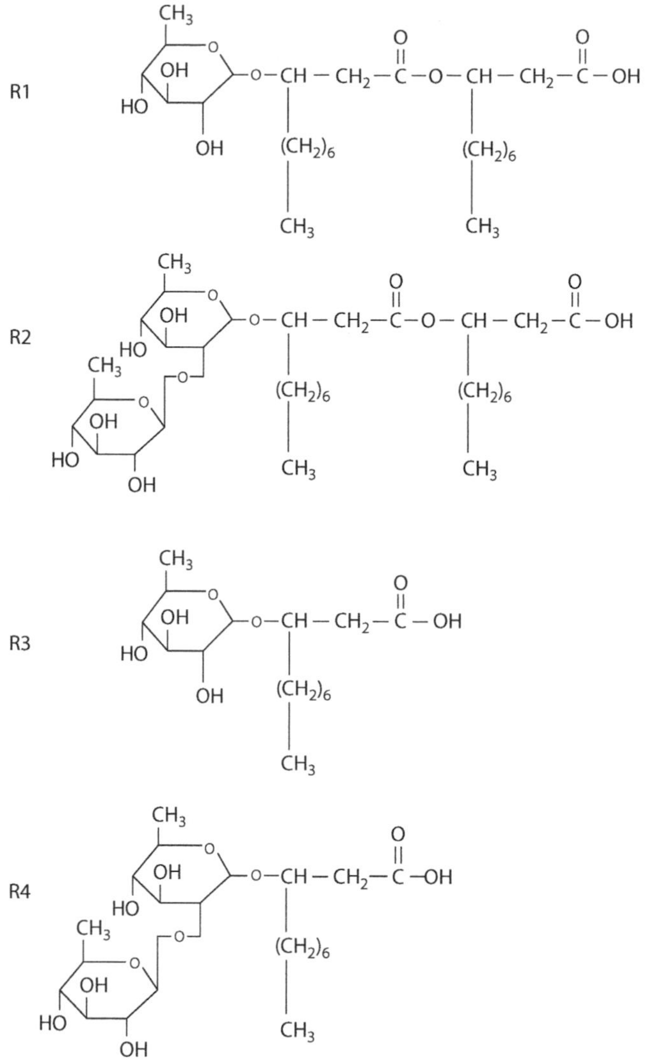

FIGURE 3.5 Structure of four different rhamnolipids. (From Mulligan, C.N., *Curr. Opin. Colloid Interface Sci.*, 14(5), 372, 2009.)

RHAMNOLIPID PROPERTIES

Numerous properties of rhamnolipids make them suitable for many applications. Many of their properties, such as their versatility and their ability to perform detergency; their foaming, wetting, emulsifying, demulsifying, solubilizing, and thickening; as well as their solubilizing, dispersing, metal sequestering, vesicle forming and phase dispersion; antimicrobial activities against both phytopathogenic fungi and bacteria; and also their thermostability and ecological compatibility, outperform other biosurfactants and synthetic surfactants. These properties make rhamnolipids an ideal candidate for numerous applications, such as cosmetics, pharmaceutical, petroleum, environmental, agricultural, and food industries (Bockmuehl, 2012; do Valle Gomes and Nitschke, 2012; Lourith and Kanlayavattanakul, 2009; Mulligan and Gibbs, 2004; Nitschke et al., 2011; Takemoto et al., 2010).

In the rhamnolipid mixture, the ratio of monorhamnolipids to dirhamnolipids significantly affects the properties and behaviors and the properties of the rhamnolipids. According to Cohen and Exerowa (2007), the ratio between monorhamnolipids and dirhamnolipids in the rhamnolipid mixture affects the thickness of the foam films and their surface electric properties, and consequently the stability of the foams. The ratio of the monorhamnolipids and dirhamnolipids in the rhamnolipid mixture is strain dependent; at the same time, some researchers such as Lotfabad et al. (2010) indicate that the ratio of dirhamnolipid to monorhamnolipids in a rhamnolipid mixture is highly substrate dependent. Therefore, by choosing an optimal substrate and strain, rhamnolipids with the desired ratio of component, and therefore with desirable properties, can be produced.

MICELLE CHARACTERISTICS

By increasing the concentration of biosurfactants, they start forming structures such as lamellae, vesicles, and micelles (Dahrazma et al., 2008). In water, the hydrophilic heads create the outside layer and hydrophobic part (tail) is positioned inside; on the other hand, in hydrophobic environment, the "inverse micelle" would be formed: the tails would be on the surface of sphere and heads are placed inside. The shape and the size of the micelle can be seen with transmission electron microscopy (Champion et al., 1995). The size and the shape of micelles depend upon the size and ionic strength of the head group and the size and shape of the hydrophobic tail. According to Dahrazma et al. (2008), data from small-angle neutron scattering confirm that the morphology of these aggregates is pH dependent. They concluded that as a result of this pH dependency, the release of the metals or drugs from these pH-sensitive vesicles could be controlled, so that they can be successfully used for drug delivery or other applications that require a controlled release. On the other hand, pH could be a controlling parameter on the efficacy of using biosurfactants in the process of bioremediation. For example, Dahrazma et al. (2008) noted that an addition of 1% NaOH increased the formation of large aggregates (>200 nm) and micelles with an average diameter of 17 Å, but in an acidic environment (pH 5.5), addition of 1% NaCl results in the formation of a mixture of large vesicles with an average radius of 600 Å. The reason for this occurrence could be that at higher pH, the charge densities on the carboxyl groups are increased, so the heads would be more repulsive, increasing the heads' tendency to increase their distance from each other, therefore, shaping a spherical structure (Ishigami and Suzuki, 1997). The other conclusion would be that the shape and the size of these micelles are not only controlled by the concentration of biosurfactants and the pH of the environment. It is also affected by the presence and concentration of ions in the solution that can affect the charge density on the carboxylic groups on the biosurfactants.

Muller (1993) studied the relation between the heat capacity of the micellization for ionic surfactants and the temperature dependence of CMC. He found out that

CMCs at any temperature are a function of the minimal value for CMC* (CMC of the ionic surfactant) and T/T^* (the ratio of temperature to the temperature at which the minimal value for the CMC of the ionic surfactant is found). It should also be noted that biosurfactant solubility is temperature dependent, and it increases by increasing the temperature. Therefore, at some temperatures, micelle can't be formed. The point where the solubility curve and CMC curve meet is called the Krafft point (Arnold and Linke, 2007; Kunieda and Shinoda, 1976).

Rhamnolipids, also like other biosurfactants, are able to form micelles at concentrations above the CMC. It has been reported that increasing the pH affects the morphology of rhamnolipids; it initially results in creating vesicles, and then by increasing more, micelles would be created (Figure 3.6) (Mulligan, 2009).

The investigations of Raza et al. (2010) showed that by changing the pH, the shape of the vesicles changes and therefore the behavior and effectiveness of the rhamnolipid solution. In their investigation, increasing pH from 5 to 8, irregular-shaped vesicles progressively turned to oligo-vesicles and, after that, to smaller spherical vesicles or micelles. These investigations agree with the previous investigations by Mulligan (2009).

Production of micelles results in increased wettability and decreased interfacial tension between solid surfaces and solutions. Micelles are increasing the mobilization of metals and organic contaminants from soils and sediments by increasing their bioavailability and also by emulsifying and mobilizing them, so that they can be easily removed from the soil matrix in a washing treatment. The removal takes place as a result of mobilizing heavy metals by chelation of ions, formation of metal complexes with the micelles, replacement of the negatively charged metal anions with anionic biosurfactants, and increasing the repulsive interfaces due to increasing the negative zeta potential of the medium (Chowdiah et al., 1998; Mulligan, 2009; Pacwa-Plociniczak et al., 2011).

As it has been noted before, the size of the micelle of rhamnolipids is highly pH dependent (Banat et al., 2010; Raza et al., 2010; Mulligan, 2009). According to

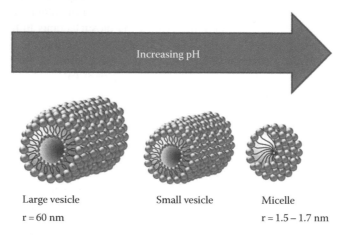

| Large vesicle | Small vesicle | Micelle |
| r = 60 nm | | r = 1.5 – 1.7 nm |

FIGURE 3.6 Effect of pH on rhamnolipid structure. (From Mulligan, C.N., *Curr. Opin. Colloid Interface Sci.*, 14(5), 372, 2009.)

Khoshdast et al. (2012), rhamnolipid has the tendency to form micelle at higher pH; therefore, rhamnolipids at the same concentration but at two different pH values act differently. At low pH, rhamnolipids have a tendency to participate; pH rise results in increasing the solubility of rhamnolipids. As pH increases, the biosurfactant activity also increases. This is as a result of micelle production and ionization of the carboxyl groups of the biosurfactants. The ionization of carboxyl groups generates and increases the binding sites for the metal cations on the biosurfactants; on the other hand, at higher pH, metal solubility increases. Furthermore, higher pH increases the solubility of the organic matter in media and results in a reduced number of absorption sites for metal ions (Chowdiah et al., 1998; Mulligan, 2009).

SURFACE-ACTIVE PROPERTIES OF RHAMNOLIPIDS

Rhamnolipid biosurfactants have fairly low CMCs and are able to decrease the surface tension of water as low as 25 mN/m. Furthermore, they are highly biodegradable, nontoxic, and are able to work at extreme pH, salinity, and various temperatures (Mulligan, 2005). Benincasa et al. (2004) reported that rhamnolipid produced by *P. aeruginosa* (RLLBI) by using soapstock as the substrate was able to create stable emulsions with hydrocarbons and vegetable oils. They also reported that the produced rhamnolipids were able to decrease the surface tension of water to 24 mN/m, with a CMC value of 120 mg/L. pH is another important factor affecting the behavior of rhamnolipids; in the investigations conducted by Mulligan (2009) and Raza et al. (2010), it was shown that by changing the pH, the shape of the vesicles changes and therefore the behavior and effectiveness of the rhamnolipid solutions. Because of the presence of carboxylic acid groups in rhamnolipids, they are expected to be anionic at higher pH (>4) (Mulligan, 2005; Penfold et al., 2012).

Costa et al. (2010) reported that in their experiments, the rhamnolipids produced by *P. aeruginosa* L2-1, by using waste cooking oil as the substrate, could achieve a surface tension of 30 mN/m and a CMC of 30 mg/L. Costa et al. (2010) also stated that L2-1 rhamnolipids are able to emulsify soybean oil completely and form stable emulsions with hydrocarbons. These rhamnolipids were also able to remove 69% of crude oil from their contaminated sand samples. Hence, they can be efficiently used in the cleanup of oil spills.

Rhamnolipids are amphiphilic molecules with both hydrophilic and hydrophobic poles, so they are located at the fluid–fluid interface with different degrees of polarity. It could be the interface between two liquids, such as oil and water, or the interface between a liquid and a gas. According to Jimenez Islas et al. (2010), properties of biosurfactants such as emulsification or their ability to work as surface-active agents depend on four primary parameters: CMC, aggregation number, hydrophilic–lipophilic balance, and cloud point (CP).

Many of the properties of rhamnolipids are concentration dependent, as mentioned before; at higher concentration, as the concentration reaches the CMC, micelles are formed spontaneously. At this point, surface tension reaches the minimum.

For rhamnolipids, pH is also a key factor that directly affects the solubility of these biosurfactants in aqueous solutions. As it has been discussed previously, the pH increase results in increasing the solubility of rhamnolipids. Experiments showed that at lower pH, the solution starts to form clouds, and at some point, it starts precipitating (Dahrazma et al., 2008). There is a need for more experiments to determine the exact CP and participation point.

Xia et al. (2011) reported that *P. aeruginosa* isolated from a petroleum reservoir was able to produce 2.66 g/L of rhamnolipids. The produced rhamnolipids were able to decrease the surface tension in aqueous as well as hydrocarbon mixtures. They were able to reduce the surface tension of supernatant separated from biomass from 71.2 to 22–30 mN/m, emulsify the crude oil and reach the emulsion index of 80%, and were able to maintain their surface-active properties at extreme conditions such as high temperature (up to 120°C), high salinity (even higher than 20 g/L), and the presence of metal ions. These biosurfactants also showed high stability at pH higher than 5, although it was observed that pH lower than 5 and the presence of monovalent and trivalent ions have negatively affected the stability of the biosurfactants.

Das and Mukherjee (2005) demonstrated that rhamnolipids showed strong emulsifying property and were able to separate significant amount of oil from an oil-saturated sand sample. Furthermore, they indicated that rhamnolipids could induce hemolysis and coagulation of platelet-poor plasma, but this surfactant didn't induce any damage to the lung, liver, heart, and kidney tissues of the subject of the experiment (chicken). Their study revealed that rhamnolipids can be used for many applications in petroleum and pharmaceutical industries, as well as environmental bioremediation.

According to Ozdemir and Malayoglu (2004), the affinity of the biosurfactants in the solution to be absorbed to the solid surfaces and a decrease in the surface tension increase the wettability and result in partially or completely wetting the solid surface by the aqueous solution. They studied the advancing contact angles of the aqueous solutions of the rhamnolipid mixture, composed of Rh1 and Rh2 with a ratio of 1:1, on solid surfaces, such as hydrophilic glass, hydrophobic polymer, polyethylene terephthalate (PET), and gold. Their experiment showed that at lower concentration, there was a change in the angle that was dependent on the affinity of the surfactant toward the solid surface that was used. Higher concentrations resulted in a decrease in the contact angle (Ozdemir and Malayoglu, 2004).

Likewise, it has been shown that lower concentrations of the rhamnolipids on hydrophobic surfaces decreased the adhesion tension, and higher concentrations caused increase in the adhesion tension. Further observations showed that on hydrophilic surfaces, there was a steady decrease in adhesion tension (Ozdemir and Malayoglu, 2004).

Moreover, Costa et al. (2009a) investigated the wetting characteristics of rhamnolipids from *P. aeruginosa* LBI strain, grown on waste oil and sodium dodecyl sulfate as substrates. They conducted their research on solid surfaces such as glass, PET, polyvinyl chloride (PVC), PCL, and a polymer blend (PVC-PCL). They noticed that by increasing the concentration of the rhamnolipids, contact angle increased, but further increasing the rhamnolipid concentration resulted in a decreasing contact angle. They noted that wetting abilities of the LBI rhamnolipids was superior in comparison with the chemical surfactant being used.

ANTIMICROBIAL AND CYTOLYTIC PROPERTIES OF RHAMNOLIPIDS

For surviving in densely populated places, such as soil, microorganisms have evolved some survival mechanisms to be able to live on and compete with their competitors. One of these mechanisms is producing the antimicrobial agent to prevent the growth of the competitors (Wecke et al., 2011). Antimicrobial properties of rhamnolipids make them a suitable candidate for use in pharmaceutical, food, cosmetic, and agricultural industries (Mulligan and Gibbs, 2004). The extent of the effectiveness of rhamnolipids is strain dependent and slightly varies by the type of the substrate being used.

According to Haba et al. (2003), rhamnolipids demonstrate excellent antimicrobial properties. In their experiments, the minimum inhibitory concentration (MIC) values of rhamnolipids for a few different strains of bacteria were measured. It varied from as little as 0.5 mg/L, for *Klebsiella pneumonia*, to 75 mg/L, for *Fusarium solani*, which is a fungus. They reported that for common bacteria such as *Bacillus subtilis*, the MIC value was 16 mg/L, and for *Staphylococcus aureus* and *S. epidermidis*, it was 32 mg/L. According to Mulligan and Gibbs (2004), Rhamnolipid antibiotics combined with its emulsifying properties make it a promising agent in cosmetic industries, and they are being used in moisturizing creams.

Likewise, Benincasa et al. (2004) reported that rhamnolipid produced by *P. aeruginosa* (RLLBI) by using soapstock as the substrate exhibited antimicrobial activities against other microorganisms. They reported much lower MIC values than what other researchers reported earlier for some of their experimental subjects (bacteria). For example, they reported that the MIC value for *B. subtilis, S. aureus*, and *Proteus vulgaris* was 8 mg/L, and for *Streptococcus faecalis*, it was 4 mg/L. They also conducted research to measure the MIC value for *P. aeruginosa*, and they found that by using a concentration of 32 mg/L, they were able to inhibit this bacterium. Furthermore, their research showed the ability of the RLLBI for inhibiting some fungal species such as *Penicillium, Alternaria, Gliocladium virens*, and *Chaetonium globosum*. They found the MIC for these species was 32 mg/L.

Abalos et al. (2001) also reported that in their experiments, rhamnolipids demonstrated excellent antifungal properties against the strains being studied. They found the MIC values of 16 µg/mL for *Aspergillus niger* and *Gliocladium virens;* 32 µg/mL for *C. globosum, P. chrysogenun*, and *A. pullulans*; and 18 µg/mL for *Botrytis cinerea* and *R. solani*. Furthermore, according to Costa et al. (2010), the rhamnolipids produced by *P. aeruginosa* L2-1 showed antimicrobial activity with the MIC value of 32 µg/mL for *B. cereus* and *Micrococcus luteus*. Also it showed an MIC value of 128 µg/mL for *S. aureus*.

Takemoto et al. (2010) have done some experiments to study the effectiveness of the rhamnolipids as a fungicide against common fungal infections of grapes. They have done their experiments on some species of fungi (*Aspergillus japonicus, Cladosporium cladosporioides, Curvularia brachyspora, Greeneria uvicola, Nigrospora sphaerica, Trichoderma* sp., *Penicillium sclerotiorum*, and *P. thomii*) isolated from a heavily infected vineyard. Their experiments showed that the rhamnolipids alone were not effective as a fungicide, but the mixture of rhamnolipids and a cyclic lipodepsinonapeptide and syringomycin E (RLs + SRE) demonstrated great inhibitory activities against those fungi in their different germination stages. Rhamnolipids not only have fungicide properties but also are able to inhibit the spore germination and mycelium growth. Moreover, as Varnier et al. (2009) noted, rhamnolipids are able to trigger the defense mechanisms of grapevines by activating the expression of defense genes and initiating a hypersensitive response. Therefore, they protect the vines from the fungus *B. cinerea*. These researchers stated that the combinations of the rhamnolipid properties make these valuable molecules able to protect the grapevines from gray mold disease.

Bockmuehl (2012) refer to rhamnolipids as antimicrobially active biosurfactants. According to them, most of the antimicrobial activities of the biosurfactants are the result of their abilities to destroy the cell membranes and prevent the adhesion of microorganisms to surfaces, in addition to their excellent detergency power. These properties are advantageous in the food industry. One of the challenges, in that sector, is to prevent and eliminate the biofilms and adhesion of the microorganisms, which are responsible for spoilage and transmission of diseases (do Valle Gomes and Nitschke, 2012). These researchers' work showed that preconditioning of surfaces by using rhamnolipids resulted in reducing the adhesion and disrupting the biofilms of microorganisms. They conducted their experiments by observing the growth of a few food-borne pathogenic bacteria such as *S. aureus, Listeria monocytogenes*, and *Salmonella enteritidis* on a polystyrene surface, in the presence and absence of rhamnolipids. Their experiments revealed that using rhamnolipids 1.0% caused a reduction in the adhesion of *L. monocytogenes* by 57.8% and adhesion of *S. aureus* by 67.8%. Their observation revealed that rhamnolipids were more effective on single strain cultures than the mixed cultures. do Valle Gomes and Nitschke (2012) also reported that 2 h of contact with the rhamnolipids at a concentration of 0.25% was able to reduce the biofilms of *S. aureus* by 58.5%, *L. monocytogenes* by 26.5%, and *S. enteritidis* by 23.0%. Also it decreased 24.0% of the biofilms in the mixed cultures.

Kaczorek et al. (2012) studied the effect of adding rhamnolipids on the cell surface of some strains of few genera of bacteria such as *Bacillus, Pseudomonas, Aeromonas, Achromobacter*, and *Flavimonas*. Their investigations showed that there were some alterations in the cell surface properties, increasing their hydrophobicity. It also has been indicated that rhamnolipids were able to affect the cell walls of both gram-positive and gram-negative bacterial strains.

Vatsa et al. (2010) stated that rhamnolipids not only work as antimicrobial agents, but are also capable of stimulating the innate immunity in plant and animal cells. These researchers stated that the combination of those two properties combined with their ecologically compatibility, their low toxicity, and biodegradability make them valuable agents for use in medical, pharmaceutical, and agricultural sectors. Furthermore, experiments of Wecke et al. (2011) proved that rhamnolipid as a single antimicrobial compound can simultaneously provoke the genes of two separate stress stimulons and trigger reactions that combine the cell envelope and secretion stress responses, such as the TCS LiaRS and CssRS to prevent the cell lysis in the presence of rhamnolipids.

The experiments of Thanomsub et al. (2009) consisted of studying the rhamnolipids produced by a thermotolerant bacterium, *P. aeruginosa* B189, which was isolated from milk factory waste. Their investigations showed that the purified rhamnolipids exhibited extensive inhibitory effect on human breast cancer cells. Therefore, they suggested that rhamnolipids produced by *P. aeruginosa* B189 is an effective compound for using in anticancer drugs.

McClure and Schiller (1992) described the cytolytic activity of rhamnolipids on human monocyte-derived macrophages (MDMs); they found out that the dirhamnolipids were much more effective than monorhamnolipids and could cause blebbing of the plasma membrane of human MDM. Although the exact mechanisms responsible for rhamnolipid activities are not well identified, they suggested that one of the possible mechanisms is through the insertion of the fatty acid tails of the rhamnolipids into the cell membrane.

Kamal et al. (2012) reported the isolation of a new strain of *Pseudomonas* from an alkaline soil from a cold desert located in the Western Himalayas. These resilient bacteria had to survive the cold temperature of the region and an extreme hot temperature during a short period in the summer and UV irradiation. By conducting analysis with NMR, FITR, and MS, they found that the *Pseudomonas* strain ICTB-745 was able to produce four different bioactive substances: monorhamnolipids, dirhamnolipids, 1-hydroxyphenazine, and phenazine-1-carboxylic acid. They reported that the produced compounds could perform as a highly effective antimicrobial against both gram-positive and gram-negative bacteria, as well as *Candida albicans*; they also demonstrated cytotoxic activities against human tumor cells.

BIODEGRADABILITY OF RHAMNOLIPIDS

In natural habitats, some bacteria could secret enzymes such as lipases, lipoyl synthase, and triacylglycerol lipase, which are able to break down the rhamnolipid (Wittgens et al., 2011). Chrzanowski et al. (2012) stated that the result of their experiments demonstrated that the microorganisms present in the soil, which are responsible for biodegradation of hydrocarbons, could easily degrade rhamnolipids as well. This biodegradation could occur in both aerobic and anaerobic conditions. It was observed that the biodegradation of rhamnolipids didn't interfere with the biodegradation rate of diesel in the soil. The observation of Chrzanowski et al. (2012) showed that the presence of rhamnolipids in soil and their biodegradation didn't favor the growth of the specific species, so it doesn't alter the natural microbial equilibrium in the soil. This proves that rhamnolipids are ecologically friendly substances. Nitschke et al. (2011) stated that one of the advantages of using rhamnolipids instead of their synthetic counterpart is their biodegradability, although there is a need for more research on the possibility of their toxicities on living organisms in the environment. Moreover, according to Wena et al. (2009), the rate of rhamnolipid degradation depends on the percentage of contamination in soil, and the highest rate of degradation was observed in contaminated sites and the lowest rate was in uncontaminated soils.

METHODS OF DETECTION AND ANALYSIS

In recent years, advancement in technology resulted in the development of new devices and new methods of detection and analysis. These methods and these novel devices enabled researchers to obtain precise and accurate information about the subjects of their study. There are many methods and devices that can be used for quantitative and qualitative detection and analysis of rhamnolipids. A few of the commonly used methods will be briefly discussed.

Measuring CMC

One of the methods that have been used commonly for the quantitative measurement of the rhamnolipid is by measuring its surface tension. A tensiometer measures the force needed to pull up a thin metal ring (platinum ring) out of the surface of the solution. To determine the concentration of the rhamnolipid in solution, the solution is sequentially diluted. At each step of dilution, the surface tension is measured until the CMC (the concentration at which surface tension starts to increase) is found. Then the data from the surface tension at each concentration are plotted. The CMC of rhamnolipid is the intercept of two straight lines from the concentration-dependent and concentration-independent sections of the curve. From the value of CMC, the quantity of the rhamnolipid in the original solution can be calculated (Ballot, 2009; Mulligan and Gibbs, 2004). Although this method is one of the easiest methods of measuring the quantity of the rhamnolipids, it should be noted that the presence of other tensioactive components in the solution can cause errors in the results (Abdel-Mawgoud et al., 2010).

Rhamnolipid Hydrolysis

Abdel-Mawgoud et al. (2011) described rhamnolipid hydrolysis test as a useful method for measuring the quantity of the produced rhamnolipid. In these tests, the solution contains rhamnolipids mixed with sulfuric acid and either orcinol (1,3-dihydroxy-5-methylbenzene) or anthrone (9,10-dihydro-9-oxoanthracene), and then heated until a greenish blue color appears. This change in color is the result of hydrolysis of the rhamnose groups in rhamnolipids. Then the measurement is done by using a spectrophotometer and result is compared with a standard curve that had been prepared by standard rhamnolipid mixtures.

Spectroscopic Methods

During recent decades, there have been many advances in the field of spectroscopy and many devices have been developed. By studying the interaction of energy radiated from a source with the matter, the structures and bonds could be identified. The energy used depends on the application and could be visible light, gamma, ultraviolet, infrared (IR), or x-ray. Describing all the methods of spectroscopy is more than the scope of this chapter. Therefore, this is a summary of some of the common spectrometry methods that have been used for measuring rhamnolipids.

Spectrophotometer

A spectrophotometer is a device for measuring the quantity of the chemicals in the solution by measuring the reflection or transmission properties of the chemical depending on the wavelength of radiated light; either in the ultraviolet, visible, or near-IR spectral regions (250–2500 nm) (Allen et al., 2012), the intensity of the light passing through the solution can be an accurate indicator of the concentration. For measuring the rhamnolipids using spectrophotometer, some researchers introduced a method consisting of using orcinol as a solvent.

The orcinol method consists of extracting and separating the extract from the cells and the heating the mixture of supernatant, solvent, and sulfuric acid and orcinol (1,3-dihydroxy-5-methylbenzene) until a blue–green color is observed. According to Abdel-Mawgoud et al. (2011), this blue–green color is the result of the hydrolyzing of rhamnose groups into methyl furfural. After the appearance of this green–blue color, the intensity of the color is measured by using a spectrophotometer at 421 nm (Abdel-Mawgoud et al., 2011), then comparing the result with the standard curve at various concentrations of the rhamnolipids for calibration (Abdel-Mawgoud et al., 2011; Ballot, 2009; Wang et al., 2007).

Mass Spectrometry

MS can be used to measure both the quantity and the quality of the materials. It can be used to determine the composition of the materials, to detect and to identify the unidentified compounds, and to identify bonds and the structure of the material. Mass spectrometers are made of three main parts: ion source, mass analyzer, and detector (Downard, 2004). In this method, the mass to charge ratio of the material is measured, and this is done through ionizing the chemical compounds in order to create charged particles. For this method, the sample loaded onto mass spectrometer is vaporized; volatile samples can be introduced directly. However, nonvolatile samples are solved in a volatile solvent.

After the vaporization of the samples, they go through an ionizing chamber, where an electron beam is focused and the particles bombarded to ionize them. Next, these ionized particles are accelerated through an electric field toward an electromagnetic field and finally a detector. These charged particles can easily react with other gases on their path. Therefore, to prevent the collision of these ionized particles with gaseous contaminants and air molecules, the operation takes place under vacuum. When these vaporized ions go through electromagnetic fields in the analyzer, they are separated according to their mass and charge. After separation, ions pass onto an ion detector. Depending on their mass and charge, these ions are producing some electrical currents. The electronic signal is amplified and recorded; then the detected values are sent to a computer. By studying the peaks of the generated graphs, quantity of each ion could be measured. Mass spectroscopy (MS) can detect as low as 10^{-12} mol (Downard, 2004; Verbeck et al., 2002). By coupling gas chromatography with mass spectroscopy (GC–MS) and coupling inductively coupled plasma source with mass spectrometer (ICP–MS), the detection level could be lowered and the need for using carrier gases (for GC) eliminated.

The investigations of Pereira et al. (2012) on rhamnolipids, produced by *P. aeruginosa* strains isolated from Brazilian crude oil, showed that MS coupled with electrospray ionization (ESI–MS) analysis provided a rapid and accurate characterization of biosurfactants. Likewise, tandem mass spectrometry (MS/MS) provided accurate information about the structure of the biosurfactants. Furthermore, GC–ICP–MS and HPLC–ICP–MS have been introduced as new generation of analyzers that could be easily used for speciation of molecules (Bird, 1998; Hata et al., 2007).

Infrared Spectroscopic Methods

The IR spectroscopy method is being used for identifying the structure of compounds by identifying the present bonds in the structure (Messmer et al., 2012; Pavel et al., 2012). Depending on the type of the bond (strong or weak) and the mass of the atoms in the bond, each bond could absorb a particular frequency of the radiated IR light. The frequency of the vibration of the bond is dictated by the type of the bond and the mass of the atoms (Hemmer et al., 1999). For example, lighter atoms result in higher frequencies, and the bond between heavier atoms moves slower and therefore causes lower frequencies. Likewise, stronger bonds result in higher frequencies and weaker bonds result in lower frequencies.

IR spectrophotometer can be used in order to quantify the complex mixtures of rhamnolipid congeners (Gartshore et al., 2000). Many hydroxyl, ester, and carboxylic groups of rhamnolipids have the ability to absorb a broad range of IR waves, and depending on their structure, they absorb specific frequencies of IR. Although this method appears to be able to accurately measure the quantity of different chemicals, it should be noted that the changes in pH and also the presence of other constituents can affect the result.

Fourier Transform Infrared Spectroscopy

Fourier transform infrared spectroscopy (FTIR) is another spectroscopy method. According to Swann and Patwardhan (2011), the advantages of this method are its rapidity, the need for a small amount of sample, and it is nondestructive. In FTIR, the interferogram that passes through the sample and reaches the detector is being decoded by a mathematical technique called Fourier transformation. Fourier transformation, which is handled by a computer, gives information about the intensity of the IR radiation in each frequency separately. It produces a graph of percentage of transmission against wave number. The radiated frequency that has been absorbed in the sample is interpreted as the presence of the bonds responsible for absorbing that frequency in the sample. FTIR is an accurate method for analyzing the structure of rhamnolipids and their interaction with other materials (Guo et al., 2009; Kumar and Mamidyala, 2011). Leitermann et al. (2008) introduced attenuated total reflectance Fourier transform infrared spectroscopy (ATR-FTIR) as a rapid and convenient method for characterizing, identifying, and quantifying substances. They indicate that this method is suitable and accurate for analyzing and quantifying rhamnolipids. They indicated that unlike HPLC, using ATR-FTIR is not time consuming and took 20 min to analyze the samples of rhamnolipids.

Nuclear Magnetic Resonance

Nuclear magnetic resonance (NMR) spectroscopy has been known as the preeminent technique for determining the structure of organic compounds. It measures the absorption of radio frequencies by the atoms of the samples and, by interpreting the data, gives accurate and complete detailed information about the sample. In the past, there was a need for larger samples, but new instruments are able to get data and analyze samples that are even less than 1 mg. Numerous NMR techniques have been developed for complete analyses of different organic compounds. According to Gustafson (2012), the usage of chiral anisotropic reagents in conjunction with

NMR analyses helps to determine the absolute configuration of secondary alcohols, α-substituted primary amines, and α-substituted carboxylic acids possible (Gustafson, 2012). Haba et al. (2003) and Monteiro et al. (2007) acknowledged NMR as an accurate tool for the structural analysis of purified rhamnolipid congeners.

^{1}H and ^{13}C NMR techniques are used for identifying structures of rhamnolipid homologues and congeners, and to determine the types of rhamnolipids in the mixture (Christova et al., 2011; Lotfabad et al., 2010). According to Lotfabad et al. (2010), ^{1}H and ^{13}C NMR spectra are measured by setting the spectrometer at 500 MHz and by using chloroform as a solvent. They reported that in the ^{13}C NMR spectrum of rhamnolipids, the signals from carboxylic groups was at δ 172 and δ 174.5; for CH_2, it was from δ 22.1 to 34.1, and for CH_3, it was δ 13.9. They indicated that using NMR techniques showed that rhamnolipid produced by *P. aeruginosa* MR01 was a mixture of monorhamnolipids and dirhamnolipids.

Similarly, de Sousa et al. (2011) used the NMR spectrum of H^1 and C^{13} and mass spectra to identify the structure of rhamnolipids. They found out that the purified mixture of rhamnolipids produced by *P. aeruginosa* MSIC02 was composed of two types of rhamnolipids: L-rhamnosyl-β-hydroxydecanoyl-β-hydroxydecanoate (RL1) and L-rhamnosyl L-rhamnosyl-β-hydroxydecanoyl-β-hydroxydecanoate (RL2). They also reported that for their experiments, they used CD_3OD and $CDCl_3$ as solvents for ^{1}H and ^{13}C NMR spectra measurement.

Monteiro et al. (2007) reported that their analysis of rhamnolipids by using NMR spectroscopy, by using chloroform-deuterated methanol (MeOD) as the solvent. Their analysis showed that the mixture of rhamnolipids produced by *P. aeruginosa* DAUPE 614 predominantly composed of rhamnosyl-β-hydroxydecanoyl-β-hydroxydecanoic acid (Rha–C_{10}–C_{10}) and β-α rhamnosyl (1 → 2) rhamnosyl-β-hydroxydecanoyl-β-hydroxydecanoic acid (Rha$_2$–C_{10}–C_{10}).

CHROMATOGRAPHIC METHODS

For measuring rhamnolipid quantity, a number of chromatographic procedures are available. In this section, the summary of a few of them will be discussed.

Thin Layer Chromatography

One of the oldest methods for separating and identifying different compounds in mixtures is thin layer chromatography (TLC). TLC is made of two parts: the stationary phase, which is an absorbent such as silica gel, which is covering the surface of a glass, aluminum, or plastic sheet. The second phase or mobile phase is a solvent such as chloroform or methanol. The solvent and the mixture move through the stationary phase, and depending on the types of compounds, they move with different speeds (Jork et al., 1994).

TLC has been used in the past for the measurement of different congeners in the rhamnolipid mixture. Rhamnolipids congeners are acidic and are solubilized in a basic solution and can be extracted by using nonpolar solvents. For quantifying the congeners in the extracted mixture, TLC is being used; in this method, silica gel is used as the stationary phase and a moderately polar solvent is being used as the mobile phase. With this method, monorhamnolipids are separated from their more polar counterpart, dirhamnolipids (Abdel-Mawgoud et al., 2011).

For separating the congeners in the mixture, Koch et al. (1991) used a mixture of methanol, water, and trifluoroacetic acid with ratios of 90, 10, and 0.25, respectively. With their method, rhamnolipid congeners with different lengths of alkyl chains were separated. After that, with the orcinol test, using reagents such as the ceric dip, they can be analyzed further (Abdel-Mawgoud et al., 2011). The original TLC method has been evolved to a new generation of chromatography method, the high-performance thin layer chromatography.

Liquid Chromatography

Liquid chromatography (LC) is another chromatographic method that uses a mobile phase (solvent) moving through the stationary phase, which could be silica gel particles in a column or a flat surface. The advantage of this method is accuracy, and it needs only a small amount of sample (Allwooda and Goodacre, 2010; Ardrey, 2003). The more advanced generation of this method is the high-pressure liquid chromatography (HPLC), which has been used extensively in laboratories around the world.

High-Pressure Liquid Chromatography

High-pressure liquid chromatography or high-performance liquid chromatography (HPLC) is one of the most used analytical methods in the laboratory today. The accuracy, the ease of use, and its ability to perform analysis with a small amount of sample make it one of the most useful tools in laboratories. In HPLC, the sample and mobile phase are moved with high pressure through the stationary phase column, a column packed with spherical particles. In reversed-phase liquid chromatography, mobile phase is a polar solvent like water and a less polar solvent like methanol and stationary phase is made by a column packed with octadecylsilyl (C_{18}). In the normal-phase liquid chromatography, the mobile phase is a less polar solvent such as toluene, and the stationary phase column is a column packed with spherical particles such as silica gel particles. Through penetration of the mobile phase through stationary phase, molecules in the samples are separated based on their hydrophobicity. For example, nonpolar functional groups are attracted by the silica particles and are bonding with them (Ardrey, 2003; TDMU, 2012).

HPLC is one of the most useful methods for analyzing rhamnolipids. Mata-Sandoval et al. (1999) reported that they developed an HPLC method for quantifying rhamnolipid species. According to Abdel-Mawgoud et al. (2011), for analyzing the rhamnolipids with HPLC, a column packed with C_8 or C_{18} was used as the stationary phase and water and acetonitrile as the mobile phase or para-bromoacetophenone. The detection was at a wavelength of 265 nm. Mata-Sandoval et al. (1999) stated that to confirm the data from the HPLC, they used MS for further analysis.

Liquid Chromatography Coupled to Mass Spectrometry

According to Van Bramer (1997), MS is an accurate technique for analyzing the molecular structure of materials. This technique helps in accurately identifying and quantifying the component of a mixture. In MS, sample is under vacuum in the source region, where they are ionized. Charged particles of the sample pass through a magnetic field to be separated based on their mass-to-charge ratio. After separating ions, they go through a detector and then data are sent for analysis.

For the accurate analysis of rhamnolipid, HPLC coupled with a mass spectrometer (HPLC–MS) has been proven to be a reliable method. HPLC–MS enables researchers to accurately detect and identify all the congeners in the rhamnolipid mixture. Costa et al. (2010) reported that in their experiments, they used HPLC–MS for characterizing the rhamnolipids. For the stationary phase of the TLC, they used silica gel 60 plates, and the carrier phase was a mixture of chloroform (65%), methanol (15%), and water (2%). They were able to identify and measure 16 different rhamnolipid congeners. Part of the Gunther et al. (2005) experiment consisted of comparing the compositions of rhamnolipids produced by *P. chlororaphis* with the ones from *P. aeruginosa* by using HPLC–MS. They found out that rhamnolipids produced by *P. chlororaphis* were only monorhamnolipids. And in rhamnolipids secreted by *P. chlororaphis,* the majority of molecules were had one monounsaturated hydroxy fatty acid of 12 carbons and one saturated hydroxy fatty acid of 10 carbons, unlike the *rhamnolipids* from *P. aeruginosa,* which had long chains of fatty acid glycosylated esterified to short fatty acids and the sugar rings.

Gas Chromatography

GC is one of the accurate methods of chromatography. The mobile phase in this method is a gas and the stationary phase is a nonmoving liquid or solid (Snow and Slack, 2000). According to Abdel-Mawgoud et al. (2011), GC cannot be used for direct measuring of rhamnolipids as they have high molecular weights. However, there are some reports on measuring rhamnolipids by using GC-MS. Abdel-Mawgoud et al. (2011) indicated that for measuring rhamnolipids with GC, they should be hydrolyzed prior to analysis by using a strong acid or base.

Gas Chromatography Coupled to Mass Spectrometry

Jadhav et al. (2011) reported on the analysis of biosurfactant produced by *Enterobacter* sp. In their experiments, the sugar composition of biosurfactants was determined by GC–MS. They described that this indirect analysis was done by hydrolyzing the biosurfactant with 2M trifluoroacetic acid (CF_3COOH); the process of hydrolyzing was done at a temperature of 120°C and took 4 h to be completed. After that, the residue was separated and washed with methanol and then was mixed with 1M sodium borohydride ($NaBH_4$); then it was acetylated with potassium acetate and acetic anhydride. The acetylation was done at a temperature of 120°C and took 2 h to be completed. Once the excess reagent was evaporated and after washing the sample with ethanol, ethyl acetate was used to separate alditol acetates and was analyzed by GC–MS. Gerwig et al. (1978) likewise used GC–MS for identifying the D or L configuration of rhamnose groups. Again, there was some modification such as methanolysis of Rha–Rha–FA–FAs using hydrochloric acid in methanol.

CONCLUSION

The outstanding properties of rhamnolipids such as high surface activity, high antimicrobial activity, degradability, renewability, low toxicity, and stable emulsions with hydrocarbons make these environmentally friendly biosurfactants excellent candidates for use in bioremediation, petroleum, pharmaceutical, cosmetic, and food industries.

Although they have been demonstrated to be advantageous substances for many applications, currently, their main market is in environmental and petroleum industries, where they had been used in enhanced oil recovery, in cleanup of oil spills, and in many more applications. Nevertheless, research has shown that this valuable material can be used as an environmentally friendly fungicide and for decomposing agricultural wastes, and thus there is a large potential market in the agricultural sector. There has been significant research in medical and pharmaceutical fields, suggesting the use of these active agents for vast applications in fighting and antiproliferative activity against human breast cancer, to be used as an additive in nanoparticle production, or functioning as an antimicrobial agent. In cosmetic and food industries, emulsifying and antimicrobial activity of these substances with their compatibility made them an ideal additive to be used in many applications, from an emulsifier in the process of food production to an efficient component in moisturizing creams.

The main source of rhamnolipid is the metabolic activities of *P. aeruginosa*, a gram-negative opportunistic pathogen. These bacteria depend on the limitations of some nutrients, and the type of carbon source is able to synthesize different homologues and congeners of rhamnolipids with varying ratios of the components. These properties help researchers to customize the composition ratio depending on what the product is intended for. Nowadays, the focus of many researchers is to find and isolate microorganisms capable of producing rhamnolipids or to genetically tailor some nonpathogenic microorganisms in order to be able to reduce the production complications and cost.

In order for rhamnolipids to be able to compete with synthetic surfactants currently in the market, the cost of the production needs to be decreased. This is possible through the selection of low-cost substrates, optimization of the fermentation environment, and isolation and creation of new strains that could produce rhamnolipids at greater rates, along with a higher ratio of the desirable congeners.

Recent advancements in analysis technology enabled researchers to obtain precise and accurate information on the molecular and structural characterization of rhamnolipids, the effect of these valuable materials on different substances, and finding new applications. Purifying the desired congeners and even initiating some changes in their structures to be more suitable became possible with the help of these new techniques. By introducing mutations on rhamnolipid-producing microorganisms, the rate of the production and the percentage of desired congeners could be increased; also by introducing some particular substrates, microorganisms would be able to produce new generations of rhamnolipids, such as deuterated rhamnolipids to be used in certain applications.

REFERENCES

Abalos, A., Pinazo, A., Infante, M. R., Casals, M., Garcia, F., and Manresa, A. 2001. Physicochemical and antimicrobial properties of new rhamnolipids produced by *Pseudomonas aeruginosa* AT10 from soybean oil refinery wastes. *Langmuir*, 17(5), 1367–1371.

Abdel-Mawgoud, A., Hausmann, R., Lépine, F., Müller, M., and Déziel, E. 2011. Rhamnolipids: Detection, analysis, biosynthesis, genetic regulation, and bioengineering of production. *Biosurfactants*, 20, 13–55.

Abdel-Mawgoud, A., Lepine, F., and Deziel, E. 2010. Rhamnolipids: Diversity of structures, microbial origins and roles. *Applied Microbiology and Biotechnology*, 86(5), 1323–1336.

Abouseoud, M., Maachi, R., Amrane, A., Boudergua, S., and Nabi, A. 2008a. Evaluation of different carbon and nitrogen sources in production of biosurfactant by *Pseudomonas fluorescens. Desalination*, 223(1–3), 143–151.

Abouseoud, M., Yataghene, A., Amrane, A., and Maachi, R. 2008b. Biosurfactant production by free and alginate entrapped cells of *Pseudomonas fluorescens. Journal of Industrial Microbiology & Biotechnology*, 35(11), 1303–1308.

Abouseoud, M., Yataghene, A., Amrane, A., and Maachi, R. 2010. Effect of pH and salinity on the emulsifying and solubilizing capacity of a biosurfactant produced by *Pseudomonas fluorescens. Journal of Hazardous Materials*, 180(1), 131–136.

Allen, D., Cooksey, C., and Tsai, B., 2012. Spectrophotometry. National Institute of Standards and Technology (NIST) [retrieved on September 10, 2012]. Retrieved from the internet. http://www.nist.gov/pml/div685/grp03/spectrophotometry.cfm.

Allwooda, J. W. and Goodacre, R. 2010. An introduction to liquid chromatography–mass spectrometry instrumentation applied in plant metabolomic analyses. *Phytochemical Analysis*, 21(1), 33–47.

Anandaraj, B. and Thivakaran, P. 2010. Isolation and production of biosurfactantproducing organism from oil spilled soil. *Bioscience Technology*, 1(3), 120–126.

Ardrey, R. E. 2003. *Liquid Chromatography—Mass Spectrometry: An Introduction* (Vol. 2). John Wiley and Sons Ltd., West Sussex, UK.

Arnold, T. and Linke, D. 2007. Phase separation in the isolation and purification of membrane proteins. *Biotechniques*, 43(4), 427.

Ballot, F. 2009. Bacterial production of antimicrobial biosurfactants. Doctoral dissertation, University of Stellenbosch, Stellenbosch, South Africa.

Banat, I., Franzetti, A., Gandolfi, I., Bestetti, G., Martinotti, M., Fracchia, L., Smyth, T., and Marchant, R. 2010. Microbial biosurfactants production, applications and future potential. *Applied Microbiology and Biotechnology*, 87(2), 427–444.

Benincasa, M., Abalos, A., Oliveira, I., and Manresa, A. 2004. Chemical structure, surface properties and biological activities of the biosurfactant produced by *Pseudomonas aeruginosa* LBI from soapstock. *Antonie Van Leeuwenhoek International Journal of General and Molecular Microbiology*, 85(1), 1–8.

Benincasa, M., Contiero, J., Manresa, M. A., and Moraes, I. O. 2002. Rhamnolipid production by *Pseudomonas aeruginosa* LBI growing on soapstock as the sole carbon source. *Journal of Food Engineering*, 54(4), 283–288.

Bergström, S. Theorell, H., and Davide, H. 1946 On a metabolic product of *Pseudomonas pyocyanea*, pyolipic acid, active against *Mycobacterium tuberculosis. Ark Kemi Mineral Geologi*, 23A, 1–12.

Bird, S. M. 1998. Elemental speciation for bioanalytical applications: HPLC-ICP-MS and CE-ICP-MS. Retrived from: http://scholarworks.umass.edu/dissertations/AAI9909148/ (Accessed on September 15, 2012).

Bockmuehl, D. 2012. Biosurfactants as antimicrobial ingredients for cleaning products and cosmetics. *Tenside Surfactants Detergents*, 49(3), 196–198.

Bondarenko, O., Rahman, P. K., Rahman, T. J., Kahru, A., and Ivask, A. 2010. Effects of rhamnolipids from *Pseudomonas aeruginosa* DS10-129 on luminescent bacteria: Toxicity and modulation of cadmium bioavailability. *Microbial Ecology*, 59(3), 588–600.

Cabrera-Valladares, N., Richardson, A.-P., Olvera, C., Trevino, L. G., Deziel, E., Lepine, F., and Soberon-Chavez, G. 2006. Monorhamnolipids and 3-(3-hydroxyalkanoyloxy)alkanoic acids (HAAs) production using *Escherichia coli* as a heterologous host. *Applied Microbiology and Biotechnology*, 73(1), 187–194.

Celik, G.Y., Aslim, B., and Beyatli, Y. 2008. Enhanced crude oil biodegradation and rhamno-lipid production by *Pseudomonas stutzeri* strain G11 in the presence of Tween-80 and Triton X-100. *Journal of Environmental Biology*, 29, 867–870.

Cha, M., Lee, N., Kim, M., Kim, M., and Lee, S. 2008. Heterologous production of *Pseudomonas aeruginosa* EMS1 biosurfactant in *Pseudomonas putida*. *Bioresource Technology*, 99(7), 2192–2199.

Champion, J. T., Gilkey, J. C., Lamparski, H., Retterer, J., and Miller, R. M. 1995. Electron-microscopy of rhamnolipid (biosurfactant) morphology—Effects of ph, cadmium, and octadecane. *Journal of Colloid and Interface Science*, 170(2), 569–574.

Chayabutra, C. and Ju, L. K. 2001a. Polyhydroxyalkanoic acids and rhamnolipids are synthesized sequentially in hexadecane fermentation by *Pseudomonas aeruginosa* ATCC 10145. *Biotechnology Progress*, 17(3), 419–423.

Chayabutra, C., Wu, J., and Ju, L. K. 2001b. Rhamnolipid production by *Pseudomonas aeruginosa* under denitrification, effects of limiting nutrients and carbon substrates. *Biotechnology and Bioengineering*, 72(1), 25–33.

Choi, M. H., Xu, J., Gutierrez, M., Yoo, T., Cho, Y. H., and Yoon, S. C. 2011. Metabolic rela-tionship between polyhydroxyalkanoic acid and rhamnolipid synthesis in *Pseudomonas aeruginosa*, Comparative C-13 NMR analysis of the products in wild-type and mutants. *Journal of Biotechnology*, 151(1), 30–42.

Chowdiah, P., Misra, B. R., Kilbane, J. J., Srivastava, V. J., and Hayes, T. D. 1998. Foam prop-agation through soils for enhanced in-situ remediation. *Journal of Hazardous Materials*, 62(3), 265–280.

Christova, N., Tuleva, B., Cohen, R., Ivanova, G., Stoev, G., Stoilova-Disheva, M., and Stoineva, I. 2011. Chemical characterization and physical and biological activities of rhamnolipids produced by *Pseudomonas aeruginosa* BN10. *Zeitschrift Fur Naturforschung Section C—Journal of Biosciences*, 66(11), 394–402.

Christova, N., Tuleva, B., Lalchev, Z., Jordanova, A., and Jordanov, B. 2004. Rhamnolipid bio-surfactants produced by *Renibacterium salmoninarum* 27BN during growth on n-hexa-decane. *Zeitschrift Fur Naturforschung C—Journal of Biosciences*, 59(1/2), 70–74.

Chrzanowski, L., Dziadas, M., Lawniczak, L., Cyplik, P., Bialas, W., Szulc, A., Lisiecki, P., and Jelen, H. 2012. Biodegradation of rhamnolipids in liquid cultures: Effect of bio-surfactant dissipation on diesel fuel/B20 blend biodegradation efficiency and bacterial community composition. *Bioresource Technology*, 111, 328–335.

Chrzanowski, L., Owsianiak, M., Szulc, A., Marecik, R., Piotrowska-Cyplik, A., Olejnik-Schmidt, A. K., Staniewski, J. et al. 2011. Interactions between rhamnolipid biosurfac-tants and toxic chlorinated phenols enhance biodegradation of a model hydrocarbon-rich effluent. *International Biodeterioration & Biodegradation*, 65(4), 605–611.

Cohen, R. and Exerowa, D. 2007. Surface forces and properties of foam films from rhamno-lipid biosurfactants. *Advances in Colloid and Interface Science*, 134(35), 24–34.

Costa, S. G., de Souza, S. R., Nitschke, M., Franchetti, S. M. M., Jafelicci, M., Lovaglio, R. B., and Contiero, J. 2009. Wettability of aqueous rhamnolipids solutions produced by *Pseudomonas aeruginosa* LBI. *Journal of Surfactants and Detergents*, 12(2), 125–130.

Costa, S. G., Deziel, E., and Lepine, F. 2011. Characterization of rhamnolipid production by *Burkholderia glumae*. *Letters in Applied Microbiology*, 53(6), 620–627.

Costa, S. G., Lepine, F., Milot, S., Deziel, E., Nitschke, M., and Contiero, J. 2009. Cassava wastewater as a substrate for the simultaneous production of rhamnolipids and polyhy-droxyalkanoates by *Pseudomonas aeruginosa*. *Journal of Industrial Microbiology & Biotechnology*, 36(8), 1063–1072.

Costa, S., Nitschke, M., Haddad, R., Eberlin, M. N., and Contiero, J. 2006. Production of *Pseudomonas aeruginosa* LBI rhamnolipids following growth on Brazilian native oils. *Process Biochemistry*, 41(2): 483–488.

Costa, S. G., Nitschke, M., Lepine, F., Deziel, E., and Contiero, J. 2010. Structure, properties and applications of rhamnolipids produced by *Pseudomonas aeruginosa* L2-1 from cassava wastewater. *Process Biochemistry*, 45(9), 1511–1516.

Costa, S. G. V. A. D., Nitschke, M., and Contiero, J. 2008. Fats and oils wastes as substrates for biosurfactant production. *Ciência e Tecnologia de Alimentos, Campinas*, 28(1), 34–38.

Dahrazma, B., Mulligan, C. N., and Nieh, M. P. 2008. Effects of additives on the structure of rhamnolipid (biosurfactant): A small-angle neutron scattering (SANS) study. *Journal of Colloid and Interface Science*, 319(2), 590–593.

Darvishi, P., Ayatollahi, S., Mowla, D., and Niazi, A. 2011. Biosurfactant production under extreme environmental conditions by an efficient microbial consortium, ERCPPI-2. *Colloids and Surfaces B: Biointerfaces*, 84(2), 292–300.

Das, K. and Mukherjee, A. K. 2005. Characterization of biochemical properties and biological activities of biosurfactants produced by *Pseudomonas aeruginosa* mucoid and non-mucoid strains isolated from hydrocarbon-contaminated soil samples. *Applied Microbiology and Biotechnology*, 69(2), 192–199.

Davis, M. L., Mounteer, L. C., Stevens, L. K., Miller, C. D., and Zhou, A. 2011. 2D motility tracking of *Pseudomonas putida* KT2440 in growth phases using video microscopy. *Journal of Bioscience and Bioengineering*, 111(5), 605–611.

de Sousa, J. R., da Costa Correia, J. A., Lima de Almeida, J. G., Rodrigues, S., Loiola Pessoa, O. D., Melo, V. M. M., and Barros Goncalves, L. R. 2011. Evaluation of a co-product of biodiesel production as carbon source in the production of biosurfactant by *P. aeruginosa* MSIC02. *Process Biochemistry*, 46(9), 1831–1839.

Desai, J. D. and Banat, I. M. 1997. Microbial production of surfactants and their commercial potential. *Microbiology and Molecular Biology Reviews*, 61(1), 47–64.

Déziel, E., Lépine, F., Dennie, D., Boismenu, D., Mamer, O. A., and Villemur, R. 1999. Liquid chromatography/mass spectrometry analysis of mixtures of rhamnolipids produced by *Pseudomonas aeruginosa* strain 57RP grown on mannitol or naphthalene. *Biochimica et Biophysica Acta*, 1440(2), 244–252.

do Valle Gomes, M. Z. and Nitschke, M. 2012. Evaluation of rhamnolipid and surfactin to reduce the adhesion and remove biofilms of individual and mixed cultures of food pathogenic bacteria. *Food Control*, 25(2), 441–447.

Downard, K. 2004. *Mass Spectrometry: A Foundation Course*. Royal Society of Chemistry, Cambridge, U.K.

Dubeau, D., Déziel, E., Woods, D. E., and Lépine, F. 2009. Burkholderia thailandensis harbors two identical rhl gene clusters responsible for the biosynthesis of rhamnolipids. *BMC Microbiology*, 9(1), 263.

Edwards, J. R. and Hayashi, J. A. 1965. Structure of a rhamnolipid from *Pseudomonas aeruginosa*. *Archives of Biochemistry and Biophysics*, 111, 415–421.

Eswari, J. S., Anand, M., and Venkateswarlu, C. 2012. Optimum culture medium composition for rhamnolipid production by *Pseudomonas aeruginosa* AT10 using a novel multi-objective optimization method. *Journal of Chemical Technology and Biotechnology*, 88(2), 271–279.

Franzetti, A., Gandolfi, I., Bestetti, G., Smyth, T. J. P., and Banat, I. M. 2010. Production and applications of trehalose lipid biosurfactants. *European Journal of Lipid Science and Technology*, 112(6), 617–627.

Gartshore, J., Lim, Y. C., and Cooper, D. G. 2000. Quantitative analysis of biosurfactants using Fourier Transform Infrared (FT-IR) spectroscopy. *Biotechnology Letters*, 22(2), 169–172.

Gerwig, G. J., Kamerling, J. P., and Vliegenthart, J. F. G. 1978. Determination of the D and L configuration of neutral monosaccharides by high-resolution capillary GLC. *Carbohydrate Research*, 62(2), 381–396.

Gunther, N. W., Nunez, A., Fett, W., and Solaiman, D. K. Y. 2005. Production of rhamnolipids by *Pseudomonas chlororaphis*, a nonpathogenic bacterium. *Applied and Environmental Microbiology*, 71(5), 2288–2293.

Gunther, N. W., Nunez, A., Fortis, L., and Solaiman, D. K. Y. 2006. Proteomic based investigation of rhamnolipid production by *Pseudomonas chlororaphis* strain NRRL B-30761. *Journal of Industrial Microbiology & Biotechnology*, 33(11), 914–920.

Gunther, N. W., Solaiman, D. K. Y., and Fett, W. F. 2007. Processes for the production of rhamnolipids. US Patent No. 7202063. U.S. Patent and Trademark Office, Washington, DC.

Guo, Y. P., Hu, Y. Y., Gu, R. R., and Lin, H. 2009. Characterization and micellization of rhamnolipidic fractions and crude extracts produced by *Pseudomonas aeruginosa* mutant MIG-N146. *Journal of Colloid and Interface Science*, 331(2), 356–363.

Gustafson, K. R. 2012. *NMR Methods for Stereochemical Assignments. Handbook of Marine Natural Products*. Springer, Dordrecht, the Netherlands, pp. 547–570.

Haba, E., Abalos, A., Jauregui, O., Espuny, M. J., and Manresa, A. 2003. Use of liquid chromatography-mass spectroscopy for studying the composition and properties of rhamnolipids produced by different strains of *Pseudomonas aeruginosa. Journal of Surfactants and Detergents*, 6(2), 155–161.

Haba, E., Espuny, M. J., Busquets, M., and Manresa, A. 2000. Screening and production of rhamnolipids by *Pseudomonas aeruginosa* 47T2 NCIB 40044 from waste frying oils. *Journal of Applied Microbiology*, 88(3), 379–387.

Hamamoto, T., Kaneda, M., Horikoshi, K., and Kudo, T. 1994. Characterization of a protease from a psychrotroph, *Pseudomonas fluorescens* 114. *Applied and Environmental Microbiology*, 60(10), 3878–3880.

Hassan, H. M. and Fridovich, I. 1980. Mechanism of the antibiotic action pyocyanine. *Journal of Bacteriology*, 141(1), 156–163.

Hata, A., Endo, Y., Nakajima, Y., Ikebe, M., Ogawa, M., Fujitani, N., and Endo, G. 2007. HPLC-ICP-MS speciation analysis of arsenic in urine of Japanese subjects without occupational exposure. *Journal of Occupational Health*, 49(3), 217–223.

Hauser, G., and Karnovsky M. L. 1954. Studies on the production of glycolipid by *Pseudomonas aeruginosa. Journal of Bacteriology*, 68, 645–654.

Haussler, S., Nimtz, M., Domke, T., Wray, V., and Steinmetz, I. 1998. Purification and characterization of a cytotoxic exolipid of *Burkholderia pseudomallei. Infection and Immunity*, 66(4), 1588–1593.

Healy, M. G., Devine, C. M., and Murphy, R. 1996. Microbial production of biosurfactants. *Resources, Conservation and Recycling*, 18(1), 41–57.

Hemmer, M. C., Steinhauer, V., and Gasteiger, J. 1999. Deriving the 3D structure of organic molecules from their infrared spectra. *Vibrational Spectroscopy*, 19(1), 151–164.

Hoermann, B., Mueller, M. M., Syldatk, C., and Hausmann, R. 2010. Rhamnolipid production by *Burkholderia plantarii* DSM 9509(T). *European Journal of Lipid Science and Technology*, 112(6), 674–680.

Hori, K., Marsudi, S., and Unno, H. 2002. Simultaneous production of polyhydroxyalkanoates and rhamnolipids by *Pseudomonas aeruginosa. Biotechnology and Bioengineering*, 78(6), 699–707.

Ishigami, Y. and Suzuki, S. (1997). Development of biochemicals—functionalization of biosurfactants and natural dyes. *Progress in Organic Coatings*, 31(1), 51–61.

Jadhav, M., Kagalkar, A., Jadhav, S., and Govindwar, S. 2011. Isolation, characterization, and antifungal application of a biosurfactant produced by *Enterobacter* sp. MS16. *European Journal of Lipid Science and Technology*, 113(11), 1347–1356.

Jarvis, F.G. and Johnson, M.J. 1949. A glyco-lipid produced by *Pseudomonas aeruginosa. Journal of American Chemical Society*, 71, 4124–4126.

Janiyani, K. L., Wate, S. R., and Joshi, S. R. 1992. Surfactant production by *Pseudomonas stutzeri. Journal of Microbial Biotechnology*, 7(1), 18–21.

JENEIL Biosurfactant Co., 2004. Material safety data sheet for JBR425, http://www. biosurfactant.com/downloads/jbr425msds.pdf, accessed on November 2009.

Jian, L. J., Chen, Y. Q., Wen, H. X., Kong, J. l., Yan, P., and Zhang, D. W. 2010. Intervention of azithromycin on *Pseudomonas aeruginosa* biofilm and virulence factors. *Zhonghua Weishengwuxue He Mianyixue Zazhi*, 30(11), 1020–1024.

Jimenez Islas, D., Medina Moreno, S. A., and Gracida Rodriguez, J. N. 2010. Biosurfactant properties, applications and production a review. *Revista Internacional De Contaminacion Ambiental*, 26(1), 65–84.

Jork, H. and Gang, J. (1994). Opportunities and limitations of modern TLC/HPTLC in the quality control of *L-tryptophan*. *L-Tryptophan. Current Prospects Medicine and Drug Safety*, 338–350.

Kaczorek, E., Jesionowski, T., Giec, A., and Olszanowski, A. 2012. Cell surface properties of *Pseudomonas stutzeri* in the process of diesel oil biodegradation. *Biotechnology Letters*, 34(5), 857–862.

Kaczorek, E., Moszyńska, S., and Olszanowski, A. 2011. Modification of cell surface properties of *Pseudomonas alcaligenes* S22 during hydrocarbon biodegradation. *Biodegradation*, 22(2), 359–366.

Kamal, A., Shaik, A. B., Kumar, C. G., Mongolla, P., Rani, P. U., Krishna, K. V., and Joseph, J. 2012. Metabolic profiling and biological activities of bioactive compounds produced by *Pseudomonas* sp. strain ICTB-745 isolated from Ladakh, India. *Journal of Microbiology and Biotechnology*, 22(1): 69–79.

Khoshdast, H., Abbasi, H., Sam, A., and Noghabi, K. A. 2012. Frothability and surface behavior of a rhamnolipid biosurfactant produced by *Pseudomonas aeruginosa* MA01. *Biochemical Engineering Journal*, 60, 127–134.

Koch, A. K., Kappeli, O., Fiechter, A., and Reiser, J. 1991. Hydrocarbon assimilation and biosurfactant production in *Pseudomonas-aeruginosa* mutants. *Journal of Bacteriology*, 173(13), 4212–4219.

Kumar, C. G. and Mamidyala, S. K. 2011. Extracellular synthesis of silver nanoparticles using culture supernatant of *Pseudomonas aeruginosa*. *Colloids and Surfaces B—Biointerfaces*, 84(2), 462–466.

Kunieda, H. and Shinoda, K. 1976. Krafft points, critical micelle concentrations, surface tension, and solubilizing power of aqueous solutions of fluorinated surfactants. *The Journal of Physical Chemistry*, 80(22), 2468–2470.

Lang, S. and Wagner, F. 1987. Structure and properties of biosurfactants. In: Kosaric, N., Gray, N. C. C., and Cairns, WL (Eds.) *Biosurfactants and Biotechnology*, Marcel Dekker, New York, pp. 21–45.

Lee, M., Kim, M., Vancanneyt, M., Swings, J., Kim, S., Kang, M., and Lee, S. 2005. *Tetragenococcus koreensis* sp. nov., a novel rhamnolipid-producing bacterium. *International Journal of Systematic and Evolutionary Microbiology*, 55(4), 1409–1413.

Leitermann, F., Syldatk, C., and Hausmann, R. 2008. Fast quantitative determination of microbial rhamnolipids from cultivation broths by ATR-FTIR Spectroscopy. *Journal of Biological Engineering*, 2(1), 1–8.

Lotfabad, T. B., Abassi, H., Ahmadkhaniha, R., Roostaazad, R., Masoomi, F., Zahiri, H. S., Ahmadian, G., Vali, H., and Noghabi, K. A. 2010. Structural characterization of a rhamnolipid-type biosurfactant produced by *Pseudomonas aeruginosa* MR01: Enhancement of di-rhamnolipid proportion using gamma irradiation. *Colloids and Surfaces B: Biointerfaces*, 81(2), 397–405.

Lourith, N. and Kanlayavattanakul, M. 2009. Natural surfactants used in cosmetics: Glycolipids. *International Journal of Cosmetic Science*, 31(4), 255–261.

Martinez-Garcia, E. and de Lorenzo, V. 2011. Engineering multiple genomic deletions in Gram-negative bacteria: Analysis of the multi-resistant antibiotic profile of *Pseudomonas putida* KT2440. *Environmental Microbiology*, 13(10), 2702–2716.

Martinez-Toledo, A. and Rodriguez-Vazquez, R. 2011. Response surface methodology (Box-Behnken) to improve a liquid media formulation to produce biosurfactant and phenanthrene removal by *Pseudomonas putida*. *Annals of Microbiology*, 61(3), 605–613.

Mata-Sandoval, J. C., Karns, J., and Torrents, A. 1999. High-performance liquid chromatography method for the characterization of rhamnolipid mixtures produced by *Pseudomonas aeruginosa* UG2 on corn oil. *Journal of Chromatography A*, 864(2), 211–220.

McClure, C. D. and Schiller, N. L. 1992. Effects of *Pseudomonas aeruginosa* rhamnolipids on human monocyte-derived macrophages. *Journal of Leukocyte Biology*, 51(2), 97–102.

Messmer, A. T., Lippert, K. M., Steinwand, S., Lerch, E.-B. W., Hof, K., Ley, D., Gerbig, D., Hausmann, H., Schreiner, P. R., and Bredenbeck, J. 2012. Two-dimensional infrared spectroscopy reveals the structure of an Evans auxiliary derivative and its SnCl4 Lewis acid complex. *Chemistry—A European Journal*, 18(47), 14989–14995.

Meyer, J. and Geoffroy, V. 2002. Siderophore Typing, a Powerful Tool for the Identification of Fluorescent and Nonfluorescent *Pseudomonads*. *Applied Environmental Microbiology*, 68(6), 2745–2753.

Molina, L., Ramos, C., Duque, E., Ronchel, M. C., Garcia, J. M., Wyke, L., and Ramos, J. L. 2000. Survival of *Pseudomonas putida* KT2440 in soil and in the rhizosphere of plants under greenhouse and environmental conditions. *Soil Biology and Biochemistry*, 32(3), 315–321.

Monteiro, S. A., Sassaki, G. L., de Souza, L. M., Meira, J. A., de Araujo, J. M., Mitchell, D. A., Ramos, L. P., and Krieger, N. 2007. Molecular and structural characterization of the biosurfactant produced by *Pseudomonas aeruginosa* DAUPE 614. *Chemistry and Physics of Lipids*, 147(1), 1–13.

Morris, J. D., Hewitt, J. L., Wolfe, L. G., Kamatkar, N. G., Chapman, S. M., Diener, J. M., & Shrout, J. D. 2011. Imaging and analysis of *Pseudomonas aeruginosa* swarming and rhamnolipid production. *Applied and Environmental Microbiology*, 77(23): 8310–8317.

Mueller, M. M., Hoermann, B., Syldatk, C., and Hausmann, R. 2010. *Pseudomonas aeruginosa* PAO1 as a model for rhamnolipid production in bioreactor systems. *Applied Microbiology and Biotechnology*, 87(1), 167–174.

Muller, N. 1993. Temperature dependence of critical micelle concentrations and heat capacities of micellization for ionic surfactants. *Langmuir*, 9(1), 96–100.

Mulligan, C. N. 2005. Environmental applications for biosurfactants. *Environmental Pollution*, 133(2), 183–198.

Mulligan, C. N. 2009. Recent advances in the environmental applications of biosurfactants. *Current Opinion in Colloid & Interface Science*, 14(5), 372–378.

Mulligan, C. N. and Gibbs, B. F. 1989. Correlation of nitrogen metabolism with biosurfactant production by *Pseudomonas aeruginosa*. *Applied and Environmental Microbiology*, 55(11), 3016–3019.

Mulligan, C. N., and Gibbs, B. F. 1990. Recovery of biosurfactants by ultrafiltration. *Journal of Chemical Technology and Biotechnology*, 47(1), 23–29.

Mulligan, C. N. and Gibbs, B. F. 1993. Factors influencing the economics of biosurfactants in Biosurfactants, Production, Properties Applications, pp. 329–371, ed. N. Kosaric (Marcel Dekker, New York).

Mulligan, C. N. and Gibbs, B. F. 2004. Types, production and applications of biosurfactants. *Proceedings of the Indian National Science Academy Part B*, 70(1), 31–56.

Mulligan, C. N., Mahmourides, G., and Gibbs, B. F. 1989a. The influence of phosphate metabolism on biosurfactant production by *Pseudomonas aeruginosa*. *Journal of Biotechnology*, 12(3), 199–209.

Mulligan, C. N., Mahmourides, G., and Gibbs, B. F. 1989b. Biosurfactant production by a chloramphenicol tolerant strain of *Pseudomonas aeruginosa*. *Journal of Biotechnology*, 12(1), 37–43.

Mulligan, C. N., Chow, T. Y. K., and Gibbs, B. F. 1989c. Enhanced biosurfactant production by a mutant *Bacillus subtilis* strain. *Applied Microbiology and Biotechnology*, 31(5–6), 486–489.

Mulligan, C. N., Yong, R. N., and Gibbs, B. F. 2001a. Heavy metal removal from sediments by biosurfactants. *Journal of Hazardous Materials*, 85(1), 111–125.

Nakahara, H., Ishikawa, T., Sarai, Y., Kondo, I., Kozukue, H., and Silver, S. 1977. Linkage of mercury, cadmium, and arsenate and drug resistance in clinical isolates of *Pseudomonas aeruginosa*. *Applied and Environmental Microbiology*, 33(4), 975–976.

Nelson, K. E., Weinel, C., Paulsen, I. T., Dodson, R. J., Hilbert, H., Martins dos Santos, V. A. P., Fouts, D. E. et al. 2002. Complete genome sequence and comparative analysis of the metabolically versatile *Pseudomonas putida* KT2440. *Environmental Microbiology*, 4(12), 799–808.

Nguyen, T. and Sabatini, D. 2011. Characterization and emulsification properties of rhamnolipid and sophorolipid biosurfactants and their applications. *International Journal of Molecular Sciences*, 12(2), 1232–1244.

Nie, M., Yin, X., Ren, C., Wang, Y., Xu, F., and Shen, Q. 2010. Novel rhamnolipid biosurfactants produced by a polycyclic aromatic hydrocarbon-degrading bacterium *Pseudomonas aeruginosa* strain NY3. *Biotechnology Advances*, 28(5), 635–643.

Nitschke, M., Costa, S. G., and Contiero, J. 2005. Rhamnolipid surfactants: An update on the general aspects of these remarkable biomolecules. *Biotechnology Progress*, 21(6), 1593–1600.

Nitschke, M., Costa, S. G., and Contiero, J. 2011. Rhamnolipids and PHAs: Recent reports on *Pseudomonas*-derived molecules of increasing industrial interest. *Process Biochemistry*, 46(3), 621–630.

Ochsner, U. A., Fiechter, A., and Reiser, J. 1994. Isolation, characterization, and expression in *Escherichia coli* of the *Pseudomonas aeruginosa* rhlAB genes encoding a rhamnosyltransferase involved in rhamnolipid biosurfactant synthesis. *Journal of Biological Chemistry*, 269(31), 19787–19795.

Ochsner, U. A., Reiser, J., Fiechter, A., and Witholt, B. 1995. Production of *Pseudomonas aeruginosa* rhamnolipid biosurfactants in heterologous hosts. *Applied and Environmental Microbiology*, 61(9), 3503–3506.

Oliveira, F. J. S., Vazquez, L., de Campos, N. P., and de Franca, F. P. 2009. Production of rhamnolipids by a *Pseudomonas alcaligenes* strain. *Process Biochemistry*, 44(4), 383–389.

Ozdemir, G. and Malayoglu, U. 2004. Wetting characteristics of aqueous rhamnolipids solutions. *Colloids and Surfaces B—Biointerfaces*, 39(1–2), 1–7.

Pacwa-Płociniczak, M., Płaza, G. A., Piotrowska-Seget, Z., and Cameotra, S. S. 2011. Environmental applications of biosurfactants: Recent advances. *International Journal of Molecular Sciences*, 12(1), 633–654.

Pajarron, A. M., Dekoster, C. G., Heerma, W., Schmidt, M., and Haverkamp, J. 1993. Structure identification of natural rhamnolipid mixtures by fast-atom-bombardment tandem mass-spectrometry. *Glycoconjugate Journal*, 10(3), 219–226.

Pantazaki, A. A., Dimopoulou, M. I., Simou, O. M., and Pritsa, A. A. 2010. Sunflower seed oil and oleic acid utilization for the production of rhamnolipids by *Thermus thermophilus* HB8. *Applied Microbiology and Biotechnology*, 88(4), 939–951.

Pantazaki, A. A., Papaneophytou, C. P., and Lambropoulou, D. A. 2011. Simultaneous polyhydroxyalkanoates and rhamnolipids production by *Thermus thermophilus* HB8. *AMB Express*, 1(1), 1–13.

Pathema, 2006. *Burkholderia pseudomallei*. Bioinformatics Resource Center [retrieved on September 10, 2012]. Retrieved from the internet. http://pathema.jcvi.org/pathema/b_pseudomallei.shtml.

Pavel, J. T., Hyde, E. C., and Bruch, M. D. 2012. Structure determination of unknown organic liquids using NMR and IR spectroscopy: A general chemistry laboratory. *Journal of Chemical Education*, 89(11), 1450–1453.

Penfold, J., Thomas, R. K., and Shen, H. H. 2012. Adsorption and self-assembly of biosurfactants studied by neutron reflectivity and small angle neutron scattering: Glycolipids, lipopeptides and proteins. *Soft Matter*, 8(3), 578–591.

Pereira, J. F. B., Gudina, E. J., Doria, M. L., Domingues, M. R., Rodrigues, L. R., Teoxeira, J. A., and Coutinho, J. A. P. 2012. Characterization by electrospray ionization and tandem mass spectrometry of rhamnolipids produced by two *Pseudomonas aeruginosa* strains isolated from Brazilian crude oil. *European Journal of Mass Spectrometry*, 18(4), 399–406.

Pham, T. H., Webb, J. S., and Rehm, B. H. 2004. The role of polyhydroxyalkanoate biosynthesis by *Pseudomonas aeruginosa* in rhamnolipid and alginate production as well as stress tolerance and biofilm formation. *Microbiology*, 150(10), 3405–3413.

Pimienta, A. L., Diaz, M. P. M., Carvajal, F. G. S., and Grossov, J. L. 1997. Production of biosurfactants (Rhamnolipids) by *Pseudomonas aeruginosa* isolated from colombian sludges. *Ciencia ,Technologia and Futuro*, 1(3), 95–108.

Pinzon, N. and Ju, L.-K. 2009. Improved detection of rhamnolipid production using agar plates containing methylene blue and cetyl trimethylammonium bromide. *Biotechnology Letters*, 31(10), 1583–1588.

Pornsunthorntawee, O., Wongpanit, P., and Rujiravanit, R. 2010. Rhamnolipid biosurfactants: Production and their potential in environmental biotechnology. *Biosurfactants*, 672, 211–221.

Propst, C. and Lubin, L. 1979. Light-mediated changes in pigmentation of *Pseudomonas aeruginosa* cultures. *Journal of General Microbiology,* 113(2): 261–266.

Rahman, K. S. M., Rahman, T. J., McClean, S., Marchant, R., and Banat, I. M. 2002. Rhamnolipid biosurfactant production by strains of *Pseudomonas aeruginosa* using low-cost raw materials. *Biotechnology Progress*, 18(6), 1277–1281.

Rahman, P., Dusane, D., Zinjarde, S., Venugopalan, V., McLean, R., and Weber, M. 2010a. Quorum sensing: Implications on rhamnolipid biosurfactant production. *Biotechnology and Genetic Engineering Reviews*, 27(1), 159–184.

Rahman, P. K. S. M., Pasirayi, G., Auger, V., and Ali, Z. 2010b. Production of rhamnolipid biosurfactants by *Pseudomonas aeruginosa* DS10–129 in a microfluidic bioreactor. *Biotechnology and Applied Biochemistry*, 55(1), 45–52.

Rashedi, H., Assadi, M. M., Jamshidi, E., and Bonakdarpour, B. 2006. Optimization of the production of biosurfactant by *Pseudomonas aeruginosa* HR isolated from an Iranian southern oil well. *Iranian Journal of Chemistry & Chemical Engineering—International English Edition*, 25(1), 25–30.

Rashedi, H., Jamshidi, E., Assadi, M. M., and Bonakdarpour, B. 2005. Isolation and production of biosurfactant from *Pseudomonas aeruginosa* isolated from Iranian southern wells oil. *International Journal of Environmental Science and Technology*, 2(2), 121–127.

Raya, A., Sodagari, M., Pinzon, N. M., He, X., Newby, B.-M. Z., and Ju, L.-K. 2010. Effects of rhamnolipids and shear on initial attachment of *Pseudomonas aeruginosa* PAO1 in glass flow chambers. *Environmental Science and Pollution Research*, 17(9), 1529–1538.

Raza, Z. A., Khalid, Z. M., Khan, M. S., Banat, I. M., Rehman, A., Naeem, A., and Saddique, M. T. 2010. Surface properties and sub-surface aggregate assimilation of rhamnolipid surfactants in different aqueous systems. *Biotechnology Letters*, 32(6), 811–816.

Raza, Z. A., Khan, M. S., and Khalid, Z. M. 2007. Evaluation of distant carbon sources in biosurfactant production by a gamma ray-induced *Pseudomonas putida* mutant. *Process Biochemistry*, 42(4), 686–692.

Rezanka, T., Siristova, L., and Sigler, K. 2011. Rhamnolipid-producing thermophilic bacteria of species *Thermus* and *Meiothermus*. *Extremophiles*, 15(6), 697–709.

Rhodes, M. E. 1959. The characterization of *Pseudomonas fluorescens*. *General Microbiology*, 21(1), 221–265.

Rooney, A. P., Price, N. P., Ray, K. J., and Kuo, T. M. 2009. Isolation and characterization of rhamnolipid-producing bacterial strains from a biodiesel facility. *FEMS Microbiology Letters,* 295(1), 82–87.

Salleh, S. M., Noh, N. A. M., and Yahya, A. R. M. 2011. Comparative study: Different recovery techniques of rhamnolipid produced by *Pseudomonas aeruginosa*. USMAR-2 *International Proceedings of Chemical, Biological and Environmental Engineering (IPCBEE)*, 18, 132–135.

Sastoque-Cala, L., Cotes-Prado, A.M., Rodríguez-Vázquez, R., and Pedroza Rodríguez, A.M. 2010. Effect of nutrients and fermentation conditions on the production of biosurfactants using rhizobacteria isolated from fique plants. *Universitas Scientiarum*, 15, 251–264.

Siegmund, I. and Wagner, F. 1991. New method for detecting rhamnolipids excreted by *Pseudomonas* species during growth on mineral agar. *Biotechnology Techniques*, 5(4), 265–268.

Sim, L., Ward, O. P., and Li, Z. Y. 1997. Production and characterisation of a biosurfactant isolated from *Pseudomonas aeruginosa* UW-1. *Journal of Industrial Microbiology & Biotechnology*, 19(4), 232–238.

Singh, B. K. and Walker, A. 2006. Microbial degradation of organophosphorus compounds. *FEMS Microbiology Reviews*, 30(3), 428–471.

Snow, N. H. and Slack, G. C. 2002. Chromatography, gas chromatography-GC. Kirk-Othmer encyclopedia of chemical technology.

Stover, C. K., Pham, X. Q., Erwin, A. L., Mizoguchi, S. D., Warrener, P., Hickey, M. J., et al. 2000. Complete genome sequence of *Pseudomonas aeruginosa* PAO1, an opportunistic pathogen. *Nature.*, 406, 959–964.

Swann, G. E. A. and Patwardhan, S. V. 2011. Application of Fourier Transform Infrared Spectroscopy (FTIR) for assessing biogenic silica sample purity in geochemical analyses and palaeoenvironmental research. *Climate of the Past*, 7, 65–74.

Takemoto, J. Y., Bensaci, M., De Lucca, A. J., Cleveland, T. E., Gandhi, N. R., and Skebba, V. P. 2010. Inhibition of fungi from diseased grape by syringomycin E-rhamnolipid mixture. *American Journal of Enology and Viticulture*, 61(1), 120–124.

TDMU (I.Horbachevsky Ternopil State Medical University, Ukraine). High-performance liquid and thin-layer chromatography. [retrieved on September 2012]. Retrieved from the internet. http://intranet.tdmu.edu.ua/data/kafedra/internal/pharma_2/lectures_stud/%D0%B0%D0%BD%D0%B0%D0%BB%D1%96%D1%82%D0%B8%D1%87%D0%BD%D0%B0%20%D1%85%D1%96%D0%BC%D1%96%D1%8F/English/21%20High-performance%20liquid%20and%20thin-layer%20chromatography.htm.

Thanomsub, B., Pumeechockchai, W., Limtrakul, A., Arunrattiyakorn, P., Petchleelaha, W., Nitoda, T., and Kanzaki, H. 2009. Erratum to chemical structures and biological activities of rhamnolipids produced by *Pseudomonas aeruginosa* B189 isolated from milk factory waste. *Bioresource Technology*, 100(23), 6141–6141.

Timmis, K. N. 2002. *Pseudomonas putida*: A cosmopolitan opportunist par excellence. *Environmental Microbiology*, 4(12), 779–781.

Tindall, B. J., Kämpfer, P., Euzéby, J. P., and Oren, A. 2006. Valid publication of names of prokaryotes according to the rules of nomenclature: Past history and current practice. *International Journal of Systematic and Evolutionary Microbiology*, 56(11), 2715–2720.

Todar, K. 2004. *Todar's Online Textbook of Bacteriology: Pseudomonas Aeruginosa.* University of Wisconsin-Madison Department of Bacteriology, Madison, WI.

Tombolini, R., Van Der Gaag, D. J., Gerhardson, B., and Jansson, J. K. 1999. Colonization pattern of the biocontrol strain pseudomonas chlororaphis MA 342 on barley seeds visualized by using green fluorescent protein. *Applied and Environmental Microbiology*, 65(8), 3674–3680.

Toribio, J., Escalante, A. E., and Soberon-Chavez, G. 2010. Rhamnolipids: Production in bacteria other than *Pseudomonas aeruginosa*. *European Journal of Lipid Science and Technology*, 112(10), 1082–1087.

Tuleva, B. K., Ivanov, G. R., and Christova, N. E. 2002. Biosurfactant production by a new *Pseudomonas putida* strain. *Zeitschrift Fur Naturforschung C—A Journal of Biosciences*, 57(3–4), 356–360.

U.S. *Environmental Protection Agency* (EPA). 2001. *Pseudomonas chlororaphis* strain 63–28 (006478) Fact Sheet, [retrieved on September 30, 2012]. Retrieved from the internet. http://www.epa.gov/opp00001/chem_search/reg_actions/registration/fs_PC-006478_01-Apr-01.pdf.

Van Bramer, S. E. 1997. An Introduction to Mass Spectrometry. Lecture notes. Widener University. Retrived from: http://science.widener.edu/svb/massspec/massspec.pdf (Accessed on November 2012).

Van Gennip, M., Christensen, L. D., Alhede, M., Phipps, R., Jensen, P. Ø., Christophersen, L., and Bjarnsholt, T. 2009. Inactivation of the rhlA gene in *Pseudomonas aeruginosa* prevents rhamnolipid production, disabling the protection against polymorphonuclear leukocytes. *Apmis,* 117(7), 537–546.

van Rij, E. T., Wesselink, M., Chin-A-Woeng, T. F., Bloemberg, G. V., and Lugtenberg, B. J. 2004. Influence of environmental conditions on the production of Phenazine-1-carboxamide by *Pseudomonas chlororaphis* PCL1391. *Molecular Plant-Microbe Interactions*, 17(5), 557–566.

Varnier, A.-L., Sanchez, L., Vatsa, P., Boudesocque, L., Garcia-Brugger, A., Rabenoelina, F., Sorokin, A. et al. 2009. Bacterial rhamnolipids are novel MAMPs conferring resistance to *Botrytis cinerea* in grapevine. *Plant Cell and Environment*, 32(2), 178–193.

Vatsa, P., Sanchez, L., Clement, C., Baillieul, F., and Dorey, S. 2010. Rhamnolipid biosurfactants as new players in animal and plant defense against microbes. *International Journal of Molecular Sciences*, 11(12), 5096–5109.

Verbeck, G., Ruotolo, B., Sawyer, H., Gillig, K., Russell, D., Ruotolo, G, B., Sawyer, H., Gillig, K., and Russell, D. 2002. A fundamental introduction to ion mobility mass spectrometry applied to the analysis of biomolecules. *Journal of Biomolecular Techniques*, 13(2), 56.

Wang, Q., Fang, X., Bai, B., Liang, X., Shuler, P. J., Goddard III, W. A., and Tang, Y. 2007. Engineering bacteria for production of rhamnolipid as an agent for enhanced oil recovery. *Biotechnology and Bioengineering*, 98(4), 842–853.

Wang, S. and Mulligan, C. N. 2009. Rhamnolipid biosurfactant-enhanced soil flushing for the removal of arsenic and heavy metals from mine tailings. *Process Biochemistry*, 44(3), 296–301.

Wecke, T., Bauer, T., Harth, H., Maeder, U., and Mascher, T. 2011. The rhamnolipid stress response of *Bacillus subtilis. FEMS Microbiology Letters*, 323(2), 113–123.

Wena, J., Staceya, S. P., McLaughlina, M. J., and Kirbyb, J. K. 2009. Biodegradation of rhamnolipid, EDTA and citric acid in cadmium and zinc contaminated soils. *Soil Biology and Biochemistry*, 41(10), 2214–2221.

Wittgens, A., Tiso, T., Arndt, T. T., Wenk, P., Hemmerich, J., Mueller, C., Wichmann, R. et al. 2011. Growth independent rhamnolipid production from glucose using the non-pathogenic *Pseudomonas putida* KT2440. *Microbial Cell Factories*, 10(1), 80.

Xia, W. J., Dong, H. P., Yu, L., and Yu, D. F. 2011. Comparative study of biosurfactant produced by microorganisms isolated from formation water of petroleum reservoir. *Colloids and Surfaces A: Physicochemical and Engineering Aspects*, 392(1), 124–130.

Zhu, Y., Gan, J. J., Zhang, G. L., Yao, B., Zhu, W. j., and Meng, Q. 2007. Reuse of waste frying oil for production of rhamnolipids using *Pseudomonas aeruginosa* zju.u1M. *Journal of Zhejiang University-Science A*, 8(9), 1514–1520.

4 Sophorolipids
Characteristics, Production, and Applications

Vivek K. Morya and Eun-Ki Kim

CONTENTS

INTRODUCTION

Biosurfactants or surface active biopolymers display a specific range of structures and have been well known for their ability to cause emulsification. Based on their chemical composition, microbe-originated biosurfactants have been broadly grouped into glycolipids, lipopeptides, lipoproteins, phospholipids, fatty acids (hydroxylated and crossed-linked), polymeric surfactants, and particulate surfactants. The glycolipids (e.g., trehalose lipids, rhamnolipids, sophorolipids, mannosylerythritol lipids, lipopeptides, surfactin, iturin, fengycin, lichenysin, emulsan, biodispersan, and liposan) have been one of the prior choices for the researcher for exploring the biosurfactants. Among all, rhamnolipids and sophorolipids are two most explored and commercialized sophorolipids. Here, we are going to focus on some fundamentals and applications of sophorolipids. Gorin et al. (1961) described about sophorolipid for the first time in 1961. After that, numerous manuscripts have been published. Like other surfactants, they also facilitated the uptake of hydrophobic substrates such as triglycerides or alkanes by the microorganisms.

The sophorolipids receive an important position in both academic and commercial areas because of various kinds of bioactivities, which include the properties of detergency, bioremediation, enhanced oil recovery, and its medicinal application. Various numbers of nonpathogenic organisms are known as sophorolipid producer, with a significantly high yield (Desai and Banat, 1997; Kitamoto et al., 2002; van Bogaert et al., 2007, 2011; Adamezak and Bednarski, 2000).

In 1961, the ascomycetous yeast species *Torulopsis magnoliae* was reported as the producer of sophorolipids (Gorin et al., 1961), which was further identified as *T. apicola* by Tulloch and Spencer in 1968, and now the current nomenclature of this fungus is *Candida apicola*. After this important finding, two other organisms have also been identified as sophorolipid producer: *C. gropengiesseri* (Jones, 1967; Jones and Howe, 1968) and *C. bombicola* (Spencer et al., 1970, 1979). The *C. bombicola* (ATCC 22214) has been highly explored as a producer as well as a model organism in research, thus this strain has received the most attention in the study of sophorolipids (Develter et al., 2003; Develter and Fleurackers, 2010). Recently, the genetic engineering has been approached to identify for a metabolically more competent strain for the production enhancement, but still this research is on a preliminary stage, which needs to be explored more (van Bogaert et al., 2009a,b).

Chemically, this molecule consists of a hydroxy fatty acid, which is glycosidically bound with sophorose (a diglucose). The sophorose moiety may contain acetyl groups at the 6'- and/or 6"-positions. This allows lactonification of the molecule by itself or may form a lactone through linking the carboxyl group of the hydroxy fatty acid and the 4"-hydroxyl of the sophorose.

Another reason for the sophorolipid being better choice in the industry is its mode of the biosynthesis. Along with *de novo* synthesis, the massive bioconversion is also possible, under suitable conditions. Still there are some limitations especially the chain length of fatty acid, as the chains below the length of C16 are completely metabolized through β-oxidation, whereas chains larger than C18 may become shortened by this mechanism and incorporated when they reach a suitable length. But these limitations have really given opportunity for the engineers and researchers to meet the challenges. Metabolic engineering, reconstruction engineering, or genetic engineering cope together, and maybe in future, this issue will also sort. The aim of this chapter is to discuss some basic issues and the current status of sophorolipid biotechnology.

STRUCTURE AND PROPERTIES

Structurally, the sophorolipids also have two major domains, like any other biosurfactant: (1) hydrophobic domain and (2) hydrophilic domain. The hydrophobic part of the sophorolipids contains a terminal or subterminal hydroxylated fatty acid, which is linked to sophorose through β-glycoside linkage (Figure 4.1). This disaccharide sophorose molecule is chemically a diglucose with an unusual β-1,2 bond, which forms the hydrophilic domain of the sophorolipid. Sophorolipids are synthesized as a mixture of slightly different molecules and the two major points of variation: acetylation and lactonization (van Bogaert et al., 2011). The presence of the acetyl groups at the 6'- and/or 6"-positions on sophorolipid may or may not be involved in internal esterification. As a result, the sophorolipid can occur in either an acidic form with a free fatty acid tail or a lactonic form with an internal esterification (Asmer et al., 1988; Davila et al., 1993; van Bogaert et al., 2007, 2008a,b,c, 2011). The overall properties of sophorolipids are influenced by the chemical properties of

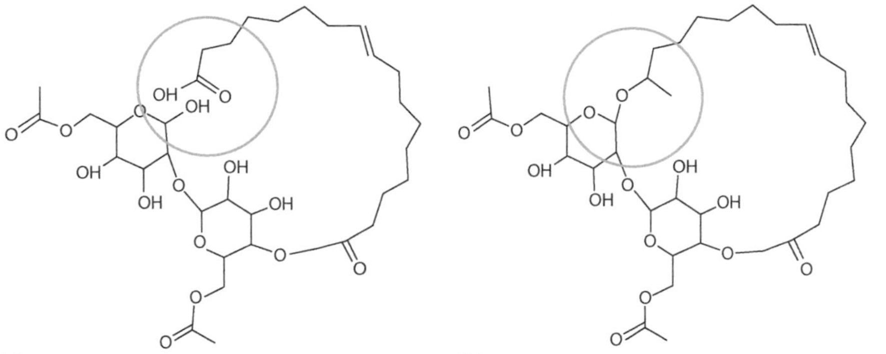

FIGURE 4.1 Diagrammatic representation of the sophorolipid: (a) acid form and (b) lactonized form.

hydroxyl fatty acid, which are affected by following parameters: (1) chain length, (2) number and position of unsaturated bonds, (3) distribution and count of hydroxyl groups, and (4) location either at the terminal (m) or at the subterminal ($m-1$) position (Asmer et al., 1988; Davila et al., 1993). These structural differences make sophorolipids more versatile and biologically active compounds than the other surfactants. Lactonized sophorolipids differ in biological and physicochemical properties while comparing with acidic forms of sophorolipids (van Bogaert et al., 2007). In general, lactonic sophorolipids have better surface activity and antimicrobial activity than acidic sophorolipids, whereas the acidic sophorolipids display a better detergency property than the lactonic sophorolipids (van Bogaert et al., 2011). Furthermore, a small change makes more differences in bioactivities, for example, acetyl groups render the molecules less water soluble, but enhance their antiviral and cytokine-stimulating effects (Kim et al., 2002; Shah et al., 2005; Sleiman et al., 2009). The di- or monoacetylation of lactonic sophorolipids shows better antibacterial activity than the nonacetylated lactonic ones or acidic forms (van Bogaert et al., 2011), and diacetylated lactonic sophorolipids possess a lower critical micelle concentration (CMC) and surface tension as compared to nonacetylated acidic molecules (Lang et al., 1989, 2000; Kim et al., 2002; van Bogaert et al., 2011). The exact mechanism and constraints or determinants of the degree of lactonization remain unclear, but the engineering of fermentation conditions can influence shifting in production of either lactonic or acidic sophorolipids (Garcia-Ochoa and Casas, 1999; Kim et al., 2009; Shin et al., 2010). The surface-active properties of sophorolipids were unaffected by heat treatment even at a higher temperature it retained the activity (van Bogaert et al., 2011; Hirata et al., 2009a,b). The structural diversity creates a range of the physiochemical properties, and hence researchers have a freedom of opportunity to screen sophorolipids against various bioactivities. For example, the solubility of sophorolipids depends on pH as at pH lower than 5.0, the sophorolipid shows dispersion, while at pH higher than 6, it perfectly dissolves. Their solubility is better in polar solvents, such as ethanol, methanol, ethyl acetate, acetonitrile, and DMSO. The emulsification ability of sophorolipids is vast enough to use as a better surfactant than any other (Hu and Ju, 2001, 2003; Bluth et al., 2006; Nguyen et al., 2010; van Bogaert et al., 2011). In general, sophorolipids lowers the surface tension from 72.80 mN/m down to 40–30 mN/m with a CMC of 11–250 mg/L (van Bogaert et al., 2007, 2011). The extreme biodegradability and relatively low toxicity (10-fold less) of the sophorolipids when compared to conventional surfactant make them a better choice over available surfactants (van Bogaert et al., 2011).

Till date, no direct study has been done for the elucidation of biosynthetic pathways of sophorolipid synthesis. Based on indirect studies, a theoretically proposed pathway has been given by van Bogaert and coworkers (2007). Figure 4.2 provides a schematic outline of the steps involved in sophorolipid biosynthesis. These studies were mainly performed on different species of *Candida*. It is well known that the building blocks for sophorolipid biosynthesis are glucose and a fatty acid. In an ideal condition, both the substrates, that is, fatty acid and glucose, must be used as such, but the fatty acid can disturb the homeostasis of the electrons in microbial cells. Thus usually methyl esters or ethyl esters or triglycerides are used instead of fatty acid. Literatures on the production of sophorolipid using unusual substrates like alkanes revealed the fact that the sophorolipid-producing *C. apicola* and *C. bombicola* possess the enzymes required for the terminal oxidation of alkanes, thereby generating fatty acids for

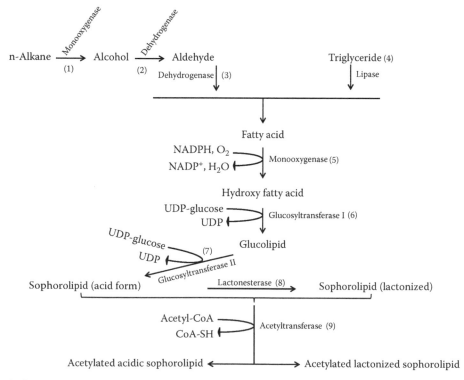

FIGURE 4.2 Proposed pathway for sophorolipid biosynthesis given by Bogaert and coworkers. (From van Bogaert, I.N. et al., *Appl. Microbiol. Biotechnol.*, 76(1), 23, 2007.)

further β-oxidation (van Bogaert et al., 2007, 2011). While in the absence of the hydrophobic substrate in the medium, the fatty acids are formed *de novo* starting from acetyl-CoA derived from glycolysis (van Bogaert et al., 2007).

In Figure 4.2, steps 1–3 are involved in sophorolipid production on lipophilic substrates other than fatty acids or esters thereof such as alkenes. While using the n-alkane as a substrate for the production sophorolipid, the cytochrome P450 monooxygenase-mediated oxidation of the *n*-alkenes to alcohol is the primary step, which is further converted into a fatty acid by successive oxidation mediated by alcohol- and aldehyde-dehydrogenase. Two genes of cytochrome P450 monooxygenase from *C. apicola* (European Molecular Biology Laboratory/GenBank accession numbers X76225 and X87640) have been reported by Lottermoser et al. (1996) and classified into the CYP52 family using amino acid similarity analysis. The cytochrome P450 enzymes of yeasts are capable of hydroxylating alkanes and/or fatty acids (Nelson, 1999; Franzetti et al., 2010; Kang et al., 2010). However, there is not so much literature available, which verify whether the mentioned gene products were involved in sophorolipid production or alkane assimilation and if they were expressed at all (van Bogaert et al., 2007). From *C. bombicola* ATCC 22214, five different cytochrome P450 monooxygenase genes has been identified which belongs to CYP52 family. One of them exposes very high similarity (91% AA identity) to the CYP52E2 gene of *C. apicola*, whereas the others probably belong to one or more new CYP52 subfamilies (van Bogaert et al. 2007). Alternatively, extracellular lipases convert any fatty acid esters into free fatty acids, which are utilized as a substrate. In the absence of lipophilic substrate, the *de novo* synthesis of fatty acids occurs to form sophorolipid using acetyl-CoA, derived from the glycolytic pathway. These fatty acids convert into sophorolipid after a series of reactions, which is initiated by hydroxylation of fatty acids in the presence of molecular oxygen and nicotinamide adenine dinucleotide phosphate—reduced form—at the ω or ω−1 position, again through the action of cytochrome P450 monooxygenase. Two glucose units are added in the resulting hydroxy fatty acid by two different glucosyltransferases. Glucose is glycosidically coupled (position C1) to the hydroxyl group of the fatty acid through the action of a specific glycosyltransferase I. Experiments with ^{13}C-labeled glucose pointed out that the bulk of the added glucose first passed through glycolysis, in this way supplementing trioses for the gluconeogenesis of glucose for sophorolipid synthesis (Hommel et al., 1987, 1994). The transferase reaction requires nucleotide-activated glucose (uridine diphosphate-glucose) as glucosyl donor (Breithaupt and Light, 1982). In a subsequent step, a second glucose is glycosidically coupled to the C2'-position of the first glucose moiety by glycosyltransferase II. Both glycosyltransferases are involved in sophorolipid synthesis. The two enzyme activities could however not be separated, and highly purified samples exhibit a single major band of 52 kDa on sodium dodecyl sulfate polyacrylamide gel electrophoresis (Esders and Light, 1972; Breithaupt and Light, 1982). Therefore, it remains open for discussion whether the consecutive glucose transfers are carried out by two different (but copurified) enzymes or by one and the same (multi)enzyme. It is supposed that sophorolipid synthesis in *C. bombicola* involves analogous enzymes. The sophorolipids obtained after the action of glucosyltransferase II are as such detected in the sophorolipid mixture as the acidic nonacetylated molecules. The majority of the sophorolipids are however further modified by both internal esterification (lactonization) and acetylation of the carbohydrate head. Lactonic sophorolipids are formed by an esterification reaction of the carboxyl group of the hydroxy fatty acid with a hydroxyl group of sophorose (Figure 4.1). The vast majority of lactones are esterified at the 4″- position, whereas a small percentage is esterified at either the 6′- or 6″-position (Asmer et al., 1988). Since several commercial lipases are able to introduce such a 6′- or 6″-ester linkage, it is suggested that those bonds are formed by cellular lipases, whereas esterification at the 4″-position is believed to be catalyzed by a specific lactone esterase. Both esterases have been identified neither in *C. bombicola* nor in other sophorolipid-producing species. The acetylation at the 6′- and/or 6″-position is carried out by an acetyl-coenzyme A (CoA)-dependent acetyl transferase. The transferase from *Rhodotorula bogoriensis* has been partially purified (Esders and Light, 1972; Bucholtz and Light, 1976), but the corresponding enzyme has not yet been identified in *C. bombicola*. The sophorolipid can be excreted as such or can be modified further. These modifications consist of lactonization by linking the carboxylic function to the 4″-hydroxyl and acetylation of the 6′- and/or 6″-hydroxyls (Figure 4.2).

SOPHOROLIPID-PRODUCING MICROBES

The sophorolipids are produced by a selected number of yeasts such as *C. bombicola*, *C. apicola, C. batistae, Wickerhamiella domericqiae*, and *R. bogoriensis* (Gorin et al., 1961; Tulloch and Spencer, 1968; Tulloch et al., 1968; Spencer et al., 1970; Chen et al., 2006a,b; Konishi et al., 2008). Without any doubt, the *C. bombicola* has been the most studied and exploited microorganism in the field of sophorolipid. This microbe was initially identified as *T. bombicola*, which was later coined as *C. bombicola*. Although several *C. bombicola* isolates are available in culture collections, the ATCC 22214 isolate is the strain of choice as this one is the most efficient sophorolipid producer (van Bogaert et al., 2011). *R. bogoriensis*, a member of Basidiomycota phylum, was identified as sophorolipid (predominantly, C22 hydroxy fatty acid, and even small amounts of C24)-producing organism (Nunez et al., 2004; van Bogaert et al., 2007). Although this organism has been known for several decades, not much attention has been given to the optimization of the production. Another well-known organism, *C. apicola*, which is phylogenetically very close to *C. bombicola*, is also a choice for some research groups for sophorolipid production (Tulloch and Spencer, 1968; Hommel et al., 1987). In a quest for novel producer of the sophorolipid, Chen and coworkers have isolated the *W. domercqiae* Y2A from oil-containing wastewater sample (Chen et al., 2006b), and surprisingly, the yield was reported up to 320 g/L, which was quite impressive (Song, 2006). The production of sophorolipids is not restricted to a single yeast species, but a number of related microorganisms are also capable to synthesize some sort of sophorolipids. Identification of new sophorolipid-producing organism *C. batistae* has been reported with a relatively higher production of the acidic sophorolipid, than the lactoic ones (Konishi et al., 2008). Some sophorolipid-producing organisms reported from various sources are listed in Table 4.1.

TABLE 4.1

Sophorolipid-Producing Organisms

Microbes	References
Candida apicola	Tulloch and Spencer (1968), Hommel et al. (1987)
Candida batistae	Konishi et al. (2008)
Candida bombicola	Rosa and Lachance (1998), Spencer et al. (1970), Rosa et al. (2003), Sugita and Nakase (1999)
Candida sp. NRRL Y-27208	Price et al. (2012)
Pichia anomala	Thaniyavarn et al. (2008)
Rhodotorula bogoriensis	Nunez et al. (2004), Zhang et al. (2011)
Sphingomonas sp. NM05	Manickam et al. (2012)
Starmerella bombicola	Kurtzman et al. (2010), Ribeiro et al. (2012), Takahashi et al. (2011)
Starmerella bombicola (ATCC 22214)	Wadekar et al. (2012)
Wickerhamiella domercqiae	Chen et al. (2006a,b), Song (2006)
Wickerhamiella domercqiae var. sophorolipid CGMCC 1576	Li et al. (2012)

PRODUCTION STRATEGIES AND MEDIA DESIGN FOR SOPHOROLIPID PRODUCTION

The ability of utilizing alkenes for the sophorolipid production by *C. bombicola* and *C. apicola* indicates that they possess enzymes that are required for modifying such kinds of substrates into fatty acids. This is one of the ways they can consume these alternative carbon sources for the energy production, via the process of β-oxidation or can be converted to sophorolipids. When culture media for these microbes are supplemented by fatty acids as a lipophilic substrate, in the presence of adequate amount of glucose or sugar content, the maximum production of sophorolipids will be occurred. However, fatty acids themselves cause a disruption in the electron balance of the cells. Due to this reason, the triglycerides or (*m*) ethyl esters are a better choice as the substrate. The extracellular lipases will gradually release free fatty acids to the culture media. This approach is predominantly used when the substrate is added in large quantities (van Bogaert et al., 2007). The basic constituent of media for the production of sophorolipids is augmented with a nitrogen source, small amounts of various minerals, and over 10% of glucose as the main source of energy and carbon. Yeast extract is mainly used as the nitrogen source on a laboratory scale, while processed waste products such as corn steep liquor are used on an industrial scale. Some researchers also used mineral nitrogen (Kim et al., 2009). Table 4.2 summarizes an account with respect to the production of sophorolipids using various types of substrates and microorganisms.

Production of sophorolipid starts with the initiation of stationary phase and is probably triggered by a high carbon-to-nitrogen ratio (C:N) (Davila et al., 1992,

TABLE 4.2

Production Optimization of Sophorolipid Using Various Substrates and Microbes: A Comparative Account

Microbes	Carbon Source (g/L)	Nitrogen Source (g/L)	Production (g/L)	Time (h)	Operation Mode	References
Candida bombicola ATCC22214	Glucose (100) Animal fat (100	Corn steep liquor (4) Urea (1.5)	120	68	Batch	Singh and Desai (1989)
Candida sp. SY 16	Glucose (15) Soybean oil (15)	Peptone (1)	95	200	Feed-batch	Kim et al. (2006)
Candida bombicola NRRL Y-17069	Glucose (100) Sunflower oil (100)	Yeast extract (1)	120	192	Batch-resting cell	Lin (1996)
Candida lipolytica UCP0988	Glucose (100) Canola oil (100)	Yeast extract (2)	8	48	Batch	Sarubbo et al. (2007)
Starmerella bombicola	Deproteinized whey (90), glucose (10), and oleic acid (100)	Yeast extract (2)	23.3	144	Feed-batch	Daverey and Pakshirajan (2009a,b)
Starmerella bombicola	Oleic acid (40)	Yeast extract (10)	38	168	Batch	Shah et al. (2007b), Ribeiro et al. (2012)
Starmerella bombicola NRRL Y-17069	Glucose (100) Glucose (100) Oleic acid (100)	Urea (1) Yeast extract (10) Urea (1)	54.16	144	Feed-batch	Bajaj et al. (2012)
Wickerhamiella domercqiae var. sophorolipid CGMCC 1576	Sunflower oil and palm (100) Glycerol (150)	Yeast extract (4) Urea (1)	6.6	144	Batch	Wadekar et al. (2012)
Rhodotorula bogoriensis	Glucose (80) Rapeseed oil (35)	Yeast extract (4)	1.26	168	Feed-batch	Nuñez et al. (2004)
Pichia anomala	Glucose (80) Soybean oil (40)	Yeast extract (1) Sodium nitrate (4)	0.2	168	Batch	Thaniyavarn et al. (2008)

1993, 1997; van Bogaert, et al., 2007). Several strategies can be applied in order to develop the cost-effective method. As a concern, to maintain the stationary phase of the microbes for better production of the sophorolipid, the process itself should be optimized like using appropriate medium, culture conditions, product recovery techniques, etc. The cost-effective substrate replacement and their optimization are a secondary approach to have an industrial-scale production. The conventional engineering is not enough to achieve the goal; thus, the overproducing mutant and recombinant strains should also be created to obtain maximum productivity. The approaches mentioned earlier should take place together to resolve the issues of the commercialization and making them as a market-compatible product. The yeast cells can be maintained in stationary phase up to 10 days while producing sophorolipid (van Bogaert et al., 2011). Sophorolipid production media were designed and optimized for the carbon and nitrogen sources, and it was found that around 100 g/L glucose was used as the carbon source with variable concentrations of yeast extract or corn steep liquor as the nitrogen source. Supplementation of lipophilic or hydrophobic carbon as carbon source, along with additional glucose feedings, can help to increase the yield either in batch or fed-batch (Shin et al., 2010).

One of the other most important factors is temperature optimization to achieve the good production of sophorolipids. The optimum growth of *C. bombicola* ATCC 22214 was reported at 28.8°C, while the production of sophorolipid was favored at 21°C (Gobbert et al., 1984). But most of the fermentation was carried out at 25°C or 30°C, as the higher temperature favors high biomass production and relatively lower glucose consumption (Casas and Garcia-Ochoa, 1999). The pH of medium also plays a crucial role in the production. During the exponential growth phase, pH drops till 3.5 or below. But it should ideally be kept at 3.5 for optimal sophorolipid production (Gobbert et al., 1984). Oxygen concentration is one of the important constraints during the entire fermentation process. The culture broth should be supplied with an adequate amount of oxygen because the yeast cells are very sensitive to oxygen and good aeration conditions are important for sophorolipid production as the cytochrome P450 monooxygenase uses molecular oxygen (Guilmanov et al., 2002; van Bogaert et al., 2007). The optimal aeration for high sophorolipid yield expressed in terms of oxygen transfer rate lies between 50 and 80 mM O_2/(Lh) (Shin et al., 2010).

In order to achieve a high amount of sophorolipid, the supplementation of hydrophobic carbon substrate is an important concern. In the last few decades, a number of substrates have been screened to design the production media such as alkanes, oils, fatty acids, and their corresponding esters. In most cases, relatively pure substrates are used. But one can go further and exploit waste streams such as biodiesel by-product streams, soybean dark oil, waste frying oil, corn oil, and molasses (Solaiman et al., 2004, 2007; Kim et al., 2005; Pekin et al., 2005; Fleurackers, 2006). An overview of various experiments in relation to different substrates or feeding strategies is given in Table 4.2. It has been reported previously that the vegetable oils are rich in oleic acid and promote sophorolipid production. Consequently, rapeseed oil is one of the substrates of choice (Davila et al., 1994; Daniel et al., 1998a,b; Kim et al., 2009; Shin et al., 2010).

Extraction of sophorolipids from the culture broth can be achieved by organic solvents such as ethyl acetate, which is followed by hexane (preferred) extraction for the removal of residual lipidic carbon. Some researchers also used pentene (Cavalero and Cooper, 2003) or t-butyl methyl ether (Rau et al., 1996, 2001) for the secondary extraction. The elimination of additional water and impurities can be achieved by the addition of polyhydric alcohols and subsequent distillation (Inoue and Miyamoto, 1980; Inoue et al., 1980; Inoue and Ito, 1982).

APPLICATIONS OF SOPHOROLIPIDS

The sophorolipids are well recognized for their surface active properties and also known as green surfactants or biosurfactants because they are produced by specific microorganisms. The application of surfactants occurs in various industries including food, pharmaceutical, cosmetic, and cleaning. Mostly, these surfactants are produced by chemical means using petrochemical raw materials. Thus they are not environmentally friendly in nature, and the use of such surfactants raise serious ecological problems (Mann and Boddy, 2000; Mann and Bidwell, 2001; van Bogaert et al., 2011). The ecological concerns like ecotoxicity, bioaccumulation, and biodegradability of surfactants are encouraging the use of biosurfactants instead of the chemical one.

The Belgian company Ecover NV (http://www.ecover.com) and the Japanese company Saraya Co, LTD (http://www.saraya.com) are working in the area of biosurfactants and in manufacturing ecofriendly detergents and cleansing agents (Futura et al., 2002; Develter and Fleurackers, 2007). The French company Soliance (http://www.groupesoliance.com) and the Korean MG Intobio Co. Ltd are also producing sophorolipid-based cosmetics and skin health products. Thus not only academic research but also commercialization of sophorolipid-based products is occupying the market and shares. An intensive screening of bioactivities of native or modified sophorolipid is required to exploit this molecule for more commercial assets.

There are many reports available that suggest the potentials of sophorolipid as an alternative to laundry detergents (Hall et al., 1996; Shin et al., 2010; van Bogaert et al., 2011). Sophorolipids can be applied for *in situ* bioremediation and degradation of hydrocarbons present in porous media such as soils and groundwater tables (Ducreux et al., 1997; Kang et al., 2010). They are used in the removal of heavy metals from sediments (Mulligan et al., 2001). The sophorolipids can also be used in secondary oil recovery and in the regeneration of hydrocarbons from dregs and muds (Baviere et al., 1994; Marchal et al., 1999; Pesce, 2002). Furthermore, the emulsifying property of sophorolipids can be used in the food industry to improve the quality and storage especially in logistics (Akari and Akari, 1987; Masaru et al., 2001).

One of the most potential applications of sophorolipids is its utilization in cosmetics either as a passive or an active ingredient. The glimpse on the effectiveness and success of the sophorolipid as cosmetics ingredient limits the production and

design of the sophorolipid-based cosmetics. Currently, cost of production and purification is the crucial challenge to overcome in order to apply biosurfactants widely
in marketable products (Muthusamy et al., 2008; Williams, 2009). The cutting-edge
companies based in France called Groupe Soliance (http://www.groupesoliance.
com) and the Korean MG Intobio Co. Ltd focus on innovative natural cosmetic
developments like sophorolipid-based products for acne cure, deodorants, and hair
cleansers (Williams, 2009; van Bogaert et al., 2011). Sophorolipids support the
skin's natural physiology and, therefore, have tremendous potential for its use in
several other cosmetic products (Shete et al., 2006). It is reported that sophorolipids
exhibited a lower cytotoxicity than surfactin (commercialized cosmetic ingredient)
or other surfactants (Yoshihiko et al., 2009; Shao et al., 2012). The antimicrobial
properties particularly against *Propionibacterium acnes* and *Corynebacterium
xerosis*—causative agents of acne and dandruff, respectively (Mager et al. 1987;
Kim et al., 2002)—and other beneficial properties regarding the protection of hair
and skin make sophorolipids an attractive component in cosmetic, hygienic, and
pharmacodermatological products (van Bogaert et al., 2007, 2011). Sophorolipids
are also reported as a stimulant for the dermal fibroblast proliferation and collagen
neosynthesis by its macrophage-activating and fibrinolytic properties. They also
act as inhibitors to free radical and elastase activity, as desquamating and depig-
menting agents, and as stimulant for leptin synthesis (Hillion et al., 1998; Borzeix,
1999; Maingault, 1999; Pellecier and Andre, 2004; Yoo et al., 2005). Because of
their antimicrobial properties, sophorolipids can be applied in germicidal mixtures
for cleaning fruits and vegetables (Pierce and Heilman, 1998). The antimicrobial
action is not merely restricted toward bacteria; sophorolipids also act as antifungal
agents against plant pathogenic fungi such as *Phytophthora* spp. and *Pythium* spp.
(Yoo et al., 2005) and inhibit algal blooms (Baek et al., 2003; Gi, 2004; Sun et al.,
2004a,b,c; Lee et al., 2008). The anticancer property of the sophorolipid is revealed
by some research especially on human promyelocytic leukemia cell line HL60
(Isoda et al., 1997) and human liver cancer cells H7402 (Chen et al., 2006a,b). Its
antiviral activity against herpes-related viral infections (Gross and Shah, 2007a)
and immunodeficiency virus was also reported. Sophorolipids also have sperm-
immobilizing activities (Shah et al., 2005), can act as septic shock antagonist (Bluth
et al., 2006), having antibacterial properties (Napolitano, 2006), and also act as
anti-inflammatory or immunomodulating agents (Sleiman et al., 2009). Table 4.3
compiles a list of applications of sophorolipids.

It has been reported that the sophorolipid cap is used as good reducing and
capping agents for cobalt and silver nanoparticle synthesis (Kasture et al., 2007,
2008). The hydroxypropyl-etherified glycolipid ester formed by sophorolipid sugar
moiety substitution with hydroxyalkyl groups has been used in pencil-shaped lip
rouge, powdered compressed cosmetic material, lip cream, and eye shadow as well
as in aqueous solutions (Kawano et al. 1981a,b; van Bogaert et al., 2011). Various
modifications and derivatizations of sophorolipids were attempted to explore the
possibility of novel bioactivity (van Bogaert et al., 2007, 2011). Table 4.4 represents
a summary of some modified sophorolipids and their potential applications.

TABLE 4.3

List of Bioactive Properties of Sophorolipid

Properties	References
Antibacterial	Mager et al. (1987), Kim et al. (2002), Shah et al. (2007), Singh et al. (2009, 2010)
Anticancer	Isoda et al. (1997), Chen et al. (2006a,b), Shao et al. (2012)
Anti-ice during cold chain	Masaru et al. (2001)
Antiviral	Shah et al. (2005), Gross and Shah (2007a)
Antifungal	Yoo et al. (2005)
Anti-inflammatory or immunomodulating agents	Napolitano (2006), Sleiman et al. (2009)
Bioremediation	Ducreux et al. (1997), Kang et al. (2010)
Cellulase production	Lo and Ju (2009), Gross and Shah (2007b)
Commercial cleaning agent	Futura et al. (2002), Develter and Fleurackers (2007)
Cosmetics	Williams (2009)
Degradation of hydrocarbons	Kang et al. (2010)
Detergent	Hall et al. (1996), Shin et al. (2010)
Germicidal	Pierce and Heilman (1998)
Inhibition of algal bloom	Lee et al. (2008), Sun et al. (2004a,b,c), Gi (2004), Baek et al. (2003)
Nanotechnology	Singh et al. (2009, 2010), Kasture et al. (2007, 2008)
Perfume and fragrance industry	Inoue and Miyamoto (1980)
Quality improvement of wheat flour products	Akari and Akari (1987)
Removal of heavy metals	Mulligan et al. (2001)
Secondary oil recovery	Baviere et al. (1994), Marchal et al. (1999), Pesce (2002)
Septic shock antagonist	Bluth et al. (2006)
Spermicidal	Shah et al. (2005)
Skin bioactivity	Hillion et al. (1998), Borzeix (1999), Maingault (1999)
Stimulant for leptin synthesis	Pellecier and André (2004)
Surfactant	Mann and Bidwell (2001), Mann and Boddy (2000), van Bogaert et al. (2011)
Wetting agent	Develter and Fleurackers (2007)

TABLE 4.4

Modification of Sophorolipid and Their Application

Modified Sophorolipid	Application	References
Hydroxypropyl-etherified glycolipid ester	Cosmetics	Kawano et al. (1981a,b), van Bogaert et al. (2011)
Amide derivates	Tunable immunoregulators	Singh et al. (2003), van Bogaert et al. (2011)
Amino acid–sophorolipid conjugates	Antibacterial, anti-HIV, and spermicidal	Zerkowski et al. (2006), Azim et al. (2006)

CONCLUSION

Interest in the use of sophorolipids (also known as biosurfactants) in commercial products and, during last few years, increase in the number of publications in the area are a clear indication that sophorolipids have a very vast application in research and industries. In some specific industrial sectors such as cosmetic and pharmaceutical, biosurfactants have high application potential and will probably play a major role after a short period of time. Sophorolipid is a choice molecule because of high-valued surface properties. They also possess a high degree of biodegradability and environmental compatibility. Sophorolipids are a type of glycolipid biosurfactants that have potential to be used in the cosmetics industry as they show various properties like antiradical properties, stimulation of dermal fibroblast metabolism, and hygroscopic properties to support healthy skin physiology. Future prospects of sophorolipid-based products include several types of facial cosmetics, lotions, beauty washes, and hair products. Still the industrial use of the sophorolipid is limited because of the high costs involved in the production process. Thus an optimization of renewable energy sources and other fermentation conditions must be optimized to achieve cheaper production technology. Although optimization of the production process is the key factor to improve yield, type, and to reduce costs, the genetic and metabolic engineering using recombinant DNA techniques could be a better solution for the manipulation of biosurfactant production, chain properties, surface properties, as well as yield. A little success has been achieved to the development molecular tools of *C. bombicola*. Due to this, successful mutant strains can be created. These studies open new perspectives to increase production yields and might become the instrument to overcome the limitations for sophorolipid commercialization.

REFERENCES

Adamczak, M. and Bednarski, W. 2000. Influence of medium composition and aeration on the synthesis of biosurfactants produced by *Candida antarctica. Biotechnology Letters*, 22:313–316.

Akari, S. and Akari, Y. 1987. Method of modifying quality of wheat flour product. *Japanese Patent*, 61205449.

Asmer, H.J., Lang, S., Wagner, F., and Wray, V. 1988. Microbial-production, structure elucidation and bioconversion of sophorose lipids. *Journal of American Oil Chemical Society*, 65:1460–1466.

Azim, A., Shah, V., Doncel, G.F., Peterson, N., Gao, W., and Gross, R. 2006. Amino acid conjugated sophorolipids: A new family of biological active functionalized glycolipids. *Bioconjugate Chemistry*, 17:1523–1529.

Baek, S.H., Sun, X.X., Lee, Y.J., Wang, S.W., Han, K.N., Choi, J.K., Noh, J.H., and Kim, E.K. 2003. Mitigation of harmful algal blooms by sophorolipid. *Journal of Microbiology and Biotechnology*, 13(5):651–659.

Bajaj, V., Tilay, A., and Annapure, U. 2012. Enhanced production of bioactive sophorolipids by *Starmerella bombicola* NRRL Y-17069 by design of experiment approach with successive purification and characterization. *Journal of Oleofin Science*, 61(7):377–386.

Baviere, M., Degouy, D., and Lecourtier, J. 1994. Process for washing solid particles comprising a sophoroside solution. US Patent, 5326407.

Bluth, M.H., Kandil, E., Mueller, C.M., Shah, V., Lin, Y.Y., Zhang, H., Dresner, L. et al. 2006. Sophorolipids block lethal effects of septic shock in rats in a cecal ligation and puncture model of experimental sepsis. *Critical Care Medicine*, 34:188–195.

Borzeix, C.F. 1999. Use of sophorolipids comprising diacetyl lactones as agent for stimulating skin fibroblast metabolism. World Patent, 9962479.

Breithaupt, T.B. and Light, R.J. 1982. Affinity-chromatography and further characterization of the glucosyltransferases involved in hydroxy-docosanoic acid sophoroside production in *Candida bogoriensis*. *Journal of Biological Chemistry*, 257:9622–9628.

Bucholtz, M.L. and Light, R.J. 1976. Acetylation of 13-sophorosyloxydocosanoic acid by an acetyltransferase purified from *Candida bogoriensis*. *Journal of Biological Chemistry*, 251:424–430.

Casas, J.A. and Garcia-Ochoa, F. 1999. Sophorolipid production by *Candida bombicola*: Medium composition and culture methods. *Journal of Bioscience and Bioengineering*, 88:488–494.

Cavalero, D.A. and Cooper, D.G. 2003. The effect of medium composition on the structure and physical state of sophorolipids produced by *Candida bombicola* ATCC 22214. *Journal of Biotechnology*, 103:31–41.

Chen, J., Song, X., Zhang, H., Qu, Y.B., and Miao, J.Y. 2006a. Production, structure elucidation and anticancer properties of sophorolipid from Wickerhamiella domercqiae. *Enzyme Microbiology and Technology*, 39:501–506.

Chen, J., Song, X., Zhang, H., Qu, Y.B., and Miao, J.Y. 2006b. Sophorolipid produced from the new yeast strain Wickerhamiella domercqiae induces apoptosis in H7402 human liver cancer cells. *Applied Microbiology and Biotechnology*, 72:52–59.

Daniel, H.J., Otto, R.T., Reuss, M., and Syldatk, C. 1998a. Sophorolipid production with high yields on whey concentrate and rapeseed oil without consumption of lactose. *Biotechnology Letters*, 20:805–807.

Daniel, H.J., Reuss, M., and Syldatk, C. 1998b. Production of sophorolipids in high concentration from deproteinized whey and rapeseed oil in a two stage fed batch process using *Candida bombicola* ATCC 22214 and Cryptococcus curvatus ATCC 20509. *Biotechnology Letters*, 20:1153–1156.

Daverey, A. and Pakshirajan, K. 2009a. Production of sophorolipids by the yeast *Candida bombicola* using simple and low cost fermentative media. *Food Research International*, 42:499–504.

Daverey, A. and Pakshirajan, K. 2009b. Production, characterization, and properties of sophorolipids from the yeast *Candida bombicola* using a low cost fermentative medium. *Applied Biochemistry and Biotechnology*, 158:663–674.

Davila, A.M., Marchal, R., Monin, N., and Vandecasteele, J.P. 1993. Identification and determination of individual sophorolipids in fermentation products by gradient elution high performance liquid chromatography with evaporative light scattering detection. *Journal of Chromatography*, 648:139–149.

Davila, A.M., Marchal, R., and Vandecasteele, J.P. 1992. Kinetics and balance of a fermentation free from product inhibition-sophorose lipid production by *Candida bombicola*. *Applied Microbiology and Biotechnology*, 38:6–11.

Davila, A.M., Marchal, R., and Vandecasteele, J.P. 1994. Sophorose lipid production from lipidic precursors predictive evaluation of industrial substrates. *Journal of Industrial Microbiology*, 13:249–257.

Davila, A.M., Marchal, R., and Vandecasteele, J.P. 1997. Sophorose lipid fermentation with differentiated substrate supply for growth and production phases. *Applied Microbiology and Biotechnology*, 47:496–501.

Desai, J.D. and Banat, I.M. 1997. Microbial production of surfactants and their commercial potential. *Microbiology and Molecular Biology Reviews*, 61:47–59.

Develter, D. and Fleurackers, S. 2007. A method for the production of short chained glycolipids. Patent EP07101581.2.

Develter, D., Renkin, M., and Jacobs, I. 2003. Detergent compositions. Patent, EP1445302A2.

Develter, D.W.G. and Fleurackers, S.J.J. 2010. Sophorolipids and rhamnolipids. In *Surfactants from Renewable Resources*. M. Kjellin and I. Johansson. (eds.) John Wiley & Sons, Ltd, Chichester, U.K. doi: 10.1002/9780470686607. Chapter 11.

Ducreux, J., Ballerini, D., Baviere, M., Bocard, C., and Monin, N. 1997. Composition containing a surface active compound and glycolipids and decontamination process for a porous medium polluted by hydrocarbons. US Patent, 5654192.

Esders, T.W. and Light, R.J. 1972. Glucosyl- and acetyltransferases involved in the biosynthesis of glycolipids from *Candida bogoriensis*. *Journal of Biological Chemistry*, 247:1375–1386.

Fleurackers, S. 2006. On the use of waste frying oil in the synthesis of sophorolipids. *European Journal of Lipid Science and Technology*, 108(1):5–12.

Franzetti, A., Gandolfi, I., Bestetti, G., Smyth, T.J.P, and Banat, I.M. 2010. Production and applications of trehalose lipid biosurfactants. *European Journal of Lipid Science and Technology*, 112:617–627.

Futura, T., Igarashi, K., and Hirata, Y. 2002. Low foaming detergent compositions. World Patent, 03002700.

Garcia-Ochoa, F. and Casas, J.A. 1999. Unstructured kinetic model for sophorolipid production by *Candida bombicola*. *Enzyme Microbiology and Technology*, 25:613–621.

Gi, K.E. 2004. Red tide inhibiting agent comprising yellow earth and biodegradable sophorolipid. Korean Patent, 0083614.

Gobbert, U., Lang, S., and Wagner, F. 1984. Sophorose lipid formation by resting cells of *Torulopsis bombicola*. *Biotechnology Letters*, 6:225–230.

Gorin, P., Spencer, J., and Tulloch, A. 1961. Hydroxy fatty acid glycosides of sophorose from *Torulopsis magnolia*. *Canadian Journal of Chemistry*, 39:846–855.

Gross, R.A. and Shah, V. 2007a. Anti herpes virus properties of various forms of sophorolipids. WO2007US63701.

Gross, R.A. and Shah, V. 2007b. Sophorolipids as protein inducers and inhibitors in fermentation medium. WO2007073371.

Guilmanov, V., Ballistreri, A., Impallomeni, G., and Gross, R.A. 2002. Oxygen transfer rate and sophorose lipid production by *Candida bombicola*. *Biotechnology and Bioengineering*, 77:489–494.

Hall, P.J., Haverkamp, J., Van Kralingen, C.G., and Schmidt, M. 1996. Laundry detergent composition containing synergistic combination of sophorose lipid and nonionic surfactant. US Patent, 5520839.

Hillion, G., Marchal, R., Stoltz, C., and Borzeix, C.F. 1998. Use of a sophorolipid to provide free radical formation inhibiting activity or elastase inhibiting activity. US Patent, 5756471.

Hirata, Y., Ryu, M., Igarashi, K., Nagatsuka, A., Furuta, T., Kanaya, S., and Sugiura, M. 2009a. Natural synergism of acid and lactone type mixed sophorolipids in interfacial activities and cytotoxicities. *Journal of Oleofin Science*, 58:565–572.

Hirata, Y., Ryu, M., Oda, Y., Igarashi, K., Nagatsuka, A., Furuta, T., and Sugiura, M. 2009b. Novel characteristics of sophorolipids, yeast glycolipid biosurfactants, as biodegradable low foaming surfactants. *Journal of Bioscience and Bioengineering*, 108:142–146.

Hommel, R., Stuwer, O., Stuber, W., Haferburg, D., and Kleber, H.P. 1987. Production of water soluble surface active exolipids by *Torulopsis apicola*. *Applied Microbiology and Biotechnology*, 26:199–205.

Hommel, R.K., Weber, L., Weiss, A., Himmelreich, U., Rilke, O., and Kleber, H.P. 1994. Production of sophorose lipid by Candida (Torulopsis) apicola grown on glucose. *Journal of Biotechnology*, 33:147–155.

Hu, Y.M. and Ju, L.K. 2001. Purification of lactonic sophorolipids by crystallization. *Journal of Biotechnology*, 87:263–272.

Hu, Y.M. and Ju, L.K. 2003. Lipase mediated deacetylation and oligomerization of lactonic sophorolipids. *Biotechnology Progress*, 19:303–311.

Inoue, S. and Ito, S. 1982. Sophorolipids from *Torulopsis bombicola* as microbial surfactants in alkane fermentations. *Biotechnology Letters*, 4:3–8.

Inoue, S., Kiruma, Y., and Kinta, M. 1980. Dehydrating purification process for a fermentation product. US Patent, 4197166.

Inoue, S. and Miyamoto, N. 1980. Process for producing a hydroxyfatty acid ester. US Patent, 4201844.

Isoda, H., Kitamoto, D., Shinmoto, H., Matsumura, M., and Nakahara, T. 1997. Microbial extracellular glycolipid induction of differentiation and inhibition of the protein kinase C activity of human promyelocytic leukemia cell line HL60. *Bioscience, Biotechnology and Biochemistry*, 61:609–614.

Jones, D. and Howe, R. 1968. Microbiological oxidation of long-chain aliphatic compounds. Part I. Alkanes and alk-1-enes. *Journal of the Chemical Society C*, 257(16):2801–2808.

Jones, D.F. 1967. Novel macrocyclic glycolipids from torulopsis gropengiesseri. *Journal of the Chemical Society (C)*, 6:479–484.

Kang, S.W., Kim, Y.B., Shin, J.D, and Kim, E.K. 2010. Enhanced biodegradation of hydrocarbons in soil by microbial biosurfactant, sophorolipid. *Applied Biochemistry and Biotechnology*, 160:780–790.

Kasture, M., Patel, P., Prabhune, A.A., Ramana, C.V., Kulkarni, A.A., and Prasad, B.L.V. 2008. Synthesis of silver nanoparticles by sophorolipids: Effect of temperature and sophorolipid structure on the size of particles. *Journal of the Chemical Society*, 120:515–520.

Kasture, M., Singh, S., Patel, P., Joy, P.A., Prabhune, A.A., Ramana, C.V., and Prasad, B.L.V. 2007. Multiutility sophorolipids as nanoparticle capping agents: Synthesis of stable and water dispersible co nanoparticles. *Langmuir*, 23:11409–11412.

Kawano, J., Suzuki, T., Inoue, S., and Hayashi, S. 1981a. Powdered compressed cosmetic material. US Patent, 4305931.

Kawano, J., Suzuki, T., Inoue, S., and Hayashi, S. 1981b. Stick shaped cosmetic material. US Patent, 4305929.

Kim, H.S., Jeon, J.W., Kim, B.H., Ahn, C.Y., Oh, H.M., and Yoon, B.D. 2006. Extracellular production of a glycolipid biosurfactant, mannosylerythritol lipid, by *Candida* sp. SY16 using fed-batch fermentation. *Applied Microbiology and Biotechnology*, 70:391–396.

Kim, H.S., Kim, Y.B., Lee, B.S., and Kim, E.K. 2005. Sophorolipid production by *Candida bombicola* ATCC 22214 from a corn oil processing byproduct. *Journal of Microbiology and Biotechnology*, 15:55–58.

Kim, K.J., Yoo, D., Kim, Y., Lee, B., Shin, D., and Kim, E.K. 2002. Characteristics of sophorolipid as an antimicrobial agent. *Journal of Microbiology and Biotechnology*, 12(2):235–241.

Kim, Y.B., Yun, H.S., and Kim, E.K. 2009. Enhanced sophorolipid production by feeding-rate-controlled fed-batch culture. *Bioresource Technology*, 100(23):6028–6032.

Kitamoto, D., Isoda, H., and Nakahara, T. 2002. Functions and potential application of glycolipids biosurfactants—From energy-saving materials to gene delivery carriers. *Journal of Bioscience and Bioengineering*, 94:187–201.

Konishi, M., Fukuoka, T., Morita, T., Imura, T., and Kitamoto, D. 2008. Production of new types of sophorolipids by *Candida batistae*. *Journal of Oleofin Science*, 57:359–369.

Kurtzman, C.P., Price, N.P.J., Ray, K.J., and Kuo, T.-M. 2010. Production of sophorolipid biosurfactants by multiple species of the Starmerella (Candida) bombicola yeast clade. *FEMS Microbiology Letters*, 311:140–146.

Lang, S., Brakemeier, A., Heckmann, R., Spockner, S., and Rau, U. 2000. Production of native and modified sophorose lipids. *Chimica Oggi (Chemistry Today)*, 18:76–79.

Lang, S., Katsiwela, E., and Wagner, F. 1989. Antimicrobial effects of biosurfactants. *Fett Wissenschaft Technologie—Fat Science Technology*, 91:363–366.

Lee, Y., Choi, J., Kim, E.K., Youn, S., and Yang, E. 2008. Field experiments on mitigation of harmful algal blooms using a sophorolipid—Yellow clay mixture and effects on marine plankton. *Harmful Algae*, 7(2):154–162.

Li, H., Ma, X., Shao, L., Shen, J., and Song, X. 2012. Enhancement of sophorolipid production of *Wickerhamiella domercqiae var. sophorolipid* CGMCC 1576 by low-energy ion beam implantation. *Applied Biochemistry and Biotechnology*, 167(3):510–523.

Lin, S.C. 1996. Biosurfactants: Recent advances. *Journal of Chemical Technology and Biotechnology*, 66:109–120.

Lo, C.M. and Ju, K.L. 2009. Sophorolipids induced cellulase production in cocultures of Hypocrea jecorina Rut C30 and *Candida bombicola*. *Enzyme Microbiology and Technology*, 44:107–111.

Lottermoser, K., Schunck, W.H. and Asperger, O. 1996. Cytochromes P450 of the sophorose lipid-producing yeast *Candida apicola*: Heterogeneity and polymerase chain reaction-mediated cloning of two genes. *Yeast*, 12:565–575.

Mager, H., Röthlisberger, R., and Wzgner, F. 1987. Use of sophorolse-lipid lactone for the treatment of dandruffs and body odour. European Patent, 0209783.

Maingault, M. 1999. Utilization of sophorolipids as therapeutically active substances or cosmetic products, in particular for the treatment of the skin. US Patent, 5981497.

Manickam, N., Bajaj, A., Saini, H.S., and Shanker, R. 2012. Surfactant mediated enhanced biodegradation of hexachlorocyclohexane (HCH) isomers by Sphingomonas sp. NM05. *Biodegradation*, 23(5):673–682.

Mann, R.M. and Bidwell, J.R. 2001. The acute toxicity of agricultural surfactants to the tadpoles of four Australian and, two exotic frogs. *Environmental Pollution*, 114:195–205.

Mann, R.M. and Boddy, M.R. 2000. Biodegradation of a nonylphenol ethoxylate by the autochthonous microflora in lake water with observations on the influence of light. *Chemosphere*, 41:1361–1369.

Marchal, R., Lemal, J., Sulzer, C., and Davila, A.M. 1999. Production of sophorolipid acetate acids from oils or esters. US Patent, 5900366.

Masaru, K., Takashi, N., Yoji, A., Kazuo, N., Tatsu, N., Sumiko, T., and Jotaro, N. 2001. Composition for high density cold storage transportation. Japanese Patent, 1131538.

Mulligan, C.N., Yong, R.N., and Gibbs, B.F. 2001. Heavy metal removal from sediments by biosurfactants. *Journal of Hazardous Materials*, 85:111–125.

Muthusamy, K., Gopalakrishnan, S., Ravi, T.K., and Sivachidambaram, P. 2008. Biosurfactants: Properties, commercial production, and application. *Current Science*, 94:736–747.

Napolitano, L.M. 2006. Sophorolipids in sepsis: Antiinflammatory or antibacterial? *Critical Care Medicine*, 34:258–259.

Nelson, D.R. 1999. Cytochrome P450 and the individuality of species. *Archives of Biochemistry and Biophysica*, 369:1–10.

Nguyen, T.T.L., Edelen, A., Neighbors, B., and Sabatini, D.A. 2010. Biocompatible lecithin based microemulsions with rhamnolipid and sophorolipid biosurfactants: Formulation and potential applications. *Journal of Colloid Interface and Science*, 348:498–504.

Nunez, A., Ashby, R., Foglia, T.A., and Solaiman, D.K.Y. 2004. LC/MS analysis and lipase modification of the sophorolipids produced by *Rhodotorula bogoriensis*. *Biotechnology Letters*, 26:1087–1093.

Pekin, G., Vardar-Sukan, F., and Kosaric, N. 2005. Production of sophorolipids from *Candida bombicola* ATCC 22214 using Turkish corn oil and honey. *Engineering and Life Sciences*, 5:357–362.

Pellecier, F. and Andre, P. 2004. Cosmetic use of sophorolipids as subcutaneous adipose cushion regulation agents and slimming application. World Patent, 2004/108063.

Pesce, L. 2002. A biotechnological method for the regeneration of hydrocarbons from dregs and muds, on the base of biosurfactants. World Patent, 02/062495.

Pierce, D. and Heilman, T.J. 1998. Germicidal composition. World Patent, 9816192.

Price, N.P. Ray, K.J., Vermillion, K.E., Dunlap, C.A., and Kurtzman, C.P. 2012. Structural characterization of novel Sophorolipid biosurfactants from a newly identified species of Candida yeast. *Carbohydrate Research*, 348: 33–41.

Rau, U., Hammen, S., Heckmann, R., Wray, V., and Lang, S. 2001. Sophorolipids: A source for novel compounds. *Industrial Crops Production*, 13:85–92.

Rau, U., Manzke, C., and Wagner, F. 1996. Influence of substrate supply on the production of sophorose lipids by *Candida bombicola* ATCC 22214. *Biotechnology Letters*, 18:149–154.

Ribeiro, I.A., Bronze, M.R., Castro, M.F., and Ribeiro, M.H. 2012. Sophorolipids: Improvement of the selective production by *Starmerella bombicola* through the design of nutritional requirements. *Applied Microbiology and Biotechnology*, 97(5):1875–1887.

Rosa, C. and Lachance, M. 1998. The yeast genus *Starmerella gen. nov.* and *Starmerella bombicola* sp. *nov.*, the teleomorph of *Candida bombicola* (Spencer, Gorin & Tullock) Meyer & Yarrow. *International Journal of Systematic Bacteriology*, 48:1413–1417.

Rosa, C.A., Lachance, M.A., Silva, J.O.C., Teixeira, A.C.P., Marini, M.M., Antonini, Y., and Martins, R.P. 2003. Yeast communities associated with stingless bees. *FEMS Yeast Research*, 4:271–275.

Sarubbo, L.A., Farias, C.B.B., and Campos-Takaki, G.M. 2007. Co-utilization of canola oil and glucose on the production of a surfactant by *Candida lipolytica*. *Current Microbiology*, 54:68–73.

Shah, V., Badia, D., and Ratsep, P. 2007a. Sophorolipids having enhanced antibacterial activity. *Antimicrobial Agents and Chemotherapy*, 51(1):397–400.

Shah, V., Doncel, G.F., Seyoum, E.K.M., Zalenskaya, I., Hagver, R., Azim, A., and Gross R. 2005. Sophorolipids, microbial glycolipids with anti-human immunodeficiency virus and sperm-immobilizing activities. *Antimicrobial Agents and Chemotherapy*, 49:4093–4100.

Shah, V., Jurjevic, M., and Badia, D. 2007b. Utilization of restaurant waste oil as a precursor for sophorolipid production. *Biotechnology Progress*, 23:512–515.

Shao, L., Song, X., Ma, X., Li, H., and Qu, Y. 2012. Bioactivities of Sophorolipid with different structures against human esophageal cancer cells. *Journal of Surgical Research*. 173(2): 286–91.

Shete, A.M., Wadhawa, G., Banat, I.M., and Chopade, B.A. 2006. Mapping of patents on bioemulsifier and biosurfactant: A review. *Journal of Science and Industrial Research* (India), 65:91–115.

Shin, J.D., Lee, J., Kim, Y.B., Han, I.S., and Kim, E.K. 2010. Production and characterization of methyl ester sophorolipids with 22-carbon-fatty acids. *Bioresource Technology*, 101(9):3170–3174.

Singh, M. and Desai, J.D. 1989. Hydrocarbon emulsification by candida tropicalis and debaryomyces polymorphus. *Indian Journal of Experimental Biology*, 27:224–226.

Singh, S., D'Britto, V., Prabhune, A.A., Ramana, C.V., Dhawan, A., and Prasad, B.L.V. 2010. Cytotoxic and genotoxic assessment of glycolipid reduced and capped gold and silver nanoparticles. *New Journal of Chemistry*, 34:294–301.

Singh, S., Patel, P., Jaiswal, S., Prabhune, A.A., Ramana, C.V., and Prasad, B.L.V. 2009. A direct method for the preparation of glycolipid metal nanoparticle conjugates: Sophorolipids as reducing and capping agents for the synthesis of water re dispersible silver nanoparticles and their anti bacterial activity. *New Journal of Chemistry*, 33:646–652.

Singh, S.K., Felse, A.P., Nunez, A., Foglia, T.A., and Gross, R.A. 2003. Regioselective enzyme catalyzed synthesis of sophorolipid esters, amides, and multifunctional monomers. *Journal of Organic Chemistry*, 68:5466–5477.

Sleiman, J.N., Kohlhoff, S.A., Roblin, P.M., Wallner, S., Gross, R., Hammerschlag, M.R., Zenilman, M.E., and Bluth, M.H. 2009. Sophorolipids as antibacterial agents. *Annals of Clinical Laboratory Science*, 39(1):60–63.

Solaiman, D.K.Y., Ashby, R.D., Nunez, A., and Foglia, T.A. 2004. Production of sophorolipids by *Candida bombicola* grown on soy molasses as substrate. *Biotechnology Letters*, 26:1241–1245.

Solaiman, D.K.Y., Ashby, R.D., Zerkowski, J.A., and Foglia, T.A. 2007. Simplified soy molasses based medium for reduced cost production of sophorolipids by *Candida bombicola*. *Biotechnology Letters*, 29:1341–1347.

Song, X.Q. 2006. *Wickerhamiella domercqiae* Y2A for producing sophorose lipid and its uses. Chinese Patent, CN1807578.

Spencer, J., Spencer, D., and Tulloch, A. 1979. Extracellular glycolipids of yeasts. *Journal of Biological Chemistry*, 3:523–540.

Spencer, J.F.T., Gorin, P.A.J., and Tulloch, A.P. 1970. *Torulopsis bombicola* sp. n. *Antonie Leeuwenhoek*, 36:129–133.

Sugita, T. and Nakase, T. 1999. Nonuniversal usage of the leucine CUG codon in yeasts: Investigation of basidiomycetous yeast. *Journal of Genetic and Applied Microbiology*, 45:193–197.

Sun, X.X., Choi, J.K., and Kim, E.K. 2004c. A preliminary study on the mechanism of harmful algal bloom mitigation by use of sophorolipid treatment original research article. *Journal of Experimental Marine Biology and Ecology*, 304(1):35–49.

Sun, X.X., Han, K.N., Choi, J.K., and Kim, E.K. 2004a. Screening of surfactants for harmful algal blooms mitigation. *Marine Pollution Bulletin*, 48(9–10):937–945.

Sun, X.X., Lee, Y.J., Choi, J.K., and Kim, E.K. 2004b. Synergistic effect of sophorolipid and loess combination in harmful algal blooms mitigation. *Marine Pollution Bulletin*, 48(9–10):863–872.

Takahashi, M., Morita, T., Wada, K., Hirose, N., Fukuoka, T., Imura, T., and Kitamoto, D. 2011. Production of sophorolipid glycolipid biosurfactants from sugarcane molasses using *Starmerella bombicola* NBRC 10243. *Journal of Oleofin Science*, 60(5):267–273.

Thaniyavarn, J., Chianguthai, T., Sangvanich, P., Roongsawang, N., Washio, K., Morikawa, M., and Thaniyavarn, S. 2008. Production of sophorolipid biosurfactant by *Pichia anomala*. *Bioscience, Biotechnology and Biochemistry*, 72(8):2061–2068.

Tulloch, A. and Spencer, J. 1968. Fermentation of long-chain compounds by *Torulopsis apicola* IV. Products from esters and hydrocarbons with 14 and 15 carbon atoms and from methyl palmitoleate. *Canadian Journal of Chemistry*, 46:1523–1528.

Tulloch, A.P., Spencer, J.F.T., and Deinema, M.H, 1968. A new hydroxy fatty acid sophoroside from *Candida bogoriensis*. *Canadian Journal of Chemistry*, 46:345–348.

van Bogaert, I.N., Saerens, K., De Muynck, C., Develter, D., Soetaert, W., and Vandamme, E.J. 2007. Microbial production and application of sophorolipids. *Applied Microbiology and Biotechnology*, 76(1):23–34.

van Bogaert, I.N., Zhang, J., and Soetaert, W. 2011. Microbial synthesis of sophorolipids. *Process Biochemistry*, 46(4):821–833.

van Bogaert, I.N.A., De Maeseneire, S.L., Develter, D., Soetaert, W., and Vandamme, E.J. 2008b. Development of a transformation and selection system for the glycolipid producing yeast *Candida bombicola*. *Yeast*, 25:273–278.

van Bogaert, I.N.A., De Maeseneire, S.L., Develter, D., Soetaert, W., and Vandamme, E.J. 2008c. Cloning and characterisation of the glyceraldehyde 3 phosphate dehydrogenase gene of *Candida bombicola* and use of its promoter. *Journal of Industrial Microbiology and Biotechnology*, 35:1085–1092.

van Bogaert, I.N.A., De Mey, M., Develter, D., Soetaert, W., and Vandamme, E.J. 2009a. Importance of the cytochrome P450 monooxygenase CYP52 family for the sophorolipid producing yeast *Candida bombicola*. *FEMS Yeast Research*, 9:87–94.

van Bogaert, I.N.A., Develter, D., Soetaert, W., and Vandamme, E.J. 2008a. Cerulenin inhibits the de novo sophorolipid synthesis of *Candida bombicola*. *Biotechnology Letters*, 30:1829–1832.

van Bogaert, I.N.A., Sabirova, J., Develter, D., Soetaert, W., and Vandamme, E.J. 2009b. Knocking out the MFE 2 gene of *Candida bombicola* leads to improved medium chain sophorolipid production. *FEMS Yeast Research*, 9:610–617.

Wadekar, S.D., Kale, S.B., Lali, A.M., Bhowmick, D.N., and Pratap, A.P. 2012. Utilization of sweetwater as a cost-effective carbon source for sophorolipids production by *Starmerella bombicola* (ATCC 22214). *Preparative Biochemistry and Biotechnology*, 42(2):125–142.

Williams, K. 2009. Biosurfactants for cosmetic application: Overcoming production challenges. *MMG Basic Biotechnology*, 5:1.

Yoo, D.S., Lee, B.S., and Kim, E.K. 2005. Characteristics of microbial biosurfactant as an antifungal agent against plant pathogenic fungus. *Journal of Microbiology and Biotechnology*, 15:1164–1169.

Yoshihiko, H., Mizuyuki, R., Yuka, O., Keisuke, I., Asami, N., Taro, F., and Masaki, S. 2009. Novel characteristics of sophorolipids, yeast glycolipid biosurfactants, as biodegradable low-foaming surfactants. *Journal of Bioscience and Bioengineering*, 108:142–146.

Zerkowski, J.A., Solaiman. D.K.Y., Ashby, R., and Foglia, T.A. 2006. Head group modified sophorolipids: Synthesis of new cationic, zwitterionic, and anionic surfactants. *Journal of Surfactants and Detergents*, 9:57–62.

Zhang, J., Saerens, K.M., van Bogaert, I.N., and Soetaert, W. 2011. Vegetable oil enhances sophorolipid production by *Rhodotorula bogoriensis*. *Biotechnology Letters*, 33(12):2417–2423.

5 Biosurfactants and Bioemulsifiers from Marine Sources

Rengathavasi Thavasi and Ibrahim M. Banat

CONTENTS

INTRODUCTION

Marine microbes are known for their many novel extra- and intracellular products such as antibiotics, enzymes, biopolymers, pigments, and toxins. Reports suggest that so far more than 10,000 metabolites with broad-spectrum biological activities and interesting medicinal properties have been isolated from marine microbes (Kelecom, 2002). However, due to the enormity of the marine biosphere, most of the marine microbial worlds remain unexplored. It has been estimated that <0.1% of marine microbial world has been explored or investigated (Ramaiah, 2005). Among various marine bioactive compounds, microbial biosurfactants (BSs) are of great importance due to their structural and functional diversity and industrial applications (Banat et al., 1991; Banat, 1995a,b; Rodrigues et al., 2006a). Marine microbial BSs are such metabolites with many interesting properties. BSs are basically amphiphilic surface active agents produced by bacteria, fungi, and actinomycetes. They belong to various classes including glycolipids, glycolipoproteins, glycopeptides, lipopeptides, lipoproteins, fatty acids, phospholipids, neutral lipids, lipopolysaccharides (Banat et al., 2010), and glycoglycerolipids (Wicke et al., 2000). The properties/applications of BSs include detergency, emulsification, foaming, dispersion, wetting, penetrating, thickening, microbial growth enhancement (e.g., oil-degrading bacteria), antimicrobial agents, metal sequestering, and resource recovering (oil recovery). These interesting properties allow BSs to have the ability to replace some of the most versatile chemical surfactants that are now in practice. In addition, BSs are promising natural surfactants that offer several advantages over chemically synthesized surfactants, such as *in situ* production using renewable substrates, lower toxicity, biodegradability, and ecological compatibility (Marchant and Banat, 2012a,b). Interest in BS research has been on the increase during the past two decades due to their interesting properties, yet the reason behind the production of BSs by many microorganisms remains mostly unknown. Several proposed physiological roles of BSs have been put forward including (1) increasing the surface area and bioavailability of hydrophobic water-insoluble substrates (e.g., oil-degrading microbes) (Ron and Rosenberg, 1999), (2) bacterial pathogenesis and quorum sensing and biofilm formation (e.g., *Pseudomonas aeruginosa*) (Davey et al., 2003), (3) antimicrobial for self-defense (e.g., antimicrobial activity of rhamnolipids) (Stanghellini and Miller, 1997), and (4) cell proliferation in the producing bacteria (e.g., viscosinamide production by *P. fluorescens*) (Nielsen et al., 1999). To isolate BS-producing microbes, combination of various screening methodologies has been studied (Maneerat and Phetrong, 2007; Satpute et al., 2008; Thavasi et al., 2011c) and extensively reviewed (Nerurkar et al., 2009; Satpute et al., 2010). Microbial communities like *Acinetobacter, Arthrobacter, Pseudomonas, Halomonas, Bacillus, Rhodococcus, Enterobacter, Azotobacter, Corynebacterium, Lactobacillus*, and yeast have been reported to produce BSs (Schulz et al., 1991; Passeri et al., 1992; Banat, 1993; Abraham et al., 1998; Maneerat et al., 2006; Thavasi et al., 2007, 2009, 2011a; Das et al., 2008a,b; Perfumo et al., 2010a). This chapter collates and highlights data search on isolation, culture methods, and potential applications for BSs from marine microbes.

MARINE BIOSURFACTANTS

Biosurfactants from marine microbes: A detailed list of BSs from marine microbes is described (Table 5.1) and a graphic representation of the number of publications and their percentages are illustrated in Figures 5.1 through 5.4. Earlier reports on BSs of marine origin mainly focused on environmental remediation applications of BSs such as emulsifiers and dispersants or BSs from oil-degrading microbes (Rosenberg et al., 1979; Schulz et al., 1991; Yakimov et al., 1998; Thavasi and Jayalakshmi, 2003, Thavasi et al., 2006, 2009, 2008, 2011a; Peng et al., 2007, 2008) and other potential applications such as medical and industrial sectors have not been studied extensively (Rodrigues et al., 2006b; Marchant and Banat, 2012a,b). Recently the trend has changed, and scientists started focusing on other potential application/ properties of marine BSs such as antimicrobial (Mukherjee et al., 2009), biofilm disruption (Kiran et al., 2010a), and nanoparticle synthesis (Kiran et al., 2010b). A detailed description of marine microbial surfactants, their composition, and producing organism are described in the following sections.

Glycolipids and glycoglycerolipids: Glycolipids are the most studied surfactants among all other marine microbial BSs, that is, 48% of the publications on marine BSs are about glycolipid BSs (Figure 5.1). Glycolipid BSs include glycolipids, glucose lipid, trehalose lipids, trehalose tetraester, trehalose corynomycolates, rhamnolipids, mannosylerythritol lipids, sophorolipids, and extracellular polysaccharide-lipids (ESLs). Glycolipids are composed of a hydrophobic fatty acid moiety esterified to a hydrophilic carbohydrate moiety. The sugar components of the glycolipids may be one or two molecules of glucose, trehalose, mannose, sophorose, or rhamnose.

TABLE 5.1
Types of Biosurfactants and Microbial Strains

Biosurfactant Type	Microorganism	Reference
Glycolipid	Bacterial strain MM1	Passeri et al. (1992)
	Nocardioides sp.	Vasileva-Tonkawa and Gesheva (2005)
	Aeromonas sp.	Ilori et al. (2005)
	Halomonas sp. ANT-3b	Pepi et al. (2005)
	Azotobacter chroococcum	Thavasi et al. (2006)
	Pseudomonas aeruginosa	Thaniyavarn et al. (2006)
	Pantoea sp.	Vasileva-Tonkawa and Gesheva (2007)
	Rhodococcus erythropolis	Peng et al. (2007)
	Bacillus megaterium	Thavasi et al. (2008)
	Brevibacterium casei	Krian et al. (2010a)
	Nocardiopsis lucentensis MSA04	Kiran et al. (2010d)
	Bacillus pumilus	
	Serratia marcescens	Dusane et al. (2011)
	Lactobacillus delbrueckii	Thavasi et al. (2011a)
	Streptomyces sp. B3	Khopade et al. (2012)
Glycolipid and phospholipids	*Alcanivorax borkumensis*	Abraham et al. (1998)
Trehalose lipids	*Rhodococcus fascians* DSM 20669	Yakimov et al. (1999)
Trehalose tetraester	*Arthrobacter* sp. EK 1	Schulz et al. (1991)
		Passeri et al. (1991)
Trehalose corynomycolates	*Arthrobacter* sp. SI 1	Schulz et al. (1991)
Glucose lipid	*Alcaligenes* sp. MM 1	Schulz et al. (1991)
	Unidentified bacterial strain MM 1	Passeri et al. (1992)
	Alcanivorax borkumensis gen. nov., sp. nov.	Yakimov et al. (1998)
Rhamnolipids	*Aspergillus* sp. MSF1	Kiran et al. (2010c)
Mannosylerythritol lipids	*Pseudozyma hubeiensis*	Konishi (2010)
Glucoglycerolipids	*Microbacterium* sp. DSM 12583	Wicke et al. (2000)
	Bacillus pumilus strain AAS3	Ramm et al. (2004)
	Micrococcus luteus	Palme et al. (2010)
Emulsan	*Acinetobacter calcoaceticus RAG-1*	Reisfeld et al. (1972)
Extracellular polysaccharide-lipid	*Alcaligenes* sp. PHY 9L-86	Goutx et al. (1987)
Sulfated heteropolysaccharide	*Halomonas eurithalina*	Calvo et al. (1998)
Extracellular polysaccharide	*Pseudomonas putida* ML2	Bonilla et al. (2005)
	Planococcus matriensis Anita I	Kumar et al. (2007)
Glycolipopeptide	*Yarrowia lipolytica* NCIM3589	Zinjarde et al. (1997)
	Yarrowia lipolytica	Amaral et al. (2006)
	Corynebacterium kutscheri	Thavasi et al. (2007)
Glycolipoprotein	*Pseudomonas nautica*	Husain et al. (1997)
	Aspergillus ustus MSF3	Kiran et al. (2009)

(continued)

TABLE 5.1 (continued)
Types of Biosurfactants and Microbial Strains

Biosurfactant Type	Microorganism	Reference
Trehaloselipid-o-dialkyl monoglycerides–protein	*Pseudomonas fluorescence*	Desai et al. (1988)
Alasan	*Acinetobacter radioresistens*	Navon-Venezia et al. (1995)
Biodispersion	*Acinetobacter calcoaceticus*	Rosenberg and Ron (1998)
Carbohydrate–protein complex	*Rhodotorula glutinis*	Oloke and Glick (2005)
Glycoprotein	*Antarctobacter* sp. TG22	Gutiérrez et al. (2007a)
	Halomonas sp. TG39 and 67	Gutiérrez et al. (2007b)
Lipopeptide	*Bacillus licheniformis* BAS50	Yakimov et al. (1995)
	Bacillus pumilus KMM150	Kalinovskaya et al. (1995)
	Pseudomonas sp. MK90e85 and MK91CC8	Gerard et al. (1997)
	*Bacillus pumilus*1364	Kalinovskaya et al. (2002)
	Rhodococcus sp. TW53	Peng et al. (2008)
	Nocardiopsis alba MSA10	Gandhimathi et al. (2009)
	Azotobacter chroococcum	Thavasi et al. (2009)
	Brevibacterium aureum MSA13	Kiran et al. (2010b)
	Pseudomonas aeruginosa	Thavasi et al. (2011d)
	Bacillus circulans DMS-2	Sivapathasekaran et al. (2011)
Ornithine lipids	*Myroides* sp. SM1	Maneerat et al. (2006)
Surfactin	*Bacillus velezensis* H3	Liu et al. (2010)
Proline lipid	*Alcanivorax dieselolei* B-5	Qiao and Shao (2010)

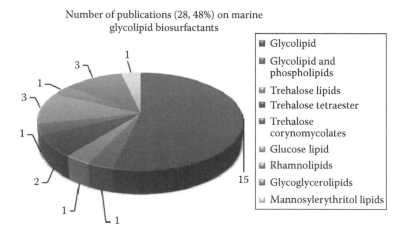

FIGURE 5.1 **(See color insert.)** Breakdown compilation of publication on marine glycolipid biosurfactants.

Number of publications (6, 10%) on marine extracellular
polysaccharide and polysaccharide lipid biosurfactants

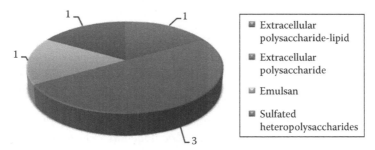

- Extracellular
  polysaccharide-lipid
- Extracellular
  polysaccharide
- Emulsan
- Sulfated
  heteropolysaccharides

FIGURE 5.2 **(See color insert.)** Breakdown compilation of publication on marine extracellular polysaccharide and polysaccharide lipid biosurfactants.

Number of publications (11, 19%) on marine glycolipopeptide
and glycolipoprotein biosurfactats

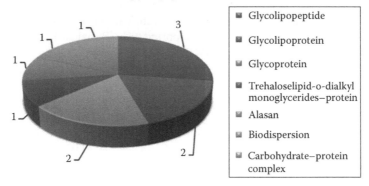

- Glycolipopeptide
- Glycolipoprotein
- Glycoprotein
- Trehaloselipid-o-dialkyl
  monoglycerides–protein
- Alasan
- Biodispersion
- Carbohydrate–protein
  complex

FIGURE 5.3 **(See color insert.)** Breakdown compilation of publication on marine glycolipopeptide, glycolipoprotein, and glycoprotein biosurfactants.

Number of publications on (13, 22%) marine lipopeptide and
lipoprotein biosurfactants

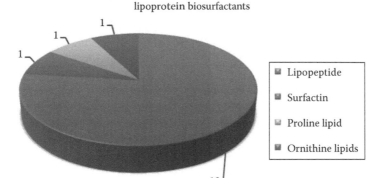

- Lipopeptide
- Surfactin
- Proline lipid
- Ornithine lipids

FIGURE 5.4 **(See color insert.)** Breakdown compilation of publication on marine lipopeptide and lipoprotein biosurfactants.

The fatty acid moiety comprises one or two fatty acids with a chain length of C_6–C_{20} with or without unsaturation depending on the source of the fatty acid and the microbial enzyme involved in processing the fatty acid part of the BS. Glycolipid-producing microbes were isolated from different marine sources such as *Halomonas* sp. ANT-3b isolated from Ross Sea, Antarctica (Pepi et al., 2005), which produced a glycolipid BS with a molecular weight of 18 kDa using *n*-hexadecane as the sole carbon source. *P. aeruginosa* A41 isolated from the gulf of Thailand was capable of utilizing wide range of carbon sources (oil, lauric acid, myristic acid, palmitic acid, stearic acid, oleic acid, and linoleic acid) and produced 6.58 g/L rhamnolipids with olive oil as the carbon substrate (Thaniyavarn et al., 2006). *Azotobacter chroococcum, Bacillus megaterium,* and *Lactobacillus delbrueckii* isolated from Tuticorin coastal waters (Tamil Nadu, India) using crude oil as the carbon source produced glycolipid BSs on crude oil, waste motor oil, and peanut oil cake carbon sources (Thavasi et al., 2006, 2008, 2011a). The highest BS production was observed with peanut oil cake as 4.6, 1.4, and 5.3 g/L, respectively, and was able to emulsify a wide range of hydrocarbons and vegetable oil in the order of waste motor lubricant oil > crude oil > peanut oil > kerosene > diesel > xylene > anthracene > naphthalene.

Among the glycolipid BSs, trehalose lipids, trehalose corynomycolates, and trehalose tetraester are the second largest group reported (9%). Trehalose lipid production by an Antarctic *Rhodococcus fascians* DSM 20669 strain was reported by Yakimov et al. (1999), and the surfactant reduced the surface tension of water from 72 to 32 mN/m. Trehalose tetraester production was found with *n*-alkanes utilizing marine bacterium, *Arthrobacter* sp. EK1 (Passeri et al., 1991). The main fraction of the purified surfactant is an anionic 2,3,4,2'-trehalose tetraester. The chain lengths of fatty acids ranged from 8 to 14. Trehalose corynomycolates producing marine *Arthrobacter* sp. SI1 was isolated from North Sea, and it produced 2 g/L of BS. The isolated BSs showed significant interfacial and emulsifying properties (Schulz et al., 1991).

A marine sponge associated *Aspergillus* sp. MSF1 strain producing rhamnolipids was also isolated from the Bay of Bengal coast of India (Kiran et al., 2010c). Rhamnolipid BS isolated from *Aspergillus* sp. MSF1 showed potential activity against pathogenic yeast, *Candida albicans* and bacteria, *Streptococcus* sp., *Micrococcus luteus*, and *Enterococcus faecalis*. Konishi et al. (2010) recently isolated a mannosyl-erythritol lipid-producing *Pseudozyma hubeiensis* from a deep sea clam *Calyptogena soyoae* at 1156 m in Sagami Bay. The major component of the BS was MEL-C (4-O-[4'-O-acetyl-2',3'-di-O-alka(e)noil-β-D-mannopyranosyl]-D-erythritol). Analysis of the lipid part of the BS revealed that major fatty acids of MEL-C were saturated fatty acids with chain lengths of C_6, C_{10}, and C_{12}. The surface tension determination of the MEL-C showed a critical micelle concentration (CMC) of 1.1×10^{-5} M.

Another class of BSs that are not in the spotlight of general research is glycoglycerolipids. There are three reports available on glycoglycerolipids from marine bacterial strains. The marine bacterium *Microbacterium* sp. DSM 12583, isolated from the Mediterranean sponge *Halichondria panicea*, was able to form a glucosylmannosyl-glycerolipid (GGL 2), 1-o-acyl-3-[α-glucopyranosyl-(1–3)-(6-o-acyl-α-mannopyranosyl)]glycerol, when grown on a complex medium with glycerol (Wicke et al., 2000). Another marine strain, *M. luteus* was isolated from the North sea and reported to produce a dimannosyl-glycerolipid (GGL 5), mannopyranosyl(1α-3)-6-acylman-nopyranosyl(*1*α-1)-3-acylglycerol on artificial seawater supplemented with glucose, yeast extract, peptone, and nitrogen/phosphate sources (Palme et al., 2010). The third marine bacterium, *B. pumilus* strain AAS3 isolated from the Mediterranean sponge *Acanthella acuta* and synthesized a diglucosyl-glycerolipid (GGL 11), 1,2-o-diacyl-3-[β-glucopyranosyl-(1–6)-β-glucopyranosyl)]glycerol, when grown on artificial medium provided with glucose, yeast extract, peptone, and nitrogen/phosphate sources (Ramm et al., 2004).

Extracellular polysaccharide-lipids (ESL) and polysaccharides (PS): ESLs and PSs are another class of BSs produced by marine microbes, and reports on ESLs and PSs contribute 10% of total publications on marine BSs (Figure 5.2). *Acinetobacter calcoaceticus* RAG-1 strain (formally known as *Arthrobacter* RAG-1) was isolated from tar balls collected in beach soil (Rosenberg et al., 1979). RAG-1 strain produced an emulsifier called emulsan, an extracellular polyanionic amphipathic heteropolysaccharide. Its amphipathic and polymeric characteristics provide it with emulsifying as well as stabilizing activity of oil/water systems. The bioemulsifier exhibits specificity toward its hydrocarbon emulsifiable substrates. It tends to concentrate at the oil/water interface of the hydrocarbon droplets preventing their coalescence and allowing the formation of concentrated emulsanosols, which were found to be useful in enhancing oil degradation, binding heavy metal cations or constituting a low-viscosity stable oil-in-water emulsion suitable for oil pipeline transportation or direct combustion. Emulsan was also found to protect its producing cells against the toxic cationic detergent cetyltrimethylammonium bromide (Shabtai and Gutnick, 1986).

Another marine bacterium *Alcaligenes* sp. PHY 9 L-86 was isolated from hydrocarbon-polluted sea-surface waters that produced surface active exopolysaccharides (extracellular carbohydrates) and lipids on a medium containing 0.1% tetradecane. Chemical analysis revealed the composition of the extracellular lipids as phospholipids, free fatty acids, triglycerides, monoglycerides, esters, and free fatty acids (73%) as the major constituent (Goutx et al., 1987). A PS BS-producing *P. putida* ML2 strain was also isolated from hydrocarbon-polluted sediment collected at the Montevideo bay (Uruguay). Chemical composition of the PS BS revealed its sugar composition as rhamnose, glucose, and glucosamine in a 3:2:1 molar ratio with a molecular weight of 10–80 kDa. Another PS surfactant-producing marine *Planococcus* strain called *P. maitriensis Anita I* was isolated from coastal area of Gujarat, India (Kumar et al., 2007). The extracellular polymeric substance produced had a chemical composition of carbohydrate (12.06%), protein (24.44%), uronic acid (11%), and sulfate (3.03%) and effectively emulsified xylene and formed stable emulsions with jatropha, paraffin, and silicone oils. Its cell-free supernatant reduced the surface tension of water from 72 to 46.07 mN/m.

Glycolipopeptides, Glycolipoproteins, and Glycoproteins (GLPs): Glycolipopeptide and glycolipoprotein BSs are made of sugar–lipid–peptides or amino acids. GLPs are the third most studied surfactants among marine BSs, and reports on GLPs contribute 19% of total publications (Figure 5.3). *Corynebacterium kutscheri,* a marine bacterium isolated by Thavasi et al. (2007) from Tuticorin harbor, India, produced a glycolipopeptide BS. The BS was composed of carbohydrate (40%), lipid (27%), and protein (29%) and was able to emulsify waste motor lubricant oil, crude oil, peanut oil, kerosene, diesel, xylene, naphthalene, and anthracene. Its emulsification activity was comparatively higher than the activity found with Triton x-100.

Amaral et al. (2006) isolated a yeast strain, *Yarrowia lipolytica*, from Guanabara Bay in Brazil, which produced a glycolipopeptide BS called Yansan in glucose-based medium. It was reported to contain 15% protein and 1% lipid. The fatty acid composition of the lipid was palmitic acid (35.8%), stearic acid (21.4%), lauric acid (8.8%), and oleic acid (6.9%). However, there was no quantitative information available on its sugar content, but its monosaccharide composition was reported as arabinose, galactose, glucose, and mannose in a ratio of 1:6:17:31. The molecular weight of the BS was approximately 20 kDa with a CMC value of 0.5 g/L. Another marine yeast strains called *Yarrowia lipolytica* NCIM 3589 isolated by Zinjarde et al. (1997) produced an emulsifier with a chemical composition of 75% lipid, 20% carbohydrate, and 5% protein. The lipid and carbohydrate part of the emulsifier comprised palmitic acid, mannose, and galactose, respectively, while the main amino acid components were aspartic acid, alanine, and threonine.

GLPs are another set of BSs produced by marine microbes composed of sugar and protein complexes. Gutiérrez et al. (2007a) reported the GLP bioemulsifier production by a marine bacterium, *Antarctobacter* sp. TG22, in a low-nutrient seawater medium supplemented with glucose. Chemical, chromatographic, and nuclear magnetic resonance spectroscopic analysis confirmed that the bioemulsifier has a high-molecular-weight (>2000 kDa) GLP with high uronic acid contents. The carbohydrate content of this emulsifier was reported as 15.4% ± 0.2%; further analysis of the carbohydrate component revealed the presence of hexoses (rhamnose, fructose, galactose, glucose, and mannose), amino sugars (galactosamine, glucosamine, and muramic acid), and uronic acids (galacturonic and glucuronic acids). The amino acid component was 5.0% ± 0.2% and mainly contained the three major amino acids, aspartic acid, glycine, and alanine.

Gutiérrez et al. (2007b) reported GLP BS production by two *Halomonas* species, TG39 and TG67. These strains produced two different GLP emulsifiers, called HE39 (strain TG39) and HE67 (strain TG67). The total carbohydrate content of HE39 and HE67 was 17.3% ± 1.0% and 22.7% ± 0.8%, respectively. The major monosaccharides identified in HE39 were rhamnose (31.7% ± 2.1%), glucuronic acid (27.9% ± 1.9%), and galactose (15.3% ± 0.5%). Carbohydrate content of HE67 showed that the main monosaccharides detected were glucuronic acid (58.8% ± 0.4%), glucosamine (10.9% ± 0.1%), and mannose (11.5% ± 0.5%), while rhamnose, galactose, galactosamine, glucose, muramic acid, and galacturonic acid, each of them present at less than 10% of the total monosaccharide composition and together they contributed only 18% to the total carbohydrate content. The total amino acid content of HE39 and HE67 was 26.6% ± 1.0% and 40.5% ± 1.6%, respectively. Amino acid analysis of hydrolyzed surfactant identified the presence of four major amino acids in both emulsifiers as aspartic acid, glutamic acid, glycine, and alanine, which in total contributed 45.1% and 50.7% to the total amino acid content of HE39 and HE67, respectively.

Lipopeptides: Lipopeptides are the second well-studied BS group from marine microorganisms that accounts for 22% of the total publications on marine BSs of which the genus *Bacillus* alone contributes 38% (Figure 5.4). Lipopeptide BSs are made of hydrophilic peptide head group and a hydrophobic lipid tail. A marine bacterium, *Azotobacter chroococcum* (Thavasi et al., 2009) isolated from seawater, grew and produced a lipopeptide BS in a medium provided with crude oil, waste motor lubricant oil, and peanut oil cake. Peanut oil cake gave the highest BS production (4.6 mg/mL). The BS product emulsified waste motor lubricant oil, crude oil, diesel, kerosene, naphthalene, anthracene, and xylene. Preliminary characterization using biochemical, Fourier transform infrared spectroscopy, and LC-MS analysis indicated that the BS was a lipopeptide with percentage lipid and protein proportion of 31.3:68.7. Sivapathasekaran et al. (2011) isolated a marine lipopeptide BS-producing *B. circulans* strain from seawater sample from Andaman and Nicobar Islands, India. Similar marine *Bacillus* strains, *B. pumilus* KMM150 isolated from an Australian marine sponge *Ircinia* sp. (Kalinovskaya et al., 1995) and *B. pumilus* KMM1364 isolated from the surface of the ascidian *Halocynthia aurantium* (Kalinovskaya et al., 2002), were also reported to produce lipopeptide BSs. Strain KMM150 produced a mixture of cyclic depsipeptides known as bacircines (1–5 fractions with different molecular weights). Bacircines 1, 2, and 3 had molecular masses of 1007, 1021, and 1021, respectively, and bacircines 4 and 5 had a molecular mass of 1035. The amino acid composition of the bacircines 1, 2, 3, 4, and 5 were similar with a combination of Leu:Val:Asp:Glu with a ratio of 4:1:1:1. Bacircines 1, 2, and 3 has C_{13}–C_{14}–hydroxyacids in their lipid part whereas bacircines 4 and 5 had 3β-hydroxypentadecanoic acid (C15-β-hydroxyacid). *B. pumilus* KMM1364 also produced a mixture of lipopeptide analogs with major components with molecular masses of 1035, 1049, 1063, and 1077. The variation in molecular weight represents changes in the number of methylene groups in the lipid and/or peptide portions of the surfactant. Structurally, these lipopeptides differ from surfactin in the substitution of the valine residue in position 4 by leucine and have been isolated as two carboxy-terminal variants, with valine or isoleucine in position 7. The lipid part of the surfactant is composed of β-hydroxy-C15-, β-hydroxy-C16-, and a high amount of β-hydroxy-C17 fatty acids. Liu et al. (2010) reported a marine *B. velezensis* strain producing nC_{14}-surfactin and *anteiso*C_{15}-surfactin. These compounds can reduce the surface tension of phosphate-buffered saline (PBS) from 71.8 to 24.8 mN/m. The CMCs of C_{14}-surfactin and C_{15}-surfactin in 0.1 M PBS (pH 8.0) were determined to be 3.06×10^{-5} and 2.03×10^{-5} M, respectively. The surface tension values at CMCs for C_{14}-surfactin and C_{15}-surfactin were 25.7 and 27.0 mM/m, respectively.

Kiran et al. (2010e) isolated a marine sponge (*Dendrilla nigra*)-associated *Brevibacterium aureum* MSA13 strain producing lipopeptide BS. The peptide part of the surfactant was made of short sequence of four amino acids including pro–leu–gly–gly coupled with a lipid moiety composed of octadecanoic acid methylester. Oil-degrading lipopeptide BS-producing *Rhodococcus* sp. was isolated from Pacific Ocean deep-sea sediments (Peng et al., 2008). The hydrophobic lipid part of the surfactant contained five types of fatty acids with chain lengths of C_{14}–C_{19} and $C_{16}H_{32}O_2$ as a major component making up 59.18% of the total. The hydrophilic fraction was composed of five kinds of amino acids with a sequence of Ala–Ile–Asp–Met–Pro.

Two strains of marine *Pseudomonas*, namely, MK90e85 and MK91CC8, produced antimycobacterial cyclic depsipeptides and viscosin (Gerard et al., 1997). The MK90e85 strain isolated from a red alga produced massetolides A, B, C, and D and the other strain MK91CC8 isolated from a marine tubeworm produced massetolides

E, F, G, H, and viscosin. Massetolide A is an optically active molecule, with a mass of 1141 Da and a molecular formula of $C_{55}H_{97}N_9O_{16}$. Another marine *Pseudomonas* strain called *P. aeruginosa* producing lipopeptide BS was also isolated from coastal waters of Tamil Nadu, India (Thavasi et al., 2011d). The BS is composed of protein (50.2%) and lipid (49.8%), and at 1 mg/mL concentration, the BS was able to emulsify waste motor lubricant oil, crude oil, peanut oil, kerosene, diesel, xylene, naphthalene, and anthracene, and the emulsification activity was higher than that achieved by Triton x-100 at similar concentration. A marine bacterium, *Myroides* sp. SM1 capable of producing ornithine lipid bioemulsifier was isolated from seawater (Maneerat et al., 2006; Maneerat and Dikit, 2007). L-Ornithine lipids were composed of L-ornithine and two different iso-3-hydroxyfatty acid (C_{15}–C_{17}) and iso-fatty acid (C_{15} or C_{16}) in a ratio of 1:1:1. Ornithine lipids exhibited emulsifying activity with crude oil in a broad range of pH, temperature, and salinity and showed high oil displacement activity.

AREAS OF POTENTIAL APPLICATIONS OF MARINE BIOSURFACTANTS

Biosurfactants and hydrocarbon degradation/remediation: Oil pollution in terrestrial and aquatic environments is a common phenomenon that causes significant ecological and social problems. The recent British Petroleum deepwater horizon oil spill at the Gulf of Mexico during the summer of 2010 is a poignant example. The traditional available treatment processes used to decontaminate polluted areas have on the main been limited in their application (Perfumo et al., 2010b). Physical collection methods such as booms, skimmers, and adsorbents typically recover no more than 10%–15% of the spilled oil, and the use of chemical surfactants as remediating agents is not favored due to their toxic effects on the existing biota in the polluted area. Therefore, despite decades of research, successful bioremediation of oil contaminated environment remains a challenge (Perfumo et al., 2010a).

Oil pollution in marine environments stimulates the indigenous community of obligate hydrocarbonoclastic bacteria to flourish, becoming the majority of the total microbial population (Yakimov et al., 2007). Microorganisms involved in oil degradation have adopted different strategies to enhance the bioavailability and gain access to hydrophobic compounds, such as hydrocarbons, including (1) BS-mediated solubilization, (2) direct access of oil drops, and (3) biofilm-mediated access (Hommel, 1990). The production of BSs and bioemulsifiers is generally involved in varying degrees, in all the strategies provided earlier. BS structural uniqueness resides in the coexistence of the hydrophilic (a sugar or peptide) and the hydrophobic (fatty acid chain) domains in the same molecule, allowing them to occupy the interfaces of mixed-phase systems (e.g., oil/water, air/water, and oil/solid/water) and consequently altering the forces governing the equilibrium conditions, which is a prerequisite for the occurrence of a broad range of surface activities including emulsification, dispersion, dissolution, solubilization, wetting, and foaming (Desai and Banat, 1997; Banat et al., 2000). Such advantage was reported by Thavasi et al. (2011a,b) investigating the effect of BS and fertilizer on biodegradation of crude oil by four marine oil-degrading bacteria, *B. megaterium*, *Corynebacterium kutscheri*, *P. aeruginosa*, and *Lactobacillus delbrueckii*. It was reported that the addition of BS and fertilizer into the lab-scale biodegradation system increased the degradation process from 19% to 37.7% as compared to the experiments where no BS and fertilizer was added. The BSs used in the earlier experiments were able to emulsify a range of different hydrocarbons.

Another example is the high-molecular exopolysaccharide-producing marine bacterium *Alcaligenes* sp. PHY 9L-86, which was able to use 0.1% tetradecane as the sole carbon and energy source and degrade 98% of the substrate within 48 h (Goutx et al., 1987). In the same way, emulsifiers from *Halomonas* sp. TG39 and TG67 showed significant emulsification activity with different edible oils as well as with hexadecane, and these emulsions remained stable for several months (Gutiérrez et al., 2007b). Low-molecular-weight BSs such as glycolipids produced by marine bacterium, *Halomonas* sp. ANT-3b, was able to degrade *n*-hexadecane and use it as the sole source of carbon to produce BSs (Pepi et al., 2005). In addition to the broad-spectrum emulsification ability to promote the biodegradation process, marine microbial BSs also exhibited higher stability at various conditions as reported by Thaniyavarn et al. (2006); the researchers isolated a *P. aeruginosa* A41 strain producing rhamnolipid BS that had good stability and activity at a wide range of temperatures (40°C–121°C), pH (2–12), and NaCl concentrations (0%–5%). Higher stability at various conditions and broad-spectrum emulsification activities found with marine BSs indicated their potential broad-spectrum application against various hydrocarbons and in different environments.

Besides the reports on the application of marine microbes and their BSs, there are few reports on the application of microbes, and their BSs isolated from nonmarine origin support the concept of the application of microbial BSs in oil bioremediation. Bioremediation of gasoline contaminated soil by bacterial consortium with poultry litter (PL), coir pith (CP), and rhamnolipid BS in an *ex situ* bioremediation system was reported by Rahman et al. (2002). In this study, the authors treated red soil (RS) with gasoline-spilled soil (GS) from a gasoline station, and different combinations of amendments were prepared using (1) mixed bacterial consortium (MC), (2) PL, (3) CP, and (4) rhamnolipid BS produced by *Pseudomonas* sp. DS10–129. The study was conducted for a period of 90 days during which bacterial growth, hydrocarbon degradation, and growth parameters of *Phaseolus aureus* RoxB (mung bean planted on the oil-polluted soil) including seed germination, chlorophyll content, and shoot and root length were measured. Results from the biodegradation experiments revealed that 67% and 78% of the hydrocarbons were effectively degraded within 60 days in soil samples amended with RS + GS + MC + PL + CP + BS at 0.1% and 1%. Maximum seed germination, shoot length, root length, and chlorophyll content of the *P. aureus* were recorded after 60 days in the earlier amendments. This study suggests that using BSs or BS-producing microbes, oil-polluted agriculture soil can be restored to its original state, and the application of BS as a growth enhancer in agriculture lands polluted with hydrocarbons. Another examples is enhanced bioremediation of *n*-alkane in petroleum sludge using bacterial consortium amended with rhamnolipid and micronutrients (Rahman et al., 2003). Results reported in this study showed that maximum hydrocarbon degradation was observed after the 56 days of treatment. Degradation of hydrocarbons with different carbon chain length showed that *n*-alkanes in the range of *n*C8–*n*C11 were degraded completely followed by *n*C12–*n*C21, *n*C22–*n*C31, and *n*C32–*n*C40 with percentage degradations of 100%,

83%–98%, 80%–85%, and 57%–73%, respectively. Addition of rhamnolipids into the biodegradation system significantly increased the microbial growth by promoting the efficient utilization of hydrocarbons.

Biosurfactants and microbially enhanced oil recovery (MEOR): BSs have extensive potential application in the petroleum industry such as emulsifiers, demulsifiers, and oil recovery agents. MEOR is a technique that either uses a crude preparation of BS or sterilized BS containing culture broth to liberate crude oil from a binding substrate (Marchant and Banat, 2012b). For example, Banat et al. (1991) carried out a crude oil sludge tank cleanup and oil recovery process in which BS-containing sterilized culture broth was used to clean up oil sludge from an oil storage tank. After 5 days of treatment involving energy addition and circulation to enhance the process of emulsification followed by a deemulsification, 91% of crude oil present in the oil storage tank was recovered. Hydrocarbon analysis for the recovered crude showed a 100% hydrocarbon content. This result indicated that MEOR process doesn't require live microorganisms or pure BSs and that sterilized BS-containing broth is sufficient to mobilize and recover significant amount of oil from oil sludge deposits.

Biosurfactants and heavy metal remediation: Microbial BSs are known for their metal-complexing activities that have been reported to be effective in the remediation of heavy metal-contaminated environments (Mulligan et al., 2001; Singh and Cameotra, 2004). The mechanisms behind metal binding are (1) anionic BSs create complexes with metals in a nonionic form by ionic bonds. These bonds are stronger than the metal's bonds with the soil/sediment and metal–BS complexes are desorbed from the soil matrix to the solution due to the lowering of the interfacial tension and (2) the cationic BSs can replace the same charged metal ions by competition for some but not all negatively charged surfaces (ion exchange). Metal ions can be removed from soil surfaces by the BS micelles. The polar head groups of micelles bind metals that mobilize the metals in water (Mulligan and Gibbs, 2004).

Applications of marine BSs in heavy metal remediation have been reported by many researchers. Das et al. (2009b) studied the bacterial cells (*B. circulans*) and BS-mediated cadmium and lead metal binding and suggested that there was no cell-mediated metal binding, but that an increase in metal binding was observed with an increase in BS concentration from 0.5 × CMC to 5 × (CMC of the BS was 40 mg/L). The percentage removal at 0.5 × CMC was 76.6%, 53.18%, 56.63%, and 42.74%, 29.72%, 23.19% for lead and cadmium, respectively, while the percentage removal was increased to 100%, 95.75%, 87.69%, and 97.66%, 88.43%, 86.35% for lead and cadmium, respectively at 5.0 × CMC of BS. A complete removal of the metals was seen at 10 × CMC.

Gnanamani et al. (2010) reported the chromium reduction and trivalent chromium tolerance behavior of marine *Bacillus* sp. MTCC5514 through its extracellular enzyme reductase and BS production. The isolate reduces 10–2000 mg/L of hexavalent chromium to trivalent chromium within 24–96 h, and the release of extracellular chromium reductase was responsible for the metal reduction. The role of the BS in this metal reduction process is to entrap the trivalent chromium in the micelle of BSs, which prevents microbial cells from exposure toward trivalent chromium. It was concluded that extracellular chromium reductase and BS mediate the remediation process and keep the cells active and provide tolerance and resistance toward high concentration of hexavalent chromium and trivalent chromium. The earlier reports on the application of marine BSs in heavy metal binding and mobilization activities clearly indicated the potential application marine BSs in metal remediation.

Antimicrobial and antifouling agents: Marine microbial surfactants have been recognized for their biological properties such as antimicrobial and antifouling/biofilm activities. Antimicrobial activity of lipopeptide BS produced by marine *B. circulans* was reported by Mukherjee et al. (2009). Significant inhibitory activity was seen against gram-positive bacteria like *M. flavus, B. pumilus,* and *Mycobacterium smegmatis,* and gram-negative bacteria like *Escherichia coli, Serratia marcescens, Proteus vulgaris, Pseudomonas* sp., and *Klebsiella aerogenes.* The BS also showed significant inhibitory action against fungal species such as *Aspergillus niger, Aspergillus flavus,* and *C. albicans.* The purified BS showed more antimicrobial activity, and its broad-spectrum activity against gram-positive and -negative and fungal cultures indicated its potential application as an antimicrobial agent in medical and household antimicrobial and disinfectant applications.

Gerard et al. (1997) reported the antimicrobial activity of cyclic depsipeptides (Massetolide A-H) and viscosin produced by *Pseudomonas* sp. MK90e85 and MK91CC8 strains. Massetolide A and viscosin showed antimicrobial action against *Mycobacterium tuberculosis* and *Mycobacterium avium-intracellulare.* Massetolide A showed a minimum inhibitory concentration (MIC) value of 5–10 μg/mL against *M. tuberculosis* and 2.5–5 μg/mL against *M. avium-intracellulare.* Viscosin had an MIC value of 10–20 μg/mL against *M. tuberculosis* and 10–20 μg/mL against *M. avium-intracellulare.* Even though these molecules are toxic to pathogenic microbial cells, massetolide A was nontoxic to mice at a dose of 10 mg/kg body weight. This nontoxic nature to mammalian cell property of the massetolide A indicates its potential application in treating infections caused by *Mycobacterium* sp. The proposed antimicrobial mechanism of BSs is membrane lipid order perturbation, which compromises the viability of microorganisms and their spores (Azim et al., 2006).

Similar to antimicrobial activity, BSs play an important role in the formation of biofilm especially attachment and detachment of cells. Das et al. (2009a) reported the antiadhesive and biofilm dislodging activities of a lipopeptide surfactant isolated from marine *B. circulans* strain. The concentration of BS used in both the experiments was 0.1–10 mg/mL, and the results suggested that maximum antiadhesive and biofilm dislodging activity was observed at 10 mg/mL concentration at which percentage antiadhesion activity was 89%, 88%, 84%, 87%, 86%, 88%, 87%, and 83% against *E. coli, M. flavus, S. marcescens, Salmonella typhimurium, P. vulgaris, Citrobacter freundii, Alcaligenes faecalis,* and *K. aerogenes,* respectively. The biofilm dislodging activity of the BS against bacterial strains was as follows: *E. coli* (59%), *M. flavus* (72%), *S. marcescens* (94%), *S. typhimurium* (89%), *P. vulgaris* (82%), *Citrobacter freundii* (77%), *Alcaligenes faecalis* (79%), and *K. aerogenes* (80%). Biofilm disruption potential of a glycolipid BS isolated from marine *B. casei* was reported by Kiran et al. (2010a). The biofilm-forming bacterial cultures used in the assay system were *Vibrio parahaemolyticus* MTCC 451, *Vibrio harveyi* MTCC 3438, *Vibrio alginolyticus* MTCC 4439, *Vibrio alcaligenes* MTCC 4442,

Vibrio vulnificus MTCC 1145, *P. aeruginosa* MTCC 2453, and *E. coli* MTCC 2339. Biofilm disruption results revealed that biofilm-forming capacity of both mixed culture and individual strains were significantly inhibited at 30 mg/mL concentration. Dusane et al. (2011) reported the antibiofilm activity of a glycolipid BS produced by a tropical marine bacterium *S. marcescens* isolated from the hard coral, *Symphyllia* sp. The reported antibiofilm activity of the glycolipid BS was 89% and 90%, 75% and 88%, 76% and 82% at 50 and 100 μg/mL concentrations against *B. pumilus, P. aeruginosa,* and *C. albicans,* respectively.

Biosurfactants in medical/therapeutic applications: Antimicrobial activities of marine BSs reviewed in the previous section of this chapter make them relevant molecules for applications in combating many diseases and as therapeutic agents. In addition, their role as antiadhesive agents against several pathogens indicates their utility as suitable antiadhesive coating agents for medical insertional materials leading to a reduction in a large number of hospital infections without the use of synthetic drugs and chemicals. Recent reports suggest that microbial surfactants also have other important medicinal and therapeutic applications. The use and potential commercial application of BSs in the medical field have increased during the past decade. There are few reports available on therapeutic applications of BSs of nonmarine origin. Example, Rodrigues et al. (2006a) reported the inhibition of microbial adhesion to silicone rubber treated with BS from *Streptococcus thermophilus* A. Pathogens used in this study were *Rothia dentocariosa* GBJ 52/2B and *Staphylococcus aureus* GB 2/1 (bacterial strains) and *C. albicans* GBJ 13/4A and *C. tropicalis* GB 9/9 (yeast strains). After 4 h of treatment of the silicon rubber with BS, a decrease in the initial cell deposition rate was observed for *Rothia dentocariosa* GBJ 52/2B and *S. aureus* GB 2/1 from 1937 ± 194 to 179 ± 21 microorganisms/cm^2/s and from 1255 ± 54 to 233 ± 26 microorganisms/cm^2/s, respectively, which was an 86% reduction of the initial deposition rate for both strains. The number of bacterial cells adhering to the silicone rubber with preadsorbed BS after 4 h was further reduced by 89% and 97% for the two strains, respectively. Antiadhesion activity observed against the two yeast strains showed 67% and 70% reduction, respectively, for *C. albicans* GBJ 13/4A and *C. tropicalis* GB 9/9.

Another example was reported by Rodrigues et al. (2006c) on antimicrobial cell adhesion activity of rhamnolipid BSs on voice prostheses to silicone rubber. After 4 h of treatment with (4 g/L concentration) rhamnolipids, an average of 66% antiadhesion activity occurred for *S. salivarius* GB 24/9 and *C. tropicalis* GB 9/9. Preformed microbial mat/biofilm had a 96% detachment of microorganisms adhered to the silicone rubber. They concluded that pretreatment with surface active compounds may be a promising strategy to reduce the microbial colonization rate of silicone rubber voice prostheses used after total laryngectomy surgery. Laryngectomy is a surgical treatment for the extensive cancer of larynx, which alters swallowing and respiration in patients, which is followed up with a surgical voice restoration procedure comprising tracheoesophageal puncture techniques with an insertion of a "voice prosthesis" to improve successful voice rehabilitation. After the surgery, microbial colonization on the silicon rubber voice prostheses is a major drawback of these devices. Antimicrobials are usually used to prevent the colonization of silicone rubber voice prostheses by microorganisms, but long-term medication may induce the development of resistant strains (Rodrigues et al., 2007). However, antiadhesion results reported by Rodrigues et al. (2006a,b) suggest an alternate way of using microbial BSs as antiadhesive agent; in this process, the BSs used not necessarily have to be an antimicrobial agent and that approach may prevent the antimicrobial resistance development among the microbes involved in adhesion.

Another interesting example for the therapeutic application of BSs is sophorolipids, a family of natural and easily chemoenzymatically modified microbial glycolipid, showed promising immune modulator activity in a rat model of sepsis. Sophorolipid administration after the induction of intra-abdominal sepsis significantly decreased the rat mortality in this model. This may be mediated in part by decreased macrophage nitric oxide production and modulation of inflammatory responses (Bluth et al., 2006).

Enhanced healing of full-thickness burn wounds using dirhamnolipid was reported by Stipcevic et al. (2006). In this study, treatment of full-thickness burn wounds with topical 0.1% dirhamnolipid accelerated the closure of wounds by 32% within 21 days of the treatment as compared to control where no rhamnolipids used. Dirhamnolipid was well tolerated by the animals (rat), which indicates the nontoxicity toward mammalian cells. This study indicated the possible potential application of dirhamnolipid in accelerating normal wound healing and perhaps in overcoming defects associated with healing failure in chronic wounds. BSs isolated from marine microbes may have similar properties, which requires further research. The earlier results clearly indicate that evaluation of marine BSs for therapeutic applications may bring more interesting properties into light and add more values to marine BSs.

Biosurfactants and nanotechnology: The biological synthesis of nanoparticles has gained considerable attention in view of their exceptional biocompatibility and low toxicity. The use of surfactants as nanoparticle stabilizing agents is an emerging field; however, synthetic surfactants are not environmentally friendly. The application of BSs as an alternate to synthetic surfactants for nanoparticle stabilization could be a green ecofriendly approach. BS-mediated nanoparticle syntheses have been reported for microbially produced rhamnolipids (Kumar et al., 2010) and sophorolipids (Kasture et al., 2008) isolated from nonmarine sources. Synthesis of silver nanoparticles by glycolipid BS isolated from marine *B. casei* MSA19 was reported by Kiran et al. (2010b). They used an *in situ* water-in-oil microemulsion phase for the particle synthesis. The glycolipid BS was used as a particle-stabilizing agent by forming reverse micelles. The silver nanoparticles synthesized in this study were uniform and stable for 2 months. Therefore, the BS-mediated nanoparticle synthesis can be considered as "green" stabilizer of nanoparticles and could be extended to other marine microbial BSs.

CONCLUSION

Although BSs have been the subject of intense investigation during the past few decades, relatively small numbers of microorganisms and research output have focused on their production from marine microorganisms. Nevertheless, some marine microbial communities, including *Acinetobacter*, *Arthrobacter*, *Pseudomonas*, *Halomonas*, *Bacillus*,

Rhodococcus, Enterobacter, Azotobacter, Corynebacterium, and *Lactobacillus,* have been explored for the production of surface active molecules both BSs and bioemulsifiers. Such biological surfactants have important potential application in different industries, and the marine ecosystems can provide an excellent opportunity to select unique and diverse producing microorganisms and chemical products. Effective screening and purification techniques are essential in order to explore and discover unique and effective BSs and bioemulsifiers able to be produced using cost-effective technology processes and at acceptable yields and quality. Promising recent biotechnological approaches coupled with highlighting of the importance of the marine resource for such novel compounds and their environmental credentials are expected to support both search and potential industrial application of BSs in many industrial applications.

ACKNOWLEDGMENT

The first author (Dr. R. Thavasi) is very much thankful to CSIR Government of India and the authorities of Annamalai University for their support to carry out the biosurfactant-related research.

REFERENCES

Abraham, W.R., Meyer, H., and Yakimov, M. 1998. Novel glycine containing glucolipids from the alkane using bacterium *Alcanivorax borkumensis. Biochimica et Biophysica Acta,* 1393:57–62.
Amaral, P.F.F., Da-Silva, J.M., Lehocky, M., Barros-Timmons, A.M.V., Coelho, M.A.Z., Marrucho, I.M., and Coutinho, J.A.P. 2006. Production and characterization of a bioemulsifier from *Yarrowia lipolytica. Process Biochemistry,* 41:1894–1898.
Azim, A., Shah, V., Doncel, G.F., Peterson, N., Gao, W., and Gross, R. 2006. Amino acid conjugated sophorolipids: A new family of biologically active functionalized glycolipids. *Bioconjugate Chemistry,* 17(6):1523–1529.
Banat, I.M. 1993. The isolation of a thermophilic biosurfactant producing *Bacillus* sp. *Biotechnology Letters,* 15:591–594.
Banat, I.M. 1995a. Biosurfactants characterization and use in pollution removal: State of the art. A review. *Acta Biotechnologica,* 15:251–267.
Banat, I.M. 1995b. Biosurfactants production and use in microbial enhanced oil recovery and pollution remediation: A review. *Bioresource Technology,* 51:1–12.
Banat, I.M., Franzetti, A., Gandolfi, I., Bestetti, G., Martinotti, M.G., Fracchia, L., Smyth, T.J., and Marchant, R. 2010. Microbial biosurfactants production, applications and future potential. *Applied Microbiology and Biotechnology,* 87:427–444.
Banat, I.M., Makkar, S.R., and Cameotra, S.S. 2000. Potential commercial application of microbial surfactants. *Applied Microbiology and Biotechnology,* 53:495–508.
Banat, I.M., Samarah, N., Murad, M., Horne, R., and Banergee, S. 1991. Biosurfactant production and use in oil tank clean-up. *World Journal of Microbiology and Biotechnology,* 7:80–88.
Bluth, M.H., Kandil, E., Mueller, C.M., Shah, V., Lin, Y.Y., Zhang, H., Dresner, L. et al. 2006. Sophorolipids block lethal effects of septic shock in rats in a cecal ligation and puncture model of experimental sepsis. *Critical Care Medicine,* 34:188–195.
Bonilla, M., Olivaro, C., Corona, M., Vazquez, A., and Soubes, M. 2005. Production and characterization of a new bioemulsifier from *Pseudomonas putida* ML2. *Journal of Applied Microbiology,* 98:456–463.

Calvo, C., Martinez-Checa, F., Mota, A., Bejar, V., and Quesada, E. 1998. Effect of cations, pH and sulfate content on the viscosity and emulsifying activity of the *Halomonas eurihalina* exopolysaccharide. *Journal of Industrial Microbiology and Biotechnology*, 20:205–209.

Das, P., Mukherjee, S., and Sen, R. 2008a. Improved bioavailability and biodegradation of a model polyaromatic hydrocarbon by a biosurfactant producing bacterium of marine origin. *Chemosphere*, 72:1229–1234.

Das, P., Mukherjee, S., and Sen, R. 2008b. Antimicrobial potentials of a lipopeptide biosurfactant derived from a marine *Bacillus circulans*. *Journal of Applied Microbiology*, 104:1675–1684.

Das, P., Mukherjee, S., and Sen, R. 2009a. Antiadhesive action of a marine microbial surfactant. *Colloids and Surfaces, B: Biointerfaces*, 71(2):183–186.

Das, P., Mukherjee, S., and Sen, R. 2009b. Biosurfactant of marine origin exhibiting heavy metal remediation properties. *Bioresource Technology*, 100(20):4887–4890.

Davey, M.E., Caiazza, N.C., and Tootle, G.A.O. 2003. Rhamnolipid surfactant production affects biofilms architecture in *Pseudomonas aeruginosa* PA-01. *Journal of Bacteriology*, 185:1027–1036.

Desai, A.J., Patel, K.M., and Desai, J.D. 1988. Emulsifier production by *Pseudomonas fluorescens* during the growth on hydrocarbons. *Current Science*, 57:500–501.

Desai, J.D. and Banat, I.M. 1997. Microbial production of surfactants and their commercial potential. *Microbiology and Molecular Biology Reviews*, 61:47–64.

Dusane, D.H., Pawar, V.S., Nancharaiah, Y.V., Venugopalan, V.P., Kumar, A.R., and Zinjarde, S.S. 2011. Anti-biofilm potential of a glycolipid surfactant produced by a tropical marine strain of *Serratia marcescens*. *Biofouling*, 27(6):645–654.

Gandhimathi, R., Kiran, G.S., Hema, T.A., Selvin, J., Rajeetha R.T., and Shanmughapriya, S. 2009. Production and characterization of lipopeptide biosurfactant by a sponge-associated marine actinomycetes *Nocardiopsis alba* MSA10. *Bioprocess and Biosystems Engineering*, 32(6):825–835.

Gerard, J., Lloyd, R., Barsby, T., Haden, P., Kelly, M.T., and Andersen, R.J. 1997. Massetolides A-H, Antimycobacterial cyclic depsipeptides produced by two *Pseudomonads* isolated from marine habitats. *Journal of Natural Products*, 60:223–229.

Gnanamani, A., Kavitha, V., Radhakrishnan, N., Suseela Rajakumar, G., Sekaran, G., and Mandal, A.B. 2010. Microbial products (biosurfactant and extracellular chromate reductase) of marine microorganism are the potential agents reduce the oxidative stress induced by toxic heavy metals. *Colloids and Surfaces B: Biointerfaces*, 79:334–339.

Goutx, M., Mutaftshiev, S., and Bertrand, J.C. 1987. Lipid and exopolysaccharide production during hydrocarbon growth of a marine bacterium from the sea surface. *Marine Ecology: Progress Series*, 40(3):259–265.

Gutiérrez, T., Mulloy, B., Bavington, C., Black, K., and Green, D.H. 2007a. Partial purification and chemical characterization of a glycoprotein (putative hydrocolloid) emulsifier produced by a marine bacterium. *Applied Microbiology and Biotechnology*, 76:1017–1026.

Gutiérrez, T., Mulloy, B., Bavington, C., Black, K., and Green, D.H. 2007b. Glycoprotein emulsifiers from two marine *Halomonas* species: Chemical and physical characterization. *Journal of Applied Microbiology*, 103:1716–1727.

Hommel, R.K. 1990. Formation and physiological role of biosurfactants produced by hydrocarbon-utilizing microorganisms. Biosurfactants in hydrocarbon utilization. *Biodegradation*, 1:107–119.

Husain, D.R., Goutx, M., Acquaviva, M., Gilewicz, M., and Bertrand, J.C. 1997. The effect of temperature on eicosane substrate uptake modes by a marine bacterium *Pseudomonas nautical* strain 617: Relationship with the biochemical content of cells and supernatants. *World Journal of Microbiology and Biotechnology*, 13:587–590.

Ilori, M.O., Amobi, C.J., and Odocha, A.C. 2005. Factors affecting biosurfactant production by oil degrading *Aeromonas* spp. isolated from a tropical environment. *Chemosphere*, 61:985–992.

Kalinovskaya, N., Kuznetsova, T., Rashkes, Ya., Mil'grom, Yu., Mil'grom, E., Willis, R., Wood, A. et al. 1995. Surfactin-like structures of five cyclic despsipeptises from the marine isolates of *Bacillus pumilus*. *Russian Chemical Bulletin* (English Translation), 44:951–955.

Kalinovskaya, N.I., Kuznetsova, T.A., Ivanova, E.P., Ludmila, A.R., Voinov, V.G., Huth, F., and Laatsch, H. 2002. Characterization of surfactin-like cyclic depsipeptides synthesized by *Bacillus pumilus* from ascidian *Halocynthia aurantium*. *Marine Biotechnology*, 4:179–188.

Kasture, M.B., Patel, P., Prabhune, A.A., Ramana, C.V., Kulkarni, A.A., and Prasad, B.L.V. 2008. Synthesis of silver nanoparticles by sophorolipids: Effect of temperature and sophorolipid structure on the size of particles. *Journal of Chemical Sciences*, 120:515–520.

Kelecom, A. 2002. Secondary metabolites from marine microorganisms. *Annals of the Brazilian Academy of Sciences*, 74:151–170.

Khopade, A., Ren, B., Liu, X.-Y., Mahadik, K., Zhang, L., and Kokare, C. 2012. Production and characterization of biosurfactant from marine *Streptomyces* species B3. *Journal of Colloid and Interface Science*, 367:311–318.

Kiran, G.S., Hema, T.A., Gandhimathi, R., Selvin, J., Thomas, T.A., Rajeetha, R.T., and Natarajaseenivasan, K. 2009. Optimization and production of a biosurfactant from the sponge-associated marine fungus *Aspergillus ustus* MSF3. *Colloids and Surfaces, B: Biointerfaces*, 73:250–256.

Kiran, G.S., Sabarathnam, B., and Selvin, J. 2010a. Biofilm disruption potential of a glycolipid biosurfactant from marine *Brevibacterium casei*. *FEMS Immunology and Medical Microbiology*, 59:432–438.

Kiran, G.S., Sabu, A., and Selvin, J. 2010b. Synthesis of silver nanoparticles by glycolipid biosurfactant produced from marine *Brevibacterium casei* MSA19. *Journal of Biotechnology*, 148:221–225.

Kiran, G.S., Thajuddin, N., Hema, T.A., Idhayadhulla, A., Kumar, R.S., and Selvin, J. 2010c. Optimization and characterization of rhamnolipid biosurfactant from sponge associated marine fungi *Aspergillus* sp. MSF1. *Desalination and Water Treatment*, 24:257–265.

Kiran, G.S., Thomas, T.A., and Selvin, J. 2010d. Production of a new glycolipid biosurfactant from marine *Nocardiopsis lucentensis* MSA04 in solid-state cultivation. *Colloids and Surfaces, B: Biointerfaces*, 78:8–16.

Kiran, G.S., Thomas, T.A., Selvin, J., Sabarathnam, B., and Lipton, A.P. 2010e. Optimization and characterization of a new lipopeptide biosurfactant produced by marine *Brevibacterium aureum* MSA13 in solid state culture. *Bioresource Technology*, 101:2389–2396.

Konishi, M., Fukuoka, T., Nagahama, T., Morita, T., Imura, T., Kitamoto, D., and Hatada, Y. 2010. Biosurfactant-producing yeast isolated from *Calyptogena soyoae* (deep-sea cold-seep clam) in the deep sea. *Journal of Bioscience and Bioengineering*, 110:169–175.

Kumar, A.S., Mody, K., and Jha, B. 2007. Evaluation of biosurfactant/bioemulsifier production by a marine bacterium. *Bulletin of Environmental Contamination and Toxicology*, 79:617–621.

Kumar, C.G., Mamidyala, S.K., Das, B., Sridhar, B., Devi, G.S., and Karuna, M.S. 2010. Synthesis of biosurfactant-based silver nanoparticles with purified rhamnolipids isolated from *Pseudomonas aeruginosa* BS-161R. *Journal of Microbiology and Biotechnology*, 20:1061–1068.

Liu, X.-Y., Ren, B., Chen, M., Wang, H.-B., Kokare, C.R., Zhou, X.-L., Wang, J.-D. et al. 2010. Production and characterization of a group of bioemulsifiers from the marine *Bacillus velezensis* strain H3. *Applied Microbiology and Biotechnology*, 87:1881–1893.

Maneerat, S., Bamba, T., Harada, K., Kobayashi, A., Yamada, H., and Kawai, F. 2006. A novel crude oil emulsifier excreted in the culture supernatant of a marine bacterium, *Myroides* sp. strain SM1. *Applied Microbiology and Biotechnology*, 70:254–259.

Maneerat, S. and Dikit, P. 2007. Characterization of cell-associated bioemulsifier from *Myroides* sp. SM1, a marine bacterium. *Songklanakarin Journal of Science and Technology*, 29:769–779.

Maneerat, S. and Phetrong, K. 2007. Isolation of biosurfactant-producing marine bacteria and characteristics of selected biosurfactant. *Songklanakarin Journal of Science and Technology*, 29:781–791.

Marchant, R. and Banat, I.M. 2012a. Biosurfactants: A sustainable replacement for chemical surfactants? *Biotechnology Letters*, 34:1597–1605.

Marchant, R. and Banat, I.M. 2012b. Microbial biosurfactants: Challenges and opportunities for future exploitation. *Trends in Biotechnology*, (In Press) doi:10.1016/j. tibtech.2012.07.003.

Mukherjee, S., Das, P., Sivapathasekaran, C., and Sen, R. 2009. Antimicrobial biosurfactants from marine *Bacillus circulans*: Extracellular synthesis and purification. *Letters in Applied Microbiology*, 48:281–288.

Mulligan, C.N. and Gibbs, B.F. 2004. Types, production and applications of biosurfactants. *Proceedings of the Indian Notational Science Academy*, 1:31–55.

Mulligan, C.N., Yong, C.N., and Gibbs, B.F. 2001. Heavy metal removal from sediments by biosurfactants. *Journal of Hazardous Materials*, 85:111–125.

Navon-Venezia, S., Zosim, Z., Gottlieb, A., Legmann, R., Carmeli, S., Ron, E.Z., and Rosenberg, E. 1995. Alasan, a new bioemulsifier from *Acinetobacter radioresistens*. *Applied and Environmental Microbiology*, 61:3240–3244.

Nerurkar, A.S., Hingurao, K.S., and Suthar, H.G. 2009. Bioemulsfiers from marine microorganisms. *Journal of Scientific and Industrial Research*, 68:273–277.

Nielsen, T.H., Christophersen, C., Anthoni, U., and Sorensen, J. 1999. Viscosinamide, a new cyclic depsipeptide with surfactant and antifungal properties produced by *Pseudomonas fluorescens* DR54. *Journal of Applied Microbiology*, 86:80–90.

Oloke, J.K. and Glick, B.R. 2005. Production of bioemulsifier by an unusual isolate of salmon/red melanin containing *Rhodotorula glutinis*. *African Journal of Biotechnology*, 4:164–171.

Palme, O., Moszyk, A., Iphöfer, D., and Lang, S. 2010. Selected microbial glycolipids: Production, modification and characterization. *Advances in Experimental Medicine and Biology*, 672:185–202.

Passeri, A., Lang, S., Wagner, F., and Wray, V. 1991. Marine biosurfactants, II. Production and characterization of an anionic trehalose tetraester from the marine bacterium *Arthrobacter* sp. EK1. *Zeitschrift fuer Naturforschung, C: Journal of Biosciences*, 46:204–209.

Passeri, A., Schmidt, M., Haffner, T., Wray, V., Lang, S., and Wagner, F. 1992. Marine biosurfactants. IV. Production, characterization and biosynthesis of an anionic glucose lipid from the marine bacterial strain MM1. *Applied Microbiology and Biotechnology*, 37:281–286.

Peng, F., Liu, Z., Wang, L., and Shao, Z. 2007. An oil-degrading bacterium: *Rhodococcus erythropolis* strain 3C-9 and its biosurfactants. *Journal of Applied Microbiology*, 102:1603–11.

Peng, F., Wang, Y., Sun, F., Liu, Z., Lai, Q., and Shao, Z. 2008. A novel lipopeptide produced by a Pacific Ocean deep-sea bacterium, *Rhodococcus* sp. TW53. *Journal of Applied Microbiology*, 105:698–705.

Pepi, M., Cesàro, A., Luit, G., and Baldi, F. 2005. An Antarctic psychrotrophic bacterium *Halomonas* sp. ANT-3b, growing on n-hexadecane, produces a new emulsifying glycolipid. *FEMS Microbiology Ecology*, 53:157–166.

Perfumo, A., Rancich, I., and Banat, I.M., 2010b. Possibilities and challenges for biosurfactants use in petroleum industry. In: *Biosurfactants Advances in Experimental Medicine and Biology*, Sen, R. (ed.), Vol. 672. Springer-Verlag, Berlin, Germany, pp. 135–157.

Perfumo, A., Smyth, T.J.P., Marchant, R., and Banat, I.M., 2010a. Production and roles of biosurfactants and bioemulsifiers in accessing hydrophobic substrates. In: *Handbook of Hydrocarbon and Lipid Microbiology*, Timmis, K.N. (ed.) Springer-Verlag, Berlin, Germany, pp. 1501–1512.

Qiao, N. and Shao, Z. 2010. Isolation and characterization of a novel biosurfactant produced by hydrocarbon-degrading bacterium *Alcanivorax dieselolei* B-5. *Journal of Applied Microbiology*, 108:1207–1216.

Rahman, K.S.M., Banat, I.M., Thahira, J., Thayumanavan, T., and Lakshmanaperumalsamy, P. 2002. Bioremediation of gasoline contaminated soil by bacterial consortium with poultry litter, coir pith and rhamnolipid biosurfactant. *Bioresource Technology*, 81:25–32.

Rahman, K.S.M., Rahman, T.J., Kourkoutas, Y., Petsas, I., Marchant, R., and Banat, I.M. 2003. Enhanced bioremediation of n-alkane in petroleum sludge using bacterial consortium amended with rhamnolipid and micronutrients. *Bioresource Technology*, 90:159–168.

Ramiah, N. 2005. Facets and opportunities. In: *Marine Microbiology*, Ramaiah, N. (ed.) National Institute of Oceanography, India, pp. 1–6.

Ramm, W., Schatton, W., Wagner-Döbler, I., Wray, V., Nimtz, M., Tokuda, H., Enjyo, F. et al. 2004. Diglucosyl-glycerolipids from the marine sponge-associated *Bacillus pumilus* strain AAS3: Their production, enzymatic modification and properties. *Applied Microbiology and Biotechnology*, 64:497–504.

Reisfeld, A., Rosenberg, E., and Gutnick, D.L. 1972. Microbial degradation of crude oil: Factors affecting the dispersion in sea water by mixed and pure cultures. *Applied Microbiology*, 24:363–368.

Rodrigues, L., Banat, I.M., Teixeira, G., and Oliveira, O. 2006a. Biosurfactants: Potential applications in medicine. *Journal of Antimicrobial Chemotherapy*, 57:609–618.

Rodrigues, L., Banat, I.M., Teixeira, J., and Oliveira, R., 2007. Strategies for the prevention of microbial biofilm formation on silicone rubber voice prostheses. *Journal of Biomedical Materials Research Part B—Applied Biomaterials*, 81B:358–370.

Rodrigues, L.R., Banat, I.M., van der Mei, H.C, Teixeira, J.A., and Oliveira, R. 2006c. Interference in adhesion of bacteria and yeasts isolated from explanted voice prostheses to silicone rubber by rhamnolipid biosurfactants. *Journal of Applied Microbiology*, 100:470–480.

Rodrigues, L., van der Mei, H.C., Banat, I.M., Teixeira, J.A., and Oliveira, R. 2006b. Inhibition of microbial adhesion to silicone rubber treated with biosurfactant from *Streptococcus thermophilus* A. *FEMS Immunology and Medical Microbiology*, 46:107–112.

Ron, E.Z. and Rosenberg, E. 1999. Natural roles of biosurfactants. *Environmental Microbiology*, 3:229–236.

Rosenberg, E., Perry, A., Gibson, D.T., and Gutnick, D.L. 1979. Emulsifier of *Arthrobacter* RAG-1: Specificity of hydrocarbon substrate. *Applied and Environmental Microbiology*, 37:409–413.

Rosenberg, E. and Ron, E.Z. 1998. Surface active polymers from the genus Acinetobacter. In: *Biopolymers from Renewable Resources*, Kaplan, D.L. (ed.) Springer-Verlag, Berlin, Germany, pp. 281–291.

Satpute, S.K., Banat, I.M., Dhakephalkar, P.K., Banpurkar, A.G., and Chopade, B.A. 2010. Biosurfactants, bioemulsifiers and exopolysaccharides from marine microorganisms. *Biotechnology Advances*, 28:436–450.

Satpute, S.K., Bhawsar, B.D., Dhakephalkar, P.K., and Chopade, B.A. 2008. Assessment of different screening methods for selecting biosurfactant producing marine bacteria. *Indian Journal of Marine Sciences*, 37:243–250.

Schulz, D., Passeri, A., Schmidt, M., Lang, S., Wagner, F., Wray, V., and Gunkel, W. 1991. Marine biosurfactants, I. Screening for biosurfactants among crude oil degrading marine microorganisms from the North Sea. *Zeitschrift fuer Naturforschung C: Journal of Biosciences*, 46:197–203.

Shabtai, Y. and Gutnick, D.L. 1986. Enhanced emulsan production in mutants of *Acinetobacter calcoaceticus* RAG-1 selected for resistance to cetyltrimethylammonium bromide. *Applied and Environmental Microbiology*, 52:146–151.

Singh, P. and Cameotra, S.S. 2004. Enhancement of metal bioremediation by use of microbial surfactants. *Biochemical and Biophysical Research Communications*, 319:291–297.

Sivapathasekaran, C., Mukherjee, S., Sen, R., Bhattacharya, B., and Samanta, R. 2011. Single step concomitant concentration, purification and characterization of two families of lipopeptides of marine origin. *Bioprocess and Biosystems Engineering*, 34:339–346.

Stanghellini, M.E. and Miller, R.M. 1997. Biosurfactants: Their identity and potential efficacy in the biological control of zoosporic plant pathogens. *Plant Disease*, 81:4–12.

Stipcevic, T., Piljac, A., and Piljac, G. 2006. Enhanced healing of full-thickness burn wounds using di-rhamnolipid. *Burns*, 32:24–34.

Thaniyavarn, J., Chongchin, A., Wanitsuksombut, N., Thaniyavarn, S., Pinphanichakarn, P., Leepipatpiboon, N., Morikawa, M., and Kanaya, S. 2006. Biosurfactant production by *Pseudomonas aeruginosa* A41 using palm oil as carbon source. *Journal of General and Applied Microbiology*, 52:215–222.

Thavasi, R. and Jayalakshmi, S. 2003. Bioremediation potential of hydrocarbonoclastic bacteria in Cuddalore harbour waters (India). *Research Journal of Chemistry and Environment*, 7:17–22.

Thavasi, R., Jayalakshmi, S., Balasubramanian, T., and Banat, I.M. 2006. Biodegradation of crude oil by nitrogen fixing marine bacteria *Azotobacter chroococcum*. *Research Journal of Microbiology*, 1:401–408.

Thavasi, R., Jayalakshmi, S., Balasubramaian, T., and Banat, I.M. 2007. Biosurfactant production by *Corynebacterium kutscheri* from waste motor lubricant oil and pea nut oil cake. *Letter in Applied Microbiology*, 45:686–691.

Thavasi, R., Jayalakshmi, S., Balasubramaian, T., and Banat, I.M. 2008. Production and characterization of a glycolipid biosurfactant from *Bacillus megaterium*. *World Journal of Microbiology and Biotechnology*, 24:917–925.

Thavasi, R., Jayalakshmi, S., Balasubramaian, T., and Banat, I.M. 2011b. Effect of biosurfactant and fertilizer on biodegradation of crude oil by marine isolates of *Bacillus megaterium*, *Corynebacterium kutscheri* and *Pseudomonas aeruginosa*. *Bioresource Technology*, 102:772–778.

Thavasi, R., Jayalakshmi, S., and Banat, I.M. 2011a. Application of biosurfactant produced from peanut oil cake by *Lactobacillus delbrueckii* in biodegradation of crude oil. *Bioresource Technology*, 102:3366–3372.

Thavasi, R., Sharma, S., and Jayalakshmi, S. 2011c. Evaluation of screening methods for the isolation of biosurfactant producing marine bacteria. *Journal of Petroleum and Environmental Biotechnology*, S1:1–6.

Thavasi, R., Subramanyam, N.V.R.M., Jayalakshmi, S., Balasubramanian, T., and Banat, I.M. 2009. Biosurfactant production by *Azotobacter chroococcum* isolated from the marine environment. *Marine Biotechnology*, 11:551–556.

Thavasi, R., Subramanyam, N.V.R.M., Jayalakshmi, S., Balasubramanian, T., and Banat, I.M. 2011d. Biosurfactant production by *Pseudomonas aeruginosa* from renewable resources. *Indian Journal of Microbiology*, 51:30–36.

Vasileva-Tonkova, E. and Gesheva, V. 2005. Glycolipids produced by Antarctic *Nocardioides* sp. during growth on n-paraffin. *Process Biochemistry*, 40:2387–2391.

Vasileva-Tonkova, E. and Gesheva, V. 2007. Biosurfactant production by Antarctic facultative anaerobe *Pantoea* sp. during growth on hydrocarbons. *Current Microbiology*, 54:136–141.

Wicke, C., Hüners, M., Wray, V., Nimtz, M., Bilitewski, U., and Lang, S. 2000. Production and structure elucidation of glycoglycerolipids from a marine sponge-associated *Microbacterium* species. *Journal of Natural Products*, 63:621–626.

Yakimov, M.M., Giuliano, L., Bruni, V., Scarfi, S., and Golyshin, P.N. 1999. Characterization of antarctic hydrocarbon-degrading bacteria capable of producing bioemulsifiers. *Microbiologica*, 22:249–256.

Yakimov, M.M., Golyshin, P.N., Lang, S., Moore, E.R.B., Abraham, W.-R., Lunsdorf, H., and Timmis, K.N. 1998. *Alcanivorax borkumensis* gen. nov., sp. nov., a new, hydrocarbon-degrading and surfactant-producing marine bacterium. *International Journal of Systematic Bacteriology*, 48:339–348.

Yakimov, M.M., Timmis, K.N., and Golyshin, P.N. 2007. Obligate oil-degrading marine bacteria. *Current Opinion in Biotechnology*, 18:257–266.

Yakimov, M.M., Timmis, K.N., Wray, V., and Fredrickson, H.L. 1995. Characterization of a new lipopeptide surfactant produced by thermotolerant and halotolerant subsurface *Bacillus licheniformis* BAS50. *Applied and Environmental Microbiology*, 61:1706–1713.

Zinjarde, S., Chinnathambi, S., Lachke, A.H., and Pant, A. 1997. Isolation of an emulsifier from *Yarrowia lipolytica* NCIM3589 using a modified mini isoelectric focusing unit. *Letters in Applied Microbiology*, 24:117–121.

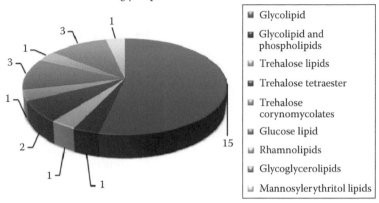

Number of publications (28, 48%) on marine glycolipid biosurfactants

- Glycolipid
- Glycolipid and phospholipids
- Trehalose lipids
- Trehalose tetraester
- Trehalose corynomycolates
- Glucose lipid
- Rhamnolipids
- Glycoglycerolipids
- Mannosylerythritol lipids

FIGURE 5.1 Breakdown compilation of publication on marine glycolipid biosurfactants.

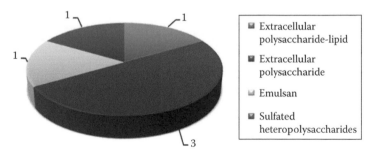

Number of publications (6, 10%) on marine extracellular polysaccharide and polysaccharide lipid biosurfactants

- Extracellular polysaccharide-lipid
- Extracellular polysaccharide
- Emulsan
- Sulfated heteropolysaccharides

FIGURE 5.2 Breakdown compilation of publication on marine extracellular polysaccharide and polysaccharide lipid biosurfactants.

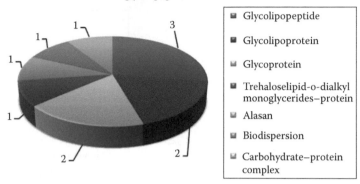

Number of publications (11, 19%) on marine glycolipopeptide and glycolipoprotein biosurfactats

- Glycolipopeptide
- Glycolipoprotein
- Glycoprotein
- Trehaloselipid-o-dialkyl monoglycerides–protein
- Alasan
- Biodispersion
- Carbohydrate–protein complex

FIGURE 5.3 Breakdown compilation of publication on marine glycolipopeptide, glycolipoprotein, and glycoprotein biosurfactants.

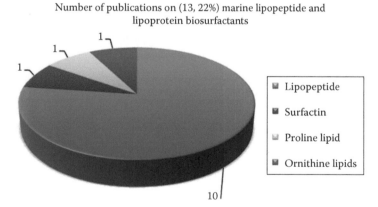

Number of publications on (13, 22%) marine lipopeptide and lipoprotein biosurfactants

- Lipopeptide
- Surfactin
- Proline lipid
- Ornithine lipids

FIGURE 5.4 Breakdown compilation of publication on marine lipopeptide and lipoprotein biosurfactants.

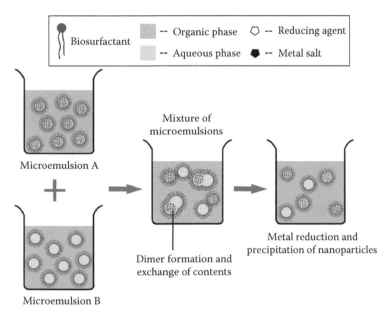

FIGURE 9.1 Mechanism for microemulsion-based nanoparticles synthesis. (Adapted from Capek, I., *Adv. Colloid Interface Sci.*, 110, 49, 2004.)

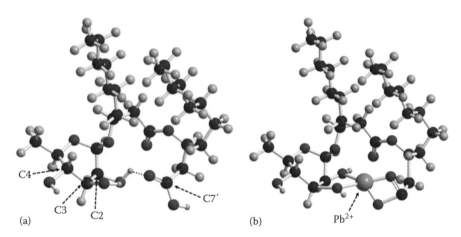

FIGURE 11.3 (a) An energy-minimized molecular mechanics model of monorhamnolipid (C10, C10) showing the oxygen-rich cavity that may serve as a cation binding pocket. (b) A model showing how a Pb^{2+} ion might interact with the binding pocket of monorhamnolipid (C10, C10).

6 Characterization, Production, and Applications of Lipopeptides

Catherine N. Mulligan

CONTENTS

INTRODUCTION

Surfactants are amphiphilic compounds that reduce the free energy of the system by replacing the bulk molecules of higher energy at an interface. They contain a hydrophobic portion with little affinity for the bulk medium and a hydrophilic group that is attracted to the bulk medium. Surfactants have been used industrially as adhesives; flocculating, wetting, and foaming agents; deemulsifiers; and penetrants (Mulligan and Gibbs, 1993). They are used for these applications based on their abilities to lower surface tensions and increase solubility, detergency power, wetting ability, and foaming capacity. Petroleum users have traditionally been the major users, as in enhanced oil removal applications by increasing the solubility of petroleum components (Falatko, 1991). They have also been used for mineral flotation and in the pharmaceutical industries. Typical desirable properties include solubility enhancement, surface tension reduction, the critical micelle concentrations (CMCs), wettability, and foaming capacity.

Surfactants are classified as cationic, anionic, zwitterionic, and nonionic and are made synthetically from hydrocarbons, lignosulfonates, or triglycerides. Some common synthetic surfactants include linear alkyl benzenesulfonates, alcohol sulfates, alcohol ether sulfates, alcohol glyceryl ether sulfonates, α-olefin sulfonates, alcohol ethoxylates, and alkylphenol ethoxylates (Layman, 1985). Surfactants have many applications industrially with multiphasic systems. Sodium dodecyl sulfate (SDS, $C_{12}H_{25}$-SO_4^- Na^+) is a widely used anionic surfactant. The effectiveness of a surfactant is determined by surface tension lowering, which is a measure of the surface free energy per unit area or the work required to bring a molecule from the bulk phase to the surface (Rosen, 1978). These amphiphilic compounds (containing hydrophobic and hydrophilic portions) concentrate at solid–liquid, liquid–liquid, or vapor–liquid interfaces. An interfacial boundary exists between two immiscible phases. The hydrophobic portion concentrates at the surface while the hydrophilic is oriented toward the solution. A good surfactant can lower the surface tension of water from 72 to 35 mN/m and the interfacial tension (tension between nonpolar and polar liquids) for water against n-hexadecane from 40 to 1 mN/m. Efficient surfactants have a low CMC (i.e., less surfactant is necessary to decrease the surface tension) as the CMC is defined as the minimum concentration necessary to initiate micelle formation (Becher, 1965). In practice, the CMC is also the maximum concentration of surfactant monomers in water and is influenced by pH, temperature, and ionic strength.

An important factor in the choice of surfactant is the product cost (Mulligan and Gibbs, 1993). In general, surfactants are used to save energy and consequently energy costs (such as the energy required for pumping or mixing). Charge type, physicochemical behavior, solubility, and adsorption behavior are some important selection criteria for surfactants.

Some surfactants, known as biosurfactants, are biologically produced from yeast or bacteria (Lin, 1996). They can be potentially as effective with some distinct advantages over the highly used synthetic surfactants due to high specificity, biodegradability, and biocompatibility (Cooper, 1986).

Biosurfactants are grouped as glycolipids, lipopeptides, phospholipids, fatty acids, and neutral lipids (Bierman et al., 1987). Most of these compounds are either anionic or neutral, with only a few cationic ones. The hydrophobic parts of the molecule are based on long-chain fatty acids, hydroxy fatty acids, or α-alkyl-β-hydroxy fatty acids. The hydrophilic portion can be a carbohydrate, amino acid, cyclic peptide, phosphate, carboxylic acid, or alcohol. A wide variety of microorganisms can produce these compounds. The CMCs of the biosurfactants generally range from 1 to 200 mg/L and their molecular weights (MWs) from 500 to 1500 amu (Lang and Wagner, 1987).

LIPOPEPTIDE BIOSURFACTANTS

Lipopeptides are produced by a variety of microorganisms, including *Bacillus*, *Lactobacillus*, *Streptomyces*, *Pseudomonas*, and *Serratia* (Cameotra and Makkar, 2004; Georgiou et al., 1992). The lipopeptides are cyclic peptides with a fatty acyl chain. Various lipopeptides include surfactin (Roongsawang et al., 2003; Youssef et al., 2007), lichenysin A (Yakimov et al., 1995) or C (Jenny et al., 1991), B (Folmsbee et al., 2006), D (Zhao et al., 2010), bacillomycin (Roongsawang et al., 2003), fengycin

(Vanittanakom and Loeffler, 1986), and iturin (Bonmatin et al., 2003). Surfactin is a cyclic heptapeptide, with antibacterial, antifungal, antiviral, and antitumor activities (Folmsbee et al., 2006; Zhao et al., 2010).

Lipopeptides have been tested in enhanced oil recovery and the transportation of crude oils (Hayes et al., 1986). They were demonstrated to be effective for antimicrobial activity and in the reduction of the interfacial tension of oil and water and the viscosity of the oil, the removal of water from the emulsions prior to processing, and the release of bitumen from oil sands. Although most biosurfactant-producing organisms are aerobic, a few anaerobic producers exist. *Bacillus licheniformis* JF-2 is an example, which would be well suited for in situ studies for enhanced oil recovery or soil decontamination (Javaheri et al., 1985).

Surfactin is the most studied lipopeptide and consists of a seven-amino acid sequence in a cyclical structure with a 13–16 carbon fatty acid (Kakinuma et al., 1969) and has two charged amino acids (glutamic and aspartic acids). In addition to surfactin, iturins and fengycins are also produced (Deleu et al., 1999). Their structures are shown in Figures 6.1 through 6.3. Iturins are cyclic peptides with seven amino acids and a ß-amino closure. Fengycin lipopeptides are ß-hydroxy fatty acids with an eight-member ring in an N-terminal decapeptide. At the C-terminal end, there is a tyrosine residue at position 3. This forms an eight-member lactone ring. Fengycin A and B vary at position 6. The A form has an Ala compared to the B form of a valine.

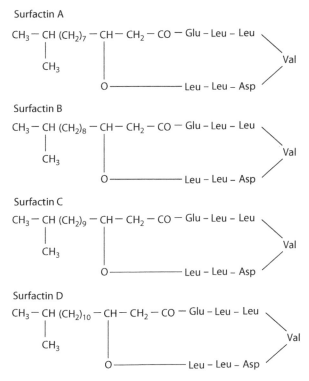

FIGURE 6.1 Structures of various forms of surfactin. (Adapted from Janek, T. et al., *Bioresour. Technol.*, 101, 6118, 2010.)

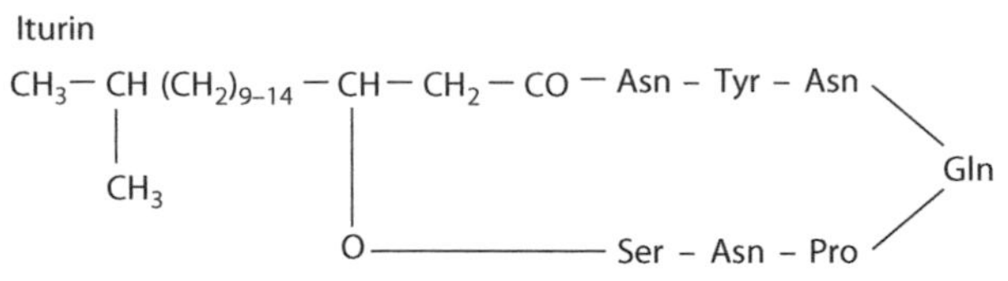

Iturin

$$CH_3 - CH\ (CH_2)_{9-14} - CH - CH_2 - CO - Asn - Tyr - Asn$$

with a branch CH_3 below the first CH, and O below the CH connecting to $Ser - Asn - Pro$, which connects to Gln at top right.

FIGURE 6.2   Structure of iturin.

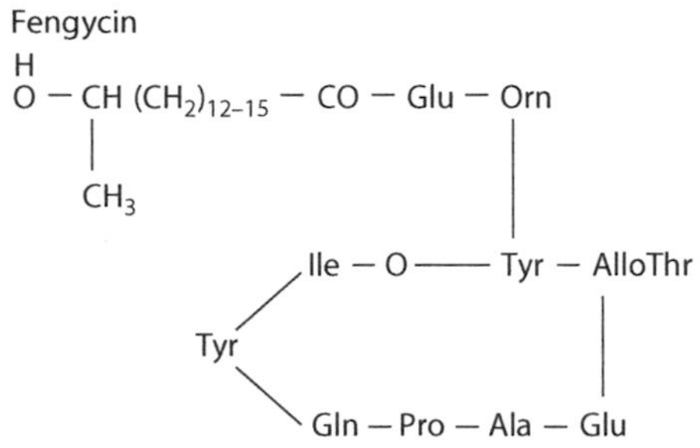

Fengycin

$$\overset{H}{O} - CH\ (CH_2)_{12-15} - CO - Glu - Orn$$

with CH_3 branch below the CH. Orn connects down to a ring:

Ile – O ——— Tyr – AlloThr

Tyr

Gln – Pro – Ala – Glu

FIGURE 6.3   Structure fengycin.

The fatty acid normally varies from 14 to 18 carbons in length (Arima et al., 1968; Matsuyama et al., 1992; Roongsawang et al., 2003). Fengycin has three charged amino acids (two glutamic acids and an ornithine). The CMC is 6.25 mg/L.

Other lipopeptides have also been studied. They include lichenysin from *B. licheniformis* (Grangemard et al., 1999; Horowitz et al., 1990), arthrofactin from *Arthrobacter* sp. (now known as *Pseudomonas* sp. MIS38) (Morikawa et al., 1993; Roongsawang et al., 2003), pumilacidin from *P. pumilus* (Naruse et al., 1990), and serrawettin from *Serratia marcescens* (Matsuyama et al., 1992). Others include massetolide A (Sen and Swaminathan, 1997), putisolvins I and II (Kuiper et al., 2004), and pumilacidin (Naruse et al., 1990). As can be seen, there is no generally accepted nomenclature based on the structure. *Pseudomonas* lipopeptides include viscosin, amphisin, tolassin, and syringomycin (Raaijmakers et al., 2006). Viscosin has nine amino acids with a 3-hydroxy fatty acid, whereas amphisin has 11 amino acids linked to a similar fatty acid (Soresen et al., 2001). A comparison of the CMC of some lipopeptides is shown in Table 6.1.

TABLE 6.1
CMC Values of Various Isolated Lipopeptides

Lipopeptide	Microbial Source	CMC (mg/L)	Reference
Fengycin	*B. circulans* DMS-2	10–13	Sivapathasekaran et al. (2009, 2010)
Surfactin	*B. subtilis*	17	Deleu et al. (1999)
Surfactin and fengycin	*B. subtilis*	10 and 11	Lin et al. (1998)
Biosurfactant	*B. licheniformis*	0.6	Barros et al. (2008)
Lipopeptide	*Rhodococcus* sp.	23.7	Peng et al. (2008)

Synthesis of the lipopeptides is performed based on a series of enzymes for each step of amino acid addition, ring closure, and acylation (Peypoux et al., 1999). This makes genetic manipulation difficult for enhanced production. Most studies are concentrated on growth optimization and isolation of overproducers (Peypoux et al., 1999) with the exception of a few studies (Gu et al., 2007; Nakayama et al., 1997; Ohno et al., 1995; Peypoux et al., 1999).

Most of the focus has been on higher-priced applications due to the low yields and high cost of the media. Higher-volume low applications including environmental remediation, enhanced oil recovery, laundry soaps, and polymerization of emulsions, need the development of low-cost substrates such as agroindustrial wastes (Makkar and Cameotra, 1999; Mukherjee et al., 2006) and isolation techniques, bioreactor design, and higher yields.

SURFACTIN

B. subtilis produces surfactin, one of the first lipopeptides found in 1968 (Figure 6.1). It has seven amino acids bonded to the carboxyl and hydroxyl groups of a 14-carbon acid (Kakinuma et al., 1969) and has blood clotting properties. Surfactin concentrations as low as 0.005% reduce the surface tension to 27 mN/m, making it a powerful biosurfactant. The CMC can be as low as 10 mg/L (Dae et al., 2006). The interfacial tension for hydrocarbon–water interfaces can be less than 1 mN/m.

The primary structure of surfactin was determined many years ago by Kakinuma et al. (1969). It is a heptapeptide with a β-hydroxy fatty acid within a lactone ring structure. The seven amino acids are bonded to the carboxyl and hydroxyl groups of a 14-carbon acid. More recently, the three-dimensional structure was determined by ^{1}H NMR techniques (Bonmatin et al., 1995). Surfactin folds into a β-sheet structure, which resembles a horse saddle both in aqueous solutions and at the air/water interface (Ishigami et al., 1995). The solubility and surface-active properties of the surfactin are dependent on the orientation of the residues. The fatty acyl chain with the hydrophobic residues formed one face, while the two carboxylic acid side chains form a claw structure enabling the chelation of heavy metals (Bonmatin et al., 1994; Gallet et al., 1999; Magetdana and Ptak, 1992). This property has been evaluated in soil remediation studies (Mulligan et al., 1999).

Mixtures of surfactin produced by *B. subtilis* have been characterized by *m*ass spectrometry (Hue et al., 2001). A combination of liquid-secondary ion mass spectrometry (LSI-MS) and high-energy tandem mass spectrometry (MS/MS) showed that the amino acid composition or length of the acyl chain can vary from 12 to 16 carbons. Leucine and isoleucine can also be differentiated. Data obtained from protonated and cationized fragments were also useful for structural characterization. They are known as A, B, C, and D forms (Figure 6.1).

SURFACTIN PRODUCTION

Most biosurfactants are produced from hydrocarbon substrates (Syldatk and Wagner, 1987). Production can be growth associated. In this case, they can either use the emulsification of the substrate (extracellular) or facilitate the passage of the substrate through the membrane (cell membrane associated). Biosurfactants, however, are also produced from carbohydrates, which are very soluble. Gram-positive and gram-negative bacteria can produce cyclic lipopeptides. The different structures can lead to different properties. The biosurfactants have been postulated to enhance the growth on hydrocarbons and, in this case, may influence the ecology of the host sponge.

Solid-state fermentation using okara, a soybean curd residue, has been performed by Ohno et al. (1995). Other substrates studied have included starch (Sandrin et al., 1990), cassava waste (Santos et al., 2000), molasses (Makkar and Cameotra, 1997b), soybean (Kim et al., 2009), and potato wastes (Fox and Bala, 2000). Surfactin yields from an autoclaved purified starch were 0.154 g/g. Low solid potato effluents exhibited a 66% lower surfactin yield than the purified starch (Thompson et al., 2001). It has been postulated that higher yields result from nutritional limitations.

Surfactin yields during production are low (0.02 g/g glucose) (de Roubin et al., 1989). Addition of iron and manganese can enhance concentrations to 0.7 g/L (Rosenberg, 1986). Further work on iron addition performed by Wei and Chu (1998) determined that addition of 1.7 mM of iron can lead to the production of up to 3.5 g/L of surfactin and enhanced biomass production. Alkaline addition is required to overcome the decrease in pH to below 5 due to acid formation. Further studies by Wei and Chu (2002) showed the effect of manganese on nitrogen utilization and subsequently surfactin production. A 0.1 mM magnesium sulfate concentration increased almost ninefold the surfactin level to 2.6 g/L. Wei et al. (2007) subsequently used the Taguchi method to optimize surfactin production with regard to the presence of Mg, K, Mn, and Fe. They found that K and Mg were critical. Kinsinger et al. (2003, 2005) further determined that concentrations of the four ions could be optimized and allowed the production of 3.34 g/L of surfactin.

Yields of 0.14 g/g sugar have been obtained using peat as a substrate after hydrolysis with 0.5% sulfuric acid for 1 h at 120°C (Sheppard and Mulligan, 1987). Citric acid addition to glucose media could also enhance production (de Roubin et al., 1989). In attempts to influence the metabolic pathway, glutamic acid, leucine, aspartic acid, and valine were added to the media but did not enhance production. Nitrogen, however, was a significant factor in surfactin production. Doubling ammonium nitrate concentrations from 0.4% to 0.8% increased yields by a factor of 1.6, while organic nitrogen addition did not have any benefit.

Other investigators (Davis et al., 1999) found that surfactin yields were highest in nitrate-limited oxygen-depleted conditions, followed by ammonium-limited (0.075 g surfactin per g biomass), oxygen-depleted conditions (0.012 g/g biomass), and carbon-limited, oxygen-depleted conditions (0.0069 g/g biomass).

A strain of B. subtilis was able to produce biosurfactant at 45°C at high NaCl concentrations (4%) and a wide pH range (4.5–10.5) (Makkar and Cameotra, 1997a,b). It was able to remove 62% of the oil in a sand pack saturated with kerosene and thus could be used for in situ oil removal and cleaning sludge from sludge tanks.

Makkar and Cameotra (2002) studied another strain of B. subtilis MTCC2423. They found it preferred sodium or potassium nitrate (3 g/L) or urea (1 g/L). Magnesium concentrations of 2.43 mM and calcium concentrations of 0.36 mM were optimal for biosurfactant yield. Unlike for previous studies for B. subtilis by de Roubin et al. (1989), aspartic acid, asparagine, glutamic acid, valine, and lysine increased biosurfactant production by 60%. While glycine and leucine addition had no affect, alanine and arginine decreased production. Production was good even at high concentrations of NaCl (up to 4%) and pH values from 4.5 to 10.5.

Solid carriers have also been evaluated for surfactin yield enhancement (Yeh et al., 2005). Activated carbon and expanded clay were added at concentrations of 133 g/L. Surfactin at concentrations of 2150 and 3300 mg/L were obtained for each carrier, respectively. Activated carbon was more appropriate for the fermentation process and seemed to increase cell growth and thus yield. A summary of the yields of surfactin can be seen in Table 6.2.

Das et al. (2009) determined that antimicrobial activity was obtained from a glucose substrate, instead of sucrose, starch, and glycerol. Emulsifying lipopeptide biosurfactants from *Azotobacter chroococcum* can be produced from oil (crude, waste motor lubricant) and peanut oil cake (Thavasi et al., 2009).

Production of another lipopeptide, brevifactin, was characterized and optimized by the marine strain *Brevibacterium aureum* MSA13 (Kiran et al., 2010).

TABLE 6.2
Production of Surfactin

B. subtilis Strain	Substrate	Surfactin Yield or Concentration	Reference
ATCC 21332	Synthetic or semisynthetic peat hydrolysate	100–250 mg/L	Arima et al. (1968); Cooper et al. (1981)
RB14	Aqueous two phase	160 mg/L 350 mg/L	Sheppard and Mulligan (1987)
	Semisynthetic	250 mg/L	Drouin and Cooper (1992)
	Solid-state okara	200–250 mg/kg wet mass	Ohno et al. (1992) Ohno et al. (1992)
MI113 (pC12)	Semisynthetic	350 mg/L	
MI113 (pC12)d	Solid-state okara	2000 mg/kg wet mass	Ohno et al. (1992)
ATCC 55033	Semisynthetic		Ohno et al. (1992)
		3500–4300 mg/L	Carrera et al. (1992)
Mutant strain of ATCC 21332	Synthetic	2000–4000 mg/L 550 mg/L	Carrera et al. (1993a,b) Mulligan et al. (1989)
C9 (KCTC 8701P) ATCC 21332	Glucose with modified salts and oxygen limitation	7.0 g/L	Kim et al. (1997)
ATCC 21332	Glucose and mineral salts with iron	3.5 g/L	Wei and Chu (1998)
ATCC 21332	Glucose with oxygen and nitrogen depletion	0.44 g/L	Davis et al. (1999)
MTCC 2423	Purified starch	0.154 g/g	Fox and Bala (2000)
SD 901	Sucrose with mineral salts	1.23 g/L	Makkar and
ATCC 21332		8,000–50,000 mg/L	Cameotra (2002)
Isolate	Bean extract		Yoneda et al. (2006)
	Solid carriers	2150–3300 mg/L	Yeh et al. (2005)
	Sucrose with foam collection	0.25 g/g	Amani et al. (2010)

Source: Adapted from Shaligram, N.S. and Singhal, R.S, *Food Technol. Biotechnol.*, 48: 119–134, 2010.

Various agro and industrial solid waste substrates including molasses, olive oil, and acrylamide were evaluated. The biosurfactant was stable over the pH range of 5–9, and up to 5% NaCl and a temperature of 121°C. The surface tension was 28.6 mN/m. The lipopeptide was characterized as an octadecanoic acid methyl ester with four amino acids pro–leu–gly–gly. This lipopeptide, thus, could have potential for microbial enhanced oil recovery and oil spill remediation.

LIPOPEPTIDE PRODUCTION REACTOR DESIGN AND OPTIMIZATION

Free and immobilized cells of *B. subtilis* ATCC 21332 were grown to produce surfactin and fengycin (Chtioui et al., 2010). Although the production of both biosurfactants was enhanced by two to four times, fengycin was particularly improved. N-heptane was used for extracting the biosurfactant. A continuous extraction with a liquid membrane called petraction was used, but the stripping was too slow. Further optimization is needed. Petraction was also employed by Dimitrov et al. (60) for surfactin. At pH 5.65, 97% recovery was achieved compared to 83% at pH 6.05 in 4 h. However, approximately 90% was removed in 30 min.

Further studies were performed using a rotating disk bioreactor (Chtioui et al., 2012). Cells were immobilized on the rotating disks. Foaming did not occur as the aeration was bubbleless. Fengycin production was favored (838 mg/L) compared to surfactin (212 mg/L). Increasing the number of disks improved the production of both products. Surfactin production was more correlated with improved oxygenation, while the fengycin production was related to more biofilm formation.

A two-phase reactor with polyethylene glycol and dextran (D-40) was evaluated for surfactin production by *B. subtilis* ATCC 21332 (Drouin and Cooper, 1992) in a cyclone reactor. Cells accumulated in the dextran phase and surfactin in the other phase. This enabled the separation of surfactin from the cells to decrease cell inhibition.

An airlift reactor in batch mode was employed to enhance aeration with a potato process effluent as the substrate (Noah et al., 2002). A 0.5 vvm air flow rate enabled surfactin removal. Conditions of a large inoculum, pH control, and the use of a pressurized reactor optimized the growth of *B. subtilis* over indigenous bacteria. Noah et al. (2005) subsequently used a chemostat and low solid potato effluents. At 0.5 vvm, a surfactin concentration increased to 1.1 g/L was obtained at high agitation rates (400 rpm).

Martinov et al. (2008) studied aeration in a stirred tank reactor with foaming. Different agitators were tested due to the decrease in aeration in the presence of surfactin. A low shear impeller Narcissus maintained stable k_La values while reducing foaming. Studies by Yeh et al. (2006) indicated however that agitation rates above 350 rpm and aeration above 2 vvm lead to higher foaming levels that caused low surfactin production and low of cells. A k_La of 0.012/s was optimal.

Sen and Swaminathan (1997, 2004) studied surfactin production by *B. subtilis* 3256. Maximal production (1.1 g/L) was at 37.4°C, pH 6.75, agitation of 140 rpm, and aeration of 0.75 vvm. Primary inoculum age (55–57 h) of 5%–6% by volume and secondary inoculation of (4–6 h) 9.5% by volume were also important for optimizing surfactin production.

Gancel et al. (2009) investigated lipopeptide production during cell immobilization on iron-enriched polypropylene particles. Immobilization improved biosurfactant production by up to 4.3 times. The amount of fengycin to surfactin varied depending on the iron content of the pellets. Highest surfactin (390 mg/L) and fengycin (680 mg/L) production was at 0.35% iron.

Guez et al. (2008) evaluated the influence of oxygen transfer rate on the production of the lipopeptide mycolysin by *B. subtilis* ATCC6633. A respiratory activity monitoring system used for the study showed that oxygen metabolism has an effect on the homologue production and that the regulatory system is complex. Chenikher et al. (2010) examined the ability to control the specific growth rate for the production of surfactin and mycosubtilin. Most feeding strategies do not take into account the loss of the biomass with the foam. This must be taken into account to enable the maintenance of the specific growth rate and subsequently production. The growth rate of 0.05/h was maintained.

An integrated foam collector was integrated for biosurfactant production to study parameters for scale-up (Amani et al., 2010). The best conditions were 300 rpm and 1.5 vvm for a surfactant yield on sucrose of 0.25 g/g. K_La of 0.01/s was achievable in shake flasks and bioreactors, and this could potentially be used for scale-up.

MEASUREMENT AND CHARACTERIZATION TECHNIQUES

Enhanced surfactin production can be determined by blood agar plate screening due to hemolysis by surfactin (Mulligan et al., 1984). To verify that the isolates are biosurfactant producers, then the cultures must be grown and the surfactin levels determined. The most common technique for determining surfactant concentration is surface tension measurement and CMC determination. HPLC is also frequently used. An assay based on hemolysis was used for the analysis of surfactin in the fermentation broth. It was determined that the method could be used as a quick low-technology method of surfactin analysis.

Huang et al. (2009) compared blood plate hemolysis, surface tension, oil spreading, and demulsification. Surface tension measurement followed by demulsification tests allowed isolation of a demulsification strain *Alcaligenes* sp. S-Xj-1, which produced a lipopeptide that was able to break O/W and W/O emulsions.

Knoblich et al. (1995) studied surfactin micelles by ice embedding and transmission electron cryomicroscopy. The micelles found were ellipsoidal with dimensions of 19, and 11 nm in width and length,respectively or spherical with a 5–9 nm in diameter, at pH 7. However, at pH 9.5, the micelles were more cylindrical with width and length dimensions of 10–14, and 40–160, or spherical with diameters of 10–20 nm. Addition of 100 mM NaCl and 20 mM $CaCl_2$ at pH 9.5 formed small spheres instead of the cylindrical micelles.

Hue et al. (2001) examined the use of a combination of LSI-MS and MS/MS for the characterization of the mixtures of surfactin produced by *B. subtilis*. Amino acid composition was determined, and the length of the acyl chain was shown to vary from 12 to 15 carbons. Leucine and isoleucine could be differentiated.

Biosurfactant proteins produced by *Lactobacillus fermentum* RC-14 have also been identified by a ProteinChip-interfaced mass spectrometer (Reid et al., 2002). Five tryptic peptide sequences by collision-induced dissociation tandem mass spectrometry were identified following on-chip digestion of collagen-binding proteins. This may lead to the determination of the factors that are responsible for antistaphylococcal activity.

^{1}H-NMR was used by Bonmatin et al. (1994) to show that surfactin can have two conformations depending on the pH. The saddle-like structure is bidentate with the two charged amino acids as sites for cation binding.

SANS studies were performed to study the characteristics of surfactin (Shen et al., 2009). At pH 7.5, the aggregation number was only 20, and the diameter of the micelles was 50 Å with a hydrophobic core of 22 Å radius. It is postulated that the leucines are in the hydrophobic core, which is consistent with its foaming characteristics. Further work (Shen et al., 2010) showed the solubilization of diphenylcarbamyl chloride.

Pecci et al. (2010) characterized the biosurfactants produced by *B. licheniformis* V9T14 strain. This strain exhibited antimicrobial activity that inhibited biofilm formation of human pathogens. LC-ESI–MS/MS analyses were used, and fengycin and surfactin homologues were determined. Fractionation was further performed by silica gel chromatography. C13, C14, and C15 surfactin homologues were found plus C17 fengycins A and B. Other C14–C16 fengycin homologues were also confirmed. Most of the surfactin (61.3%) was in the C15 form with an MW of 1035. The two most common forms of fengycin A and B, respectively, were the C17 of MW 1477 (25.1%) and 1505 (55.1%). The LC-ESI–MS/MS proved useful for the characterization of the lipopeptides.

An oil emulsification test was used to screen for biosurfactants and bioemulsifiers for strains from a sea mud (Liu et al., 2010). A *B. velezensis* H3 strain was isolated and could produce biosurfactants on starch and ammonium sulfate. C14 and C15 surfactins were discovered, which could lower the surface tension to 25.7 and 27.0 mN/m, respectively, from pH 4 to 10. CMCs were in the order of 10^{-5} mol/L. Antimicrobial properties were shown. The yield however was only 0.49 g/L. Highest yields of up to 50 g/L have been previously found by a strain on maltose and soybean flour (Yoneda et al., 2006).

GENETICS OF LIPOPEPTIDE PRODUCTION

Ultraviolet radiation mutation between argC4 and hisA1 on the genetic map led to a strain that produced 3.5 times surfactin (Mulligan et al., 1989). Another technique included random mutagenesis by N-methyl-N′ nitro-N-nitrosoguanidine of *B. licheniformis,* where an increase in surfactin production of 12-fold was obtained (Lin et al., 1998). Tsuge et al. (2001) not only found that the *yerP* gene is involved in surfactin resistance in the strain but also evaluated if this gene was involved in surfactin production. Although the *sfp* gene was inserted into the strain, production was low. Therefore, it did not appear that the *yerP* gene was linked to surfactin production.

Washio et al. (2010) analyzed the genetics of arthrofactin production by *Pseudomonas* sp. MIS38. Arthrofactin are cyclic lipopeptides that function as antibiotics, immunosuppressants, antitumor agents, siderophores, and surfactants. Schwartzer et al. (2003) postulated it to be superior as a biosurfactant to surfactin and is necessary for the swarming and biofilm formation by the bacteria. Mutants from gene insertion were isolated that did not produce the biosurfactant, gaining some info on the synthesis.

Ion beam implantation has also been utilized for generating high surfactin–producing mutant of *B. subtilis* (Liu et al., 2006). N$^+$ is implanted by this method. Gong et al. (2009) indicated that a mutant using this technique on the concentration of a crude surfactin of 12.2 g/L could be produced from 6.5 g/L of biomass. It is not known what the effect of the implantation has on the metabolism of the microorganisms.

Chelardi et al. (2012) also studied the motility of *B. subtilis*. They confirmed that *swrA* gene is needed for swarming, and surfactin increases surface wettability to allow swarming on low humidity surfaces.

Various oil reservoirs of salinities from 2.1% to 15.9% were examined to determine the presence of biosurfactant-producing strains (Simpson et al., 2011). The presence of surfactin (*srfA3*) and lichenysin (*licA3*) genes to evaluate the potential for biosurfactant production potential was confirmed. Subsequently, nutrient addition was performed to stimulate production. This confirmed the ability to biostimulate biosurfactant production in an oil reservoir for oil recovery.

EXTRACTION OF LIPOPEPTIDES

Crude extraction of lipopeptides is summarized in Table 6.3 but is usually by chemical extraction. For example, surfactin can be extracted by acid precipitation (pH 2) followed by solvent extraction by methanol (Vater et al., 2002). For *Pseudomonas* lipopeptides, multiple extraction by the ethyl acetate can be used (Kuiper et al., 2004).

Purification of lipopeptides is important for subsequent industrial application. Thin layer chromatography, HPLC, gel permeation chromatography, ion exchange chromatography, and ultrafiltration have been used. HPLC by reverse phase chromatography in particular is often used with the detection by UV absorbance or mass spectrometry to provide some information on the molecular mass of the components. Ultrafiltration with 30 kDa (UF-1) and 10 kDa (UF-II) cutoff membranes was employed in a single step (Sivapathasekaran et al., 2011). The recovery was higher with the 10 kDa membrane (89%) compared to 73% with the 30 kDa. Purity was also higher (83% compared to 78%). The product was a mixture of surfactin and fengycin.

The foaming characteristic of surfactin can be used to remove it during fermentation (Figure 6.4). At a concentration of 0.05 mg/L, it is comparable to SDS and bovine serum albumin (Razafindralambo et al., 1996). Low agitation speeds in the fermentor were beneficial for the removal of high concentrations of surfactin in the collected foam (Davis et al., 2001). Between 10 and 30 h, stirrer speeds of 146 and 166 rpm led to surfactin concentrations of 1.67 and 1.22 g/L, respectively, and agitation rates of 269 rpm produced concentrations of only 75 mg/L. Overall recovery of the produced surfactin in the foam was over 90%. Makkar and Cameotra (2001) also used foam fractionation to recover surfactin

TABLE 6.3
Recovery of Surfactin by Foam Collection

Producing Organism	Description	Foam Generation	Recovery%[a]	Enrichment Factor[b]	Reference
B. subtilis BBK 006	Recovery during batch operation	Not controlled	92.3	55.0	Chen et al. (2006a)
B. subtilis BBK 006	Recovery during continuous operation	Not controlled	28.7	50.1	Chen et al. (2006b)
B. subtilis ATCC 21331	Separate foam fractionation of cell-free broth	Controlled	97.1	2.9	Davis et al. (2001)
	Separate foam fractionation of cell broth	Controlled	97.3	1.7	
	Recovery during batch operation under oxygen depletion and low agitation (146 rpm)	Not controlled	90.0	34.0	

Source: Modified from Winterburn, J.B. and Martin, P.J, *Biotechnol. Letters*, 34(2): 187–195.

[a]   Recovery = $(C_f V_f / C_i V_i)$, where C_f, C_i are the foam and initial surfactin concentrations and V_i, V_f are the initial liquid and foam volumes.

[b]   Enrichment = C_f / C_i.

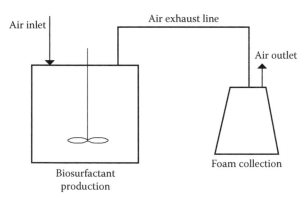

FIGURE 6.4   Recovery of surfactin by foam collection.

by *B. subtilis* MTCC 2423. Using sucrose as the substrate, yields of 4.5 g/L were obtained at 45°C. Other studies are shown in Table 6.4.

Liu et al. (2007) evaluated different conditions for adsorption for surfactin on activated carbon. pH values from 6.5 to 8.5 and 30°C were optimal. Adsorption onto activated carbon was studied to incorporate surfactin removal with production (Montastruc et al., 2008). Adsorption capacities were about 30 mg of surfactin

TABLE 6.4

Extraction Processes for Lipopeptides

Process	Biosurfactants Recovered by Each Method
Adsorption	Lipopeptides
Foam fractionation	Surfactin
Precipitation by acid	Surfactin
Ultrafiltration	Surfactin

Source: Adapted from Desai, J.D. and Banat, I.M., *Microbiol. Mol. Biol. Rev.*, 61, 47, 1997.

per gram of activated carbon and slightly lower (20 mg/g) from culture media. Ninety percent of the surfactin could then be removed by pure methanol.

Two resins (AG1-X4 and XAD-7) for the adsorption of surfactin were evaluated by Chen et al. (2008). Sorption capacities were 1.76 and 0.41 g/g, respectively. The large micelles decreased sorption in the resins.

An automated collection method was used to isolate fengycin produced by *B. subtilis* (Glazyrina et al., 2008). A flounder was used to remove the surfactant concentrated at the surface. The fraction removed was nine times higher in concentration than the bulk solution. No solvents or foam fractionation was required.

Foaming has been used by Davis et al. (2001) as a recovery process with a stirrer speed of 146 rpm lead to an enrichment of 34% and 90% recovery surfactin from the cell broth. Although the enrichment was low (1.7), the recovery was 97%. Higher speeds (204 and 269 rpm) caused high levels of foaming. With a chemostat at a dilution of 0.2/h, a high factor of enrichment was shown (Chen et al., 2007).

Chen et al. (2008a) used hexane to extract surfactin from the fermentation broth of *B. subtilis* ATCC 21332 with a microporous polyvinylidene fluoride hollow fiber module (0.2 μm pore size). The micelles did not easily pass through the pores and were sorbed onto the membrane material. Desorption by ethanol from the membranes improved the surfactin purity to 78%. Further work by Chen et al. (2008b) was performed by acid precipitation and redissolution with NaOH. Ethyl acetate was shown to be a better extractant than hexane. However, the addition of ammonium cations of Aliquat 336 could bind to surfactin and enhance the extraction to 92% for a 3 g/L concentration. Recoveries of 90% and 88% could be achieved with NaCl or ammonium sulfate addition to ethanol/water.

MEMBRANE LIPOPEPTIDE RECOVERY

Ultrafiltration membranes can be used to retain micelles of surfactin and other lipopeptides as they are larger than monomers (Mulligan and Gibbs, 1990) as shown in Figure 6.5. Sen and Swaminathan (2005) demonstrated the purification of surfactin with a polymeric membrane. Optimal flux (260 L/m^2-h) and a 166-fold concentration factor were obtained at a pH 8.5. Chen et al. (2007) used a two-step ultrafiltration process. Ultrafiltration membranes of 100 kDa were used to recover the micelles

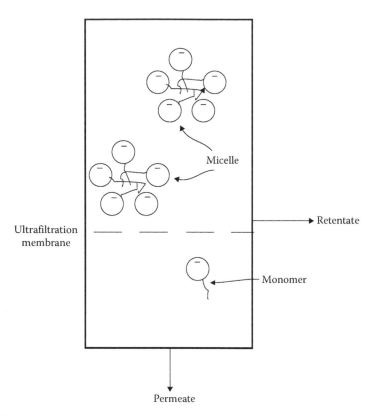

Micelle

Retentate

Ultrafiltration
membrane

Monomer

Permeate

FIGURE 6.5   Removal of metals from contaminated soil by surfactin.

while 1 kDa MW cutoff nanomembranes were used for concentrating the monomers. Recoveries of 97% were high, but purities were low (55%). Purities were increased by using a two-step ultrafiltration process with 100 kDa MW cutoff membranes. The surfactin recovered in the permeate of one membrane was then passed through the other membrane to remove salts. Recoveries were 72% with an 83% purity. Chen et al. (2008a) also evaluated two different membranes of the same MW cutoff (100 kDa). Although the cellulose ester membrane gave a higher recovery (97% compared to 88%) than the polyethersulfone membrane, it was not recommended due to the flux decrease from concentration polarization, gel formation, and amino acid sorption at the membrane. Further studies by Chen et al. (2008c) indicated that a combination of salting out and membrane filtration could increase yields and purity.

STRAIN ISOLATION

Gandhimathi et al. (2009) isolated a lipopeptide from a marine sponge-associated actinomycete, *Nocardia alba* MSA10. The lipopeptide exhibited the properties of lipase production, and demulsification, hemolytic, antibiotic, and surface activities. The solvents, ethyl acetate diethyl ether and dichloromethane, were used for

TABLE 6.5
Lipopeptide-Producing Marine Organisms

Producing Microorganism	Chemical Composition	Properties	Reference
Bacillus licheniformis BAS50	Lichenysin of MW 1006–1034	Surface tension reduction to 28 mN/m, CMC of 12 mg/L Antibacterial activity	Yakimov et al. (1995)
Bacillus circulans	Novel biosurfactant	Antimicrobial activity	Mukherjee et al. (2009)
Azotobacter chroococcum	Lipid: protein (31.3:68.7)	Emulsify ability for waste motor oil, crude oil, diesel, kerosene, naphthalene, anthracene, xylene	Thavasi et al. (2009)

Source: Adapted from Satpute et al., 2010.

extraction. The substrate glucose and peptone were used for production, which was optimal at pH 7, 30°C, and 1% salinity. It was stable between 4 and 9 and up to 80°C. Other marine lipopeptide-producing isolates are shown in Table 6.5.

Sriram et al. (2011a) isolated a lipopeptide from a metal-tolerant strain of *B. cereus* NK1. The strain was tolerant to ferrous sulfate, zinc, and lead. Biofilm inhibition by pathogens and antimicrobial activities against fungi, and gram-positive and -negative bacteria were indicated. The strain was resistant to various antibiotics including ampicillin, bacitracin, erythromycin, and rifampicin but was less resistant to others. The lipopeptide had a CMC of 45 mg/L with a surface tension of 36 mN/m. The metal resistance of the strain could enable it to be used for remediation purposes.

Another strain, *Escherichia fergusonii* KLU01 was isolated by Sriram et al. (2011b) from an oil-contaminated soil. It was able to produce a biosurfactant of CMC of 36 mg/L, with emulsification properties, and excellent stability over a range of pH (4–10), temperature (20°C–100°C), and various salts. The strain was also tolerant against manganese, lead, iron, nickel, copper, and zinc.

Jing et al. (2011) also isolated a strain of *B. subtilis* JA-1 from an oil reservoir. The strain was able to grow at temperatures of 60°C. Surface tension could be reduced to 28.3 mN/m with a CMC of 48 mg/L. The biosurfactant was stable up to pH 12, 121°C, and salt concentrations of up to 14%. The strain thus could be potentially useful for enhanced oil recovery.

Ismail et al. (2012) isolated a crude oil–emulsifying *Bacillus* sp. I-15 from oil contamination. The surface tension was reduced to 42 mN/m and the CMC was 200 mg/L. It could potentially be beneficial for natural attenuation of oil contamination.

Ghojavand et al. (2011) studied a strain of *B. mojavensis* that produced a biosurfactant. The strain could tolerate high salinities (up to 10% NaCl) and temperatures up to 55°C and could grow under anaerobic conditions. The biosurfactant could reduce the surface tension to 27 mN/m. Emulsification stability, however, was poor. Previous work by Ghojavand et al. (2008) isolated thermotolerant, halo-tolerant, and facultative biosurfactant-producing *B. subtilis* strains as shown by the 16S ribosomal deoxyribonucleic acid gene. These strains could potentially be used for enhanced oil recovery. Another strain *B. mojavensis* XH1 was studied by Li et al. (2012). A biodemulsifier was produced and isolated by ethanol extraction and then sephadex and silicon gel column chromatography. A response surface methodology was used to optimize the media for production. Biodemulsifier yield increased to 2.07 g/L.

A strain of *B. mojavensis* (PTCC 1696) (Ghojavand et al., 2012) was isolated from an oil field. The biosurfactant produced by the strain could reduce the surface tension to 26.7 mN/m. Biosurfactant was added for water flooding to enhance oil recovery from a low-permeability carbonate reservoir. Although the concentration of the surfactant was low (0.1 g/L), it showed potential for the oil removal. Costs of the purified surfactin for biomedical research are in the range of $10 per mg compared to $2–$4 per kg for emulsion formulations.

A licheniformin biosurfactant was produced by *B. licheniformis* MS3. The lipo-peptide contained the amino acids, Gly, Ala, Val, Asp, Ser, Gly, Tyr, and a lac-tone ring with a fatty acid moiety at the N-terminal amino acid residue (structure). The MW was determined as 1438 Da. The surface tension could be lowered to 38 mN/m by the isolated biosurfactant at a concentration of 15 mg/L. Isolation was per-formed using an electroflotation column. It was stable over a range of temperatures (45°C–85°C) and from pH 3 to 11.

Janek et al. (2010) isolated lipopeptides produced by *Pseudomonas fluorescens* BD 5 from the arctic. The biosurfactants were named pseudofactin I and II. They were cyclic with good emulsification abilities for plant oils and hydrocarbons, comparable to that of surfactin (Abdel-Mawgoud et al., 2008). Pseudofactin II reduced the surface tension to 31.5 mN/m with a CMC of 72 mg/L. Approximately 10 mg/L was recovered. The yield of pseudofactin I was 1/12th that of the other form. They could potentially be used for various medical and biotechnological applications (Figure 6.6).

A strain of *B. mycoides* was isolated from an oil field (Najafi et al., 2010). The isolate produced a lipopeptide derivative that could lower the surface tension to 34 mN/m. To optimize production, a response surface methodology was employed. Optimal production of surfactin of 3.3 g/L were obtained at 16.6 g/L glucose sub-strate concentration, 39°C, pH 7.4, and total salt concentration of 55.4 g/L.

A response surface methodology was also employed by Wei et al. (2010) for fengycin production. The *B. subtilis* F29-3 strain had been isolated from a potato farm. The fengycin was isolated by acid precipitation at pH 2 followed by ultrafiltra-tion and nanofiltration for purification. The media design that was optimal included mannitol, soybean meal, sodium nitrate, and magnesium sulfate and increased pro-duction by almost threefold (3.5 g/L).

Another strain *Brevibacillus brevis* (Wang et al., 2010) was isolated from an oil field. It consisted of Asp, Glu, Val, and Leu in a ratio of 1:1:1:4 like surfactin. The surface tension was 26.8 mN/m and the CMC was 9×10^{-6} M. The MWs of the various fractions varied from 1008 to 1035, depending on the C13–C15 hydrocarbon portion. This was the first time this species was shown to produce surfactin.

Pseudofactin I

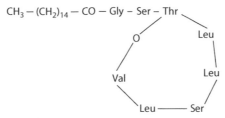

Pseudofactin II

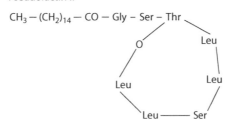

FIGURE 6.6   Structure of pseudofactin. (Adapted from Janek, T. et al., *Biores. Technol.*, 101, 6118, 2010.)

A *B. subtilis* strain was isolated from a refinery soil by Fonseca et al. (2007). A factorial design and response surface analysis indicated that the optimal C/N ratio was 3 and agitation rate was 250 rpm. The minimum surface tension was 31 mN/m. The lipopeptide on preliminary characterization was shown to be different from surfactin.

Low-cost substrates that have been used include cassava wastewater, sludge palm oil, vegetable oil refinery waste, and molasses. Raw glycerol from biodiesel production was investigated by de Faria et al. (2011). The strain *B. subtilis* LSFM-05 was grown on this substrate and produced C14/Leu7 surfactin with a CMC of 70 μM. The surfactant was collected in the foam and was produced at a concentration of 1.37 g/L. The surface tension was 29.5 mN/m. An esterified glutamic acid present in the surfactin differed from the commercial Sigma-Aldrich.

Pemmaraju et al. (2012) isolated biosurfactant-producing strains from an oily sludge using a combination of surface tension reduction, hemolytic activity, emulsification activity, drop-collapse assay, and cell surface hydrophobicity studies. Up to 6.9 g/L of surfactant (a mixture of surfactins, iturins, and fengycins) was produced by *B. subtilis* DSVP23 and is growth-associated. C12–30 hydrocarbons (saturates and aromatics) could be degraded within 5 days, indicating the potential for bioremediation by this strain.

Lipopeptides were isolated from *Paenibacillus* sp. (HRAC30) (Canova et al., 2010). The lipopeptides were characterized as a surfactin and was C15 lipopeptide of MW 1036. It was extracted using ethyl acetate and then elution through a Sephadex column. The compound was shown to be an active phytopathogen suppressor and could be used to control *Rhizoctonia solani*, a pathogen for commercial crops. The minimum inhibitory concentration was 14 μg/mL, which is almost as effective as iturin.

Another species, *Paenibacillus alvei* (Najafi et al., 2011), was isolated from an oil field. Biosurfactant production was optimized with central composite rotatable design response surface methodology. The biosurfactant reduces the surface tension to 35 mN/m. A glucose concentration of 13 g/L, temperature of 35°C, a 51 g/L total salt concentration, and pH 6.9 were the optimal conditions.

A strain of *B. amyloliquefaciens* was isolated from crude oil (Sang-Cheol et al., 2010). Lipopeptides of molecular mass of 1086.9 and 1491.2 *m/z* were determined. The higher MW corresponds to fengycin B. However, as the structure differed from fengycin A and B forms, it was designated as fengycin S. Due to its properties of emulsification, it could be used for oil spills.

Rufino et al. (2012) produced a lipopeptide from a yeast *Candida lipolytica*. A waste soybean oil residue was used as the substrate. The surface tension was 25 mN/m, and the CMC of the crude form was 0.03% and consisted of 50% protein, 20% lipid, and 8% carbohydrate.

Pseudomonas aeruginosa strains MTCC7815 and MTCC7812 were studied for solubilization and metabolism of fluorene, pyrene, and phenanthrene (Bordolai and Konwar, 2009). Pyrene and fluorene were solubilized by these two lipopeptide-producing strains, which enhanced growth. The surface tension was reduced to 35 mN/M by these strains, and the CMC were 100 and 110 mg/L for MTCC7815 and MTCC 7812, respectively. Previous studies (Bordolai and Konwar, 2008) showed that these biosurfactants were very stable (pH 2.5–1) and up to 100°C. Crude oil–saturated sand pack studies indicated that 50%–60% of the oil could be recovered from room temperature to 90°C, indicating potential for enhanced oil recovery. Glucose and glycerol were the best carbon sources.

Saimmai et al. (2013) isolated strains from mangrove sediments that produce biosurfactants. Many pollutants such as hydrocarbons are found in the sediments, and thus biosurfactant production could enhance the uptake of these pollutants by the bacteria. They identified a strain of *S. ruminantium* CT2 that produced the lipopeptide for the first time. It grew the best on molasses and could reduce the surface tension to 25.5 mN/m with a CMC of 8 mg/L. Maximum production was 5 g/L. Ethyl acetate could be used to extract the biosurfactant from the broth. The biosurfactant was characterized as a lipopeptide similar to surfactin. It showed good pH and temperature stability and could enhance motor oil removal from contaminated sand, ability to solubilize polycyclic aromatic hydrocarbons, and antimicrobial activity.

PROPERTIES AND APPLICATIONS OF LIPOPEPTIDES

Surfactin addition improved the mechanical dewatering of peat by greater than 50% at very low concentrations (0.0013 g/g wet peat) by altering the flow characteristics of the trapped water within the peat particles (Cooper et al., 1986). Surfactin has also shown the ability to inhibit blood coagulation and protein denaturation, to accelerate fibrinolysis, and to have antimyoplasmic properties (Vollenbroich et al., 1997). Mycoplamata leads to respiratory inflammation, urogenital tract diseases, and cofactors in the AIDS (Vollenbroich et al., 1997) pathogenesis. Antibiotic therapy is not effective against mycoplasmata, but surfactin can cause leakage of the plasma membrane and finally disintegration. One disadvantage is the competition with proteins. Endoflaxacin coaddition allowed a synergistic effect (Seydlová and Svobodová, 2008). Cao et al. (2009) showed that when the lipopeptide from *B. natto* TK-1 was tested against MCF-17 human breast cancer cells, it indicated antitumor behavior. The inhibition was as the G2/M phase of growth.

Other potential medical benefits for surfactin have been identified. For example, surfactin can reduce the inflammatory activity of the lipopolysaccharides against eukaryotic cells (Seydlová and Svobodová, 2008). Surfactin C was better than surfactin A, B, or D for anti-inflammatory activity, antiviral activity against herpes simplex virus (HSV-1 and 2), semliki forest virus, simian immunodeficiency virus, vesicular stomatitis virus and feline calicivirus.

Das et al. (2008) isolated at a biologically active fraction from a marine *B. circulans* by methanol extraction and HPLC fractionation. One of the fractions showed surface tension lowering to 28 mN/m and antimicrobial activity against gram-positive and gram-negative pathogenic and semipathogenic bacteria. Unlike surfactin, however, it was not hemolytic.

The adhesion of bacteria and biofilm formation are the first steps to bacterial infection.

Therefore, inhibition of this can reduce the growth of pathogenic bacteria. Lipopeptides can reduce adhesion and biofilm formation, and thus this was studied (Das et al., 2009). Higher concentrations of purified surfactant (10 mg/mL) could reduce adhesion by over 80% of several strains and over 70% for biofilm inhibition. This indicates the potential in biomedical application as bacteria in biofilms are resistant to antibiotics. The presence of *Salmonella typhimurium*, *Salmonella enterica*, *Escherichia coli*, and *Proteus mirabilis* could be reduced in PVCs and vinyl urethral catheters by running surfactin through the catheter before use (Seydlová and Svobodová, 2008).

Rivardo et al. (2010) investigated the effect of lipopeptide addition with silver on *E. coli* biofilm inhibition. The biosurfactant V9T19 was obtained from *B. licheniformis*. Silver is a well-known disinfectant. Adding the lipopeptide was able to reduce the amount of silver required by 129- to 258-fold, demonstrating its synergistic effect.

Singh and Cameotra (2004) describe the biomedical properties of surfactin and iturin A produced by *B. subtilis*. Tanaka et al. (1997) has also described the antiviral properties of surfactin. The mechanism has been postulated to be related to disruption of the virus lipid membrane.

Lima et al. (2011a) investigated the biodegradability of various surfactants. Lipopeptides were obtained from various strains *Bacillus* sp. LBBMA 111A, *B. subtilis* LBBMA 155, and *Arthrobacter oxydans* LBBMA 201. SDS, a synthetic surfactant, was also compared. Although biosurfactants are supposed to be more biodegradable, there have been few studies on this. Pure and mixed cultures were studied for the biodegradation tests, and carbon dioxide emissions were monitored. The biodegradation of the biosurfactants was significantly more than the synthetic surfactant, indicating their potential for environmental applications. Subsequent work by Lima et al. (2011b) indicated that these biosurfactants could remove phenanthrene and cadmium for contaminate soil in combination with an inorganic ligand iodide.

Reddy et al. (2009) evaluated surfactin as a stabilizing agent for the synthesis of silver nanoparticles. The stability of the nanoparticles was determined and found to be stable for a period of 2 months in the presence of surfactin. pH and temperature conditions affected particle size. Surfactin as a stabilizing agent is renewable, less toxic, and biodegradable, and thus an environmentally friendly additive.

In addition, the presence of two negative charges, one on the aspartate and the other on the glutamate residue of surfactin, enables the binding of various metals such as calcium, barium, lithium, magnesium, manganese, and rubidium (Thimon et al., 1992). Eliseev et al. (1991) also showed that a *Bacillus* species could release oil at low concentrations of 0.04 mg/L from oil-saturated columns.

A strain of *B. subtilis* isolated from contaminated sediments (Olivera et al., 2000) could produce surfactin. A crude form was then added to ship bilge waste to enhance biodegradation. Although aliphatic and aromatic compounds in a nonsterile environment were degraded more quickly in the presence of the biosurfactant, N-C17 pristane and N-C18 phytane degradation were not.

Using a technique called micellar enhanced ultrafiltration, Mulligan et al. (1999) studied the removal of various concentrations of metals from water by various concentrations of surfactin by a 50,000 Da MW cutoff ultrafiltration membrane. Cadmium and zinc rejection ratios were superior (close to 100%) at pH values of 8.3 and 11, while copper rejection ratios were the highest at pH 6.7 (about 85%). The addition of 0.4% oil as a cocontaminant slightly decreased the retention of the metals by the membrane. The ultrafiltration membranes also indicated that metals became associated with the surfactin micelles as the metals remained in the retentate and did not pass through into the permeate as illustrated in Figure 6.5. The ratio of metals to the surfactin was determined to be 1.2:1, which was only slightly different from the theoretical value of 1 mol metal: 1 mol surfactin due to the two charges on the surfactin molecule.

Batch soil washing experiments were performed to evaluate the feasibility of using surfactin from *B. subtilis* for the removal of heavy metals from a contaminated soil and sediments (Mulligan et al., 1999). Compared to minimal amounts for the control, 0.25% surfactin with 1% NaOH removed 25% of the copper and 6% of the zinc from the soil and 15% of the copper and 6% of the zinc from the sediments. A series of five washings of the soil with 0.25% surfactin with 1% NaOH removed 70% of the copper and 22% of the zinc. Ultrafiltration, octanol–water partitioning, and zeta potential measurement determined that surfactin was able to remove the metals by sorption and complexation at the soil interface, then desorption of the metal through interfacial tension lowering and fluid forces into solution and finally micellar complexation (Figure 6.5).

CONCLUSION

Surfactin has very interesting surfactant properties. Potential medical applications are related to antiinflammatory, antiviral, antibiotic, and antiadhesive activities. However, the economics are not competitive due to poor yields and the requirement for expensive and complex substrates. Portillo-Rivera et al. (2009) have postulated that biosurfactant costs can be as low as $0.50 per liter from molasses sugarcane. Low-cost purification methods are also needed as downstream costs can account for 60% of the cost (Mukherjee et al., 2006). Purity of the product is also a major consideration. Ninety-eight percent pure surfactin is sold for $10 per mg but can be reduced to $2–$4 per kg for tank cleaning or oil recovery applications (Bognolo, 1999). Although more information is available concerning the biosynthesis of surfactin, there is still a lack of information regarding the secretion, metabolic route, primary cell metabolism, and physicochemical properties of the biosurfactant. Research is thus required to accelerate the knowledge in this area and possibly will enhance the applications of the surfactant. New forms of surfactin and other lipopeptides could also become available.

REFERENCES

Abdel-Mawgoud, A.M., Aboulwafa, M.M., and Hassouna, N.A.H. 2008. Characterization of surfactin produced by *Bacillus subtilis* isolate BS5. *Applied Biochemistry and Biotechnology*, 150: 289–303.

Amani, H., Mehria, M.R., Sarrafzadeh, M.H., Haghghi, M., and Soudi, M.R. 2010. Scale up and application of biosurfactant from *Bacillus subtilis* in enhanced oil recovery. *Applied Biochemistry and Biotechnology* 162: 510–523.

Arima, K., Kakinuma, A., and Tamura, G. 1968. Surfactin, a crystalline peptide lipid surfactant produced by *Bacillus subtilis*. Isolation, characterization and its inhibition of fibrin clot formation. *Biochemical and Biophysical Research Communications* 31(3): 488–494.

Barros, F.F.C., Ponezi, A.N., and Pastore, G.M. 2008. Production of biosurfactant by *Bacillus subtilis* LB5a on a pilot scale using cassava wastewater as substrate. *Journal of Industrial Microbiology and Biotechnology* 35: 1071–1078.

Becher, P. 1965. *Emulsions, Theory and Practice,* 2nd edn. Reinhold Publishing, New York.

Biermann, M., Lange, F., Piorr, R., Ploog, U., Rutzen, H., Schindler, J., and Schmidt, R. 1987. Surfactants in consumer products. In *Theory, Technology and Application*, Falbe, J. (ed.). Springer-Verlag, Heidelberg, Germany, pp. 86–106.

Bognolo, C. 1999. Biosurfactants as emulsifying agents for hydrocarbons. *Colloids and Surfaces A. Physico Engineering Aspects* 152: 41–52.

Bonmatin, J.M., Genest, M., Labbe, H., and Ptak, M. 1994. Solution 3-dimensional structure of surfactin. A cyclic lipopeptide studied by H-1-NMR distance geometry and molecular dynamics. *Biopolymers* 34(7): 975–986.

Bonmatin, J.M., Labbe, H., Grangemard, I., Peypoux, F., Magetdana, R., Ptak, M., and Michel, G. 1995. Production, isolation and characterization of [Leu(4)]surfactins and [Ile(4)]surfactins from *Bacillus subtilis*. *Letters in Peptide Science* 2: 41– 47.

Bonmatin, J-M., Laprevote, O., and Peypoux, F. 2003. Diversity among microbial cyclic lipopeptides: iturins and surfactins. Activity-structure relationships to design new bioactive agents. *Combinatorial Chemistry and High Throughput Screening* 6: 541–556.

Bordolai, N.K. and Konwar, B.K. 2008. Microbial surfactant-enhanced mineral oil recovery under laboratory conditions. *Colloids and Surfaces B Biointerfaces*. 63(1):73–82

Bordolai, N.K. and Konwar, B.K. 2009. Bacterial biosurfactant in enhancing solubility and metabolism of petroleum hydrocarbons. *Journal of Hazardous Materials* 170: 495–505.

Cameotra, S.S. and Makkar, R.S. 2004. Recent applications of biosurfactants as biological and immunological molecules, *Current Opinion in Microbiology* 7(3): 262–266.

Canova, S.P., Petta, T., Reyes, L.F., Zucchi, T.D., Moraes, L.A.B. and Melo, I.S. 2010. Characterization of lipopeptides from *Paenibacillus sp.* (IIRAC30) suppressing *Rhizoctonia solani, World Journal of Microbiology and Biotechnology* 26: 2241–2247.

Cao, X.-H., Liao, Z.-Y., Wang, C.-L., Cai, P., Yang, W.-Y., Lu, M.-F., and Huang, G. 2009. Purification and antitumour activity of a lipopeptide biosurfactant produced by *Bacillus natto* TK-1. *Biotechnology and Biochemistry* 52: 97–106.

Carrera, P., Cosmina, P., and Grandi, G. 1992. Mutant of *B. subtilis* and production of surfactin by use of the mutant. Patent J04299981.

Carrera, P., Cosmina, P., and Grandi, G. 1993a. Mutant of *Bacillus subtilis*. US Patent 5264363.

Carrera, P., Cosmina, P., and Grandi, G. 1993b. Method of producing surfactin with the use of *Bacillus subtilis*. US Patent 5227294.

Chelardi, E., Salvetti, S., Ceraglioli, M., Gueye, S.A., Calandroni, F., and Senesi, S. 2012. Contribution of surfactin and SwrA to flagellin expression, swimming and surface mobility in *Bacillus subtilis*. *Applied and Environmental Microbiology* 78: 6540–6544.

Chen, C.Y., Baker, S.C., and Darton, R.C. 2006a. Batch production of biosurfactant with foam fractionation. *Journal of Chemical Technology and Biotechnology* 81: 1923–1931.

Chen, C.Y., Baker, S.C., and Darton, R.C. 2006b. Continuous production of biosurfactant with foam fractionation. *Journal of Chemical Technology and Biotechnology* 81: 1915–1922.

Chen, H.L., Chen, Y.S., and Juang, R.S. 2007. Separation of surfactin from fermentation broths by acid precipitation and two-stage dead-end ultrafiltration processes. *Journal of Membrane Science* 299: 114–121.

Chen, H.L., Chen, Y.S., and Juang, R.S. 2008a. Flux, decline and membrane cleaning in cross-flow ultrafiltration of treated fermentation broths for surfactin recovery. *Separation and Purification Technology* 62: 47–55.

Chen, H.L., Chen, Y.S., and Juang, R.S. 2008b. Purification of surfactin in pretreated fermentation broths by adsorptive removal of impurities. *Biochemical Engineering Journal* 40: 452–459.

Chen, H.L., Chen, Y.S., and Juang, R.S. 2008c. Recovery of surfactin from fermentation broths by a hybrid salting-out and membrane filtration process. *Separation and Purification Technology* 59: 244–255.

Chenikher, S., Guez, J.S., Coutte, F., Pekpe, M., Jacque, P., and Cassar, J.P. 2010. Control of the specific growth rate of *Bacillus subtilis* for the production of biosurfactant lipopeptides in bioreactors with foam overflow. *Process Biochemistry* 45: 1800–1807.

Chtioui, O., Dimitrov, K., Gancel, F., Dhuslter, P., and Nikov, I. 2012. Rotating discs bioreactor: A new tool for lipopeptides production. *Process Biochemistry* 47: 2020–2024.

Cooper, D.G. 1986. Biosurfactants. *Microbiological Science* 3: 145–149.

Cooper, D.G., Macdonald, C.R., Duff, J.P., and Kosaric, N. 1981. Enhanced production of surfactin from *Bacillus subtilis* by continuous product removal and metal cation additions. *Applied and Environmental Microbiology* 42: 408–412.

Cooper, D.G., Pillon, D.W., Mulligan, C.N., and Sheppard, J.D. 1986. Biological additives for improved mechanical dewatering of fuel-grade peat. *Fuel* 65: 255–259.

Dae, K.S., Cho, J.Y., Park, H.J., Lim, C.R., Lim, J.-H. Yun, H.I., Park, S.C., Kim, S.K., and Rhee, M.H. 2006. A comparison of the anti-inflammatory activity of surfactin A, B, C and D *from B. subtilis*. *Journal of Microbiology and Biotechnology* 16: 1656–1659.

Das, P., Mukjerjee, S., and Sen, R. 2008. Antimicrobial potential of a lipopeptide biosurfactant derived from a marine *Bacillus circulans*. *Journal of Applied Microbiology* 104: 1675–1684.

Das, P., Mukherjee, S., and Sen, R. 2009. Substrate dependent production of biosurfactants from a marine bacterium. *Bioresource Technology* 100: 1015–1019.

Davis, D.A., Lynch, H.C., and Varley, J. 1999. The production of surfactin in batch culture by *Bacillus subtilis* ATCC 21332 is strongly influenced by the conditions of nitrogen metabolism. *Enzyme and Microbial Technology* 25: 322–329.

Davis, D.A., Lynch, H.C., and Varley, J. 2001. The application of foaming for the recovery of surfactin from *B. subtilis* ATCC 21332 cultures. *Enzyme and Microbial Technology* 28: 346–354.

de Faria, A.F., Stéfani, D., Vaz, B.G., Silva, Í.S, Garcia, J.S., Eberlin, M.N., Grossman, M.J., Alves, O.L., and Durrant, L.R. 2011. Purification and structural characterization of fengycin homologues produced by *Bacillus subtilis* LSFM-05 grown on raw glycerol. *Journal of Industrial Microbiology and Biotechnology* 38(7): 863–871.

de Roubin, M.R., Mulligan, C.N., and Gibbs, B.F. 1989. Correlation of enhanced surfactin production with decreased isocitrate dehydrogenase activity. *Canadian Journal of Microbiology* 35: 854–859.

Deleu, M., Razafindralambo, H., Popineu, Y., Jacques, P., Thonart, P., and Paquot, M. 1999. Interfacial and emulsifying properties of lipopeptides from *Bacillus subtilis*. *Colloids and Surfaces A* 152(1–2): 3–10.

Desai, J.D. and Banat, I.M. 1997. Microbial production of surfactants and their commercial potential. *Microbiological and Molecular Biology Review* 61: 47–64.

Drouin, C.M. and Cooper, D.G. 1992. Biosurfactants and aqueous two-phase fermentation. *Biotechnology and Bioengineering* 40: 86–90.

Eliseev, S.A., Vildanova-Martsishin, R.I., Shulga, A.N., Shabo, Z.V., and Turovsky, A.A. 1991. Oil-washing bioemulsifier produced by *Bacillus* species ± potential application of bioemulsifier in oil removal for sand decontamination. *Mikrobiol Zh* 53: 61–66 (in Russian).

Falatko, D.M. 1991. Effects of biologically reduced surfactants on the mobility and biodegradation of petroleum hydrocarbons. MS thesis. Virginia Polytechnic Institute and State University, Blackburg, VA.

Folmsbee, M., Duncan, K.E., Han, S.-O., Nagle, D.P., Jennings, E., and McInerney, M. J. 2006. Re-identification of the halotolerant, biosurfactant-producing *Bacillus licheniformis* strain JF-2 as *Bacillus mojavensis* strain JF-2. *Systematic and Applied Microbiology* 29: 645–649.

Fonseca, R.R., Silva, A.J.R., De Franca, F.P., Cardosa, V.L., and Servulo, E.F.C. 2007. Optimizing carbon/nitrogen ratio for biosurfactant production by a *Bacillus subtilis* strain. *Applied Biochemistry and Biotechnology* 136–140: 471–486.

Fox, S.L. and Bala, G.A. 2000. Production of surfactant from *Bacillus subtilis* ATCC 21332 using potato substrates. *Bioresource Technology* 75: 235–240.

Gallet, X., Deleu, M., Razafindralambo, H., Jacques, P., Thonart, P., Paquot, M., and Brasseur, R. 1999. Computer simulation of surfactin conformation at a hydrophobic/hydrophilic interface. *Langmuir* 15(7): 2409–2413.

Gancel, F., Montastruc, L., Liu, T., Zhao, L., and Nikov, I. 2009. Lipopeptide overproduction by cell immobilization on iron-enriched light polymer particles. *Process Biochemistry* 44: 975–978.

Gandhimathi, R., Seghal Kiran, G., Hema, T.A., Selvin, J., Rajeetha Raviji, T., and Shamughapriya, S. 2009. Production and characterization of lipopeptide biosurfactant by a sponge-associated marine actinomycetes *Nocardiopsis alba* MSA 10. *Bioprocess and Biosystems Engineering* 32: 825–835.

Georgiou, G., Lin, S.C., and Sharma, M.M. 1992. Surface active compounds from microorganisms. *Biotechnology* 10(1): 60–65.

Ghojavand, H., Vahabzadeh, F., and Azizmohseni, F. 2011. A halotolerant, thermotolerant, and facultative biosurfactant producer: Identification and molecular characterization of a bacterium and evolution of emulsifier stability of a lipopeptide biosurfactant. *Biotechnology and Bioprocess Engineering* 16: 72–80.

Ghojavand, H., Vahabzadeh, F., Mehranian, M., Radmehr, M., Shahraki, K.A., Zolfagharian, F., Emadi, M.A., and Roayaei, E. 2008. Isolation of thermotolerant, halotolerant, facultative biosurfactant-producing bacteria *Applied Microbiology and Biotechnology* 80: 1073–1085.

Ghojavand, H., Vahabzadeh, F., and Shahraki, A.K. 2012. Enhanced oil recovery from low permeability dolomite cores using biosurfactant produced by a *Bacillus mojavensis* (PTCC 1696) isolated from Madjed-I Soleyman field. *Journal of Petroleum Science and Engineering* 81: 24–30.

Glazyrina, J., Junne, S., Thiesen, P., Lunkenheimer, K., and Goetz, P. 2008. In situ removal and purification of biosurfactants by automated surface enrichment. *Applied Microbiology and Biotechnology* 81: 23–31.

Gong, G., Zheng, Z., Chen, H., Yuan, C., Wang, P., Yao, L., and Yu, Z. 2009. Enhanced production of surfactin by *Bacillus subtilis* E8 mutant obtained by ion beam implantation. *Food Technology and Biotechnology* 47: 27–31.

Grangemard, I., Bonmatin, J.M., Bernillon, J., Das, B.C., and Peypoux, F. 1999. Lichenysins G, a novel family of lipopeptide biosurfactants from *Bacillus licheniformis* 1M 1307. Production, isolation and structural evaluation by NMR and mass spectrometry. *Journal of Antibiotics* 52(4): 363–373.

Gu, J.Q., Nguyen, K.T., Gandhi, C., Rajgarhia, V., Baltz, R.H., Brian, P., and Chu, M. 2007. Structural characterization of daptomycin analogues A21978c(1–3_(D-Asn(11)) produced by a recombinant *Streptomyces roseosporus* strain. *Journal of Natural Products* 70(2): 233–240.

Guez, J.S., Muller, C.H., Danze, P.M, Buchs, J., and Jacques, P. 2008. Respiration activity monitoring system (RAMOS), an efficient tool to study the influence of the oxygen transfer rate on the synthesis of lipopeptide by *Bacillus subtilis* ATCC6633. *Journal of Biotechnology* 134: 121–126.

Hayes, M.E., Nestau, E., and Hrebenar, K.R. 1986. Microbial surfactants. *Chemtech* 16: 239–245.

Horowitz, S., Gilbert, J.N., and Griffin, W.M. 1990. Isolation and characterization of a surfactant produced *Bacillus licheniformis*—86. *Journal of Industrial Microbiology* 6(4): 243–248.

Huang, X.F., Liu, J., Lu, L.J., Wen, Y., Xu, J.C., Yang, D.H., and Zhou, Q. 2009. Evaluation of screening methods for demulsifying bacteria and characterization of lipopeptide bio-demulsifier produced by *Alcaligenes* sp. *Bioresource Technology* 100(3): 1358–1365.

Hue, N., Serani, L., and Laprévote, O. 2001. Structural investigation of cyclic peptidolipids from *Bacillus subtilis* by high energy tandem mass spectrometry. *Rapid Communications in Mass Spectrometry* 15: 203–209.

Ishigami, Y., Osman, M., Nakahara, H., Sano, Y., Ishiguro, R., and Matsumoto, M. 1995. Significance of β-sheet formation for micellization and surface adsorption of surfactin. *Colloids and Surfaces B Biointerfaces* 4: 341–348.

Ismail, W., Al-Rowaihi, I.S., Al-Humam, A.A., Hamza, R.Y., El Nayal, A.M., and Bououdina, M. 2012. Characterization of a lipopeptide biosurfactant produced by a crude-oil-emulsifying *Bacillus* sp.1–15. *International Journal of Biodeterioration and Biodegradation* 84: 168–178.

Janek, T., Kukaszewkwicz, M., Rezanka, T., and Krasowska, A. 2010. Isolation and characterization of two new lipopeptide biosurfactants produced by *Pseudomonas fluorescens* BD5 isolated from water from the Arctic Archipelago of Svalbard. *Bioresource Technology* 101: 6118–6123.

Javaheri, M., Jenneman, G.E., McInnerey, M.J., and Knapp, R.J. 1985. Anaerobic production of a biosurfactant by *Bacillus licheniformis*. *Applied and Environmental Microbiology* 50: 698–700.

Jenny, K., Kappeli, O., and Fiechter, A. 1991. Biosurfactants from *Bacillus licheniformis* structural analysis and characterization. *Applied Microbiology and Biotechnology* 36: 5–13.

Jing, W., Guang, J., Jing, T., Hongdan, Z., Hanping, D., and Li, Y. 2011. Functional characterization of a biosurfactant-producing thermo-tolerant bacteria isolated from an oil reservoir. *Petroleum Science* 8: 353–356.

Kakinuma, A., Oachida, A., Shima, T., Sugino, H., Isano, M., Tamura, G., and Arima, K. 1969. Confirmation of the structure of surfactin by mass spectrometry. *Agricultural and Biological Chemistry* 33: 1669–1672.

Kim, K.M., Lee, J.Y., Kim, C.K., and Kang, J.S. 2009. Isolation and characterization of surfactin produced by *Bacillus polyfermenticus* KJS-2. *Archives of Pharmaceutical Research* 32: 711–715.

Kinsinger, R.F., Kearns, D.B., Hale, M., and Fall, R. 2005. Genetic requirements for potassium ion-dependent colony spreading in *Bacillus subtilis. Journal of Bacteriology* 187: 8462–8469.

Kinsinger, R.F., Shirk, M.C., and Fall, R. 2003. Rapid surface motility in *Bacillus subtilis* is dependent on extracellular surfactin and potassium ion. *Journal of Bacteriology* 185: 5627–5631.

Kiran, G.S., Thomas, T.A., Selvin, J., Sabarathnam, B., and Lipton, A.P. 2010. Optimization and characterization of a new lipopeptide biosurfactant produced by *marine Brevibacterium aureum* MSA13 in solid state culture. *Bioresource Technology* 101: 2389–2396.

Knoblich, A., Matsumoto, M., Ishiguro, R., Murata, K., Fujiyoshi, Y., Ishigami, Y., and Osman, M. 1995. Electron cryo-microscopic studies on micellar shape and size of surfactin, an anionic lipopeptide. *Colloids and Surfaces B* 5: 43–48.

Kuiper, I., Lagendijk, E.L., Pickford, R., Derrick, J.P., Lamers, G.E.M., Thomas-Oates, J.E., Lugtenberg, B.J.J., and Blowemberg, G.V. 2004. Putisolvin I and I which inhibit biofilm formation and break down existing biofilms. *Molecular Microbiology* 51(1): 97–113.

Lang, S. and Wagner, F. 1987. Structure and properties of biosurfactants. In *Biosurfactants and Biotechnology,* Kosaric, N., Cairns, W.L., and Gray, N.C.C. (eds.). Marcel Dekker, New York, pp. 21–45.

Layman, P.L. 1985. Industrial surfactants set for strong growth. *Chemical and Engineering News* 63: 23–48.

Li, X., Li, A., Liu, C., Yang, J., Ma, F., Hou, N., Xu, Y., and Ren, N. 2012. Characterization of the extracellular biodemulsifier of *Bacillus mojavensis* XH1 and the enhancement of demulsifying efficiency by optimization of the production medium composition. *Process Biochemistry* 47: 626–634.

Lima, T.M.S., Procopio, L.C., Brandao, F.D., Carvalho, A.M.X., Totola, M.R., and Borges, A.C. 2011a. Biodegradability of bacterial surfactants. *Biodegradation* 22: 585–592.

Lima, T.M.S., Procopio, L.C., Brandao, F.D., Carvalho, A.M.X., Totola, M.R., and Borges, A.C. 2011b. Simultaneous phenanthrene and cadmium removal from contaminated soil by a ligand/biosurfactant solution. *Biodegradation* 22: 1007–1015.

Lin, S.C. 1996. Biosurfactant: Recent advances. *Journal of Chemical Technology and Biotechnology* 63: 109–120.

Lin, S.C., Lin, K.C., Lo, C.C., and Lin, Y.M. 1998. Enhanced biosurfactant production by a *Bacillus licheniformis* mutant. *Enzyme and Microbial Technology* 23: 267–273.

Liu, Q., Yang, H., Wang, J., Gong, G., Zhou, W., Fan, Y., Wang, L., Yao, J., and Yu, Z. 2006. A mutant of *Bacillus subtilis* with high-producing surfactin by ion beam implantation. *Plasma Science and Technology* 8: 491–496.

Liu, T., Montastruc, L., Gancel, F., Zhao, L., and Nikov, I. 2007. Integrated process for production of surfactin. Part I: Adsorption rate of pure surfactin onto activated carbon. *Biochemical Engineering Journal* 35: 333–340.

Liu, X., Ren, B., Chen, M., Wang, H., Kokare, C.R., Zhou, X., Wang, J. et al. 2010. Production and characterization of a group of bioemulsifiers from the marine *Bacillus velezensis* strain H3. *Applied Microbiology and Biotechnology* 87: 1881–1893.

Magetdana, R. and Ptak, M. 1992. Interfacial properties of surfactin. *Journal of Colloid and Interface Science* 153(1): 285–291.

Makkar, R.S. and Cameotra, S.S. 1997a. Biosurfactant production by a thermophilic *Bacillus subtilis* strain. *Journal of Industrial Microbiology and Biotechnology* 18: 37–42.

Makkar, R.S. and Cameotra, S.S. 1997b. Utilization of molasses for biosurfactant production by two *Bacillus* strains at thermophilic conditions. *Journal of American Oil Chemical Society* 74: 887–889.

Makkar, R.S. and Cameotra, S.S. 1999. Biosurfactant production by microorganisms on unconventional carbon sources. *Journal of Surfactants and Detergents* 2(2): 237–241.

Makkar, R.S. and Cameotra, S.S. 2001. Synthesis of enhanced biosurfactant by *Bacillus subtilis* MTCC 2423 at 45°C by foam fraction. *Journal of Surfactants and Detergents* 4: 355–357.

Makkar, R.S. and Cameotra, S.S. 2002. Effects of various nutritional supplements on biosurfactant production by a strain of *Bacillus subtilis* at 45°C. *Journal of Surfactants and Detergents* 5: 11–17.

Martinov, M., Gancel, F., Jacques, P., Nikov, I., and Vlaev, S. 2008. Surfactant effect on aeration performance of stirred tank reactors. *Chemical Engineering Technology* 31: 1494–1500.

Matsuyama, T., Kaneda, K., Nakagawa, Y., Isa, K., Harahotta, H., and Yano, I. 1992. A novel extracellular cyclic lipopeptide which promotes flagellum-dependent and flagellum-independent spreading growth of *Serratia marcescens*. *Journal of Bacteriology* 174(6): 1769–1776.

Montastruc, L., Liu, T., Gancel, F., Zhao, L., and Nikov, I. 2008. Integrated process for production of surfactin Part 2: Equilibrium and kinetic study of surfactin adsorption onto activated carbon. *Biochemical Engineering Journal* 38: 349–354.

Morikawa, M., Daido, H., Takao, T., Murata, S., Shimonishi, Y., and Imanaka T. 1993. A new lipopeptide biosurfactant produced by *Arthrobacter* sp. Strain MIS38. *Journal of Bacteriology* 175(20): 6459–6466.

Mukherjee, C., Das, P., Sivapathasekaram, C., and Sen, R. 2009. Antimicrobial biosurfactants from marine *Bacillus circulans*: Extracellular synthesis and purification. *Letters of Applied Microbiology* 48: 281–288.

Mukherjee, S., Das, P., and Sen, R. 2006. Towards commercial production of microbial surfactants. *Trends in Biotechnology* 24(11): 509–515.

Mulligan, C.N., Chow, T.Y.-K., and Gibbs, B.F. 1989. Enhanced biosurfactant production by a mutant *Bacillus subtilis* strain. *Applied Microbiology and Biotechnology* 31: 486–489.

Mulligan, C.N., Cooper, D.G. and Neufeld, R.J. 1984. Selection of microbes producing biosurfactants in media without hydrocarbons. *Journal of Fermentation Technology* 62: 311–314.

Mulligan, C.N. and Gibbs, B.F. 1990. Recovery of biosurfactants by ultrafiltration. *Journal of Chemical Technology and Biotechnology* 47: 23–29.

Mulligan, C.N. and Gibbs, B.F. 1993. Factors influencing the economics of biosurfactants. In *Biosurfactants, Production, Properties, Applications*, Kosaric, N. (ed.). Marcel Dekker, New York, pp. 329–371.

Mulligan, C.N., Yong, R.N., and Gibbs, B.F. 1999. Metal removal from contaminated soil and sediments by the biosurfactant surfactin. *Environmental Science and Technology* 33(21): 3812–3820.

Najafi, A.R., Rahimpour, M.R., Jahanmiri, A.H., Roostaazad, R., Arabian, D., and Ghobadi, Z. 2010. Enhancing biosurfactant production from an indigenous strain of *Bacillus mycoides* by optimizing the growth conditions using a response surface methodology. *Chemical Engineering Journal* 163: 188–194.

Nakayama, S., Takahashi, S., Hirai, M., and Shoda, M. 1997. New variants of surfactin, by a recombinant *Bacillus subtilis*. *Applied and Microbiology and Biotechnology* 48(1): 80–82.

Naruse, N., Tenmyo, O., Kobaru, S., Kamei, H., Miyaki, T., Konishi, M., and Oki, T. 1990. Pumilacidin, a complex of new antiviral antibiotics-production, isolation, chemical properties, structure and biological activity. *Journal of Antibiotics* 43(3): 267–280.

Noah, K.S., Bruhn, D.F., and Bala, G.A. 2005. Surfactin production from potato process effluent by *B. subtilis* in a chemostat. *Applied Biochemistry and Biotechnology* 122: 465–473.

Noah, K.S., Fox, S.I., Bruhn, D.F., Thompson, D.N., and Bala, G.A. 2002. Development of continuous surfactin production from potato process effluent by *Bacillus subtilis* in an airlift reactor. *Applied Biochemistry and Biotechnology* 98–10: 803–813.

Ohno, A., Ano, T., and Shoda, M. 1992. Production of a lipopeptide antibiotic surfactin with recombinant *Bacillus subtilis*. *Biotechnology Letters* 14: 1165–1168.

Ohno, A., Ano, T., and Shoda, M. 1995. Production of a lipopeptide antibiotic surfactin by recombinant *Bacillus subtilis* in solid state fermentation. *Biotechnology and Bioengineering* 47(2): 209–214.

Olivera, N.L., Commendatore, M.G., Moran, A.C., and Esteves, J.L. 2000. Biosurfactant-enhanced degradation of residual hydrocarbons from ship bilge wastes. *Journal of Industrial Microbiology and Biotechnology* 25: 70–73.

Pecci, Y., Rivardo, F., Martinotti, M.G., and Allegrone, G. 2010. LC-ESI-MS/MS characterization of lipopeptide biosurfactants produced by the *Bacillus licheniformis* V9T14 strain. *Journal of Mass Spectrometry* 45: 772–778.

Pemmaraju, S.C., Sharma, D., Singh, N., Panwar, R., Cameotra, S.S., and Pruthi, V. 2012. Production of microbial surfactants from oily sludge-contaminated soil by *Bacillus subtilis* DSVP23. *Applied Biochemistry and Biotechnology* 167: 1119–1131.

Peng, F., Wang, Y., Sun, F., Liu, Z., Lai, Q., and Shao, Z. 2008. A novel lipopeptide produced by a Pacific Ocean deep-sea bacterium, *Rhodococcus* sp. TW53. *Journal of Applied Microbiology* 105(3): 698–705.

Peypoux, F., Bonmatin, J.M., and Wallach, J. 1999. Recent trends in biochemistry of surfactin. *Applied Microbiology and Biotechnology* 51(5): 553–563.

Portillo-Rivera, O.M., Teliez-Luis, S.J., Ramirez de Leon, J.A., and Vasquez, M. 2009. Production of microbial transglutaminase on media made from sugar cane molasses and glycerol. *Food Technology and Biotechnology* 47: 19–26.

Raaijmakers, J.M., De Bruijn, I., and De Kock, M.J.D. 2006. Cyclic lipopeptide production by plant-associated *Pseudomonas* spp.: Diversity, activity, biosynthesis, and regulation. *Molecular Plant-Microbe Interactions* 19: 699–710.

Razafindralambo, H., Paquot, M., Baniel, A., Popineau, Y., Hbid, C., Jacques, P., and Thonat, P. 1996. Foaming properties of surfactin, a lipopeptide from *Bacillus subtilis*. *JAOCS* 73: 149–151.

Reddy, A.S., Chen, C.-Y., Baker, S.C., Chen, C.-C., Jean, J.-S., Fan, C.-W., Chen, H.-R., and Wang, J.-C. 2009. Synthesis of silver nanoparticles using surfactin: A biosurfactant stabilizing agent. *Material Letters* 63: 1227–1230.

Reid, G., Gan, B.S., She, Y.-M., Ens, W., Weinberger, S., and Howard, J.C. 2002. Rapid identification of probiotic lactobacillus biosurfactant proteins by ProteinChip tandem mass spectrometry tryptic peptide sequencing. *Applied and Environmental Microbiology* 68: 977–980.

Rivardo, F., Martinotti, M.G., Turner, R.J., and Ceri, H. 2010. The activity of silver against *Escherichia coli* biofilm is increased by a lipopeptide biosurfactant. *Canadian Journal of Microbiology* 56: 272–278.

Roongsawang, N., Hase, K., Haruki, M., Imanaka, T., Morikawa, M., and Kanaya, S. 2003. Cloning and characterization of the gene cluster encoding arthrofactin synthetase from *Pseudomonas* sp. MIS38. *Chemistry and Biology* 10(9): 869–880.

Rosen, M.J. 1978. *Surfactants and Interfacial Phenomena*. John Wiley & Sons, New York.

Rosenberg, E. 1986. Microbial surfactants. *CRC Critical Reviews in Biotechnology* 1(2): 87–107.

Rufino, R.D., Luna, J.M., Campos-Takaki, G.M., Ferreira, S.R.M., and Sarubbo, L.A. 2012. Application of the biosurfactant produced by *Candida lipopolytica* in the remediation of heavy metals. *Chemical Engineering Transactions* 27: 61–66.

Saimmai, A., Onlamool, T., Sobhon, V., and Maneerat, S. 2013. An efficient biosurfactant-producing bacterium *Selenomonas ruminantium* CT2 isolated from mangrove sediment in south of Thailand. *World Journal of Microbiology and Biotechnology* 29(1): 87–102.

Sang-Cheol, L., Kim, S.-H., Park, I-H., Chung, S-Y., Chandra, M.S., and Yon-Lark, C. 2010. Isolation, purification, and characterization of novel fengycin S from LSC04 degrading-crude oil. *Biotechnology and Bioprocess Engineering* 15: 246–253.

Sandrin, C., Peypoux, F., and Michel, G. 1990. Coproduction of surfactin and iturin A lipopeptides with surfactant and antifungal properties by *Bacillus subtilis*. *Biotechnology and Applied Biochemistry* 12: 370–375.

Santos, C.F.C., Pastore, G.M., Damasceno, S., and Cereda, M.P. 2000. Producao de biosurfactantes por linhagens de *Bacillus subtilis* utilizando manipueira como susbtrato. *Revista ciencia e Tecnologia de Alimentos* 33: 157–161.

Satpute, S.K., Banat, I.M., Dhakephalkar, P.K., Banpurkar, A.G., and Chopade, B.A. 2010. Biosurfactants, bioemulsifiers and exopolysaccharides from marine microorganisms. *Biotechnology Advances*, 28: 436–450.

Sen, R. and Swaminathan, T. 1997. Application of response-surface methodology to evaluate the optimum environmental conditions for the enhanced production of surfactin. *Applied Microbiology and Biotechnology* 47: 358–363.

Sen, R. and Swaminathan, T. 2004. Response surface modeling and optimization to elucidate and analyse the effects of inoculum age and size on surfactin production. *Biochemical Engineering Journal* 21: 141–148.

Sen, R. and Swaminathan, T. 2005. Characterization of concentration and purification parameters and operating conditions for the small-scale recovery of surfactin. *Process Biochemistry* 40: 2853–2958.

Seydlová, G. and Svobodová, J. 2008. Review of surfactin chemical properties and the potential biomedical application. *Central European Journal of Medicine* 3: 123–133.

Shaligram, N.S. and Singhal, R.S. 2010. Surfactin—A review, on biosynthesis, fermentation, purification and applications. *Food Technology and Biotechnology* 48: 119–134.

Shen, H.-H., Thomas, R.K., Chen, C.-Y., Darton, R.C., Baker, S.C., and Penfold, J. 2009. Aggregation of the naturally occurring lipopeptide, surfactin at interfaces and in solution: An unusual type of surfactant? *Langmuir* 25: 4211–4218.

Shen, H.H., Thomas, R.K., Penfold, J., and Fragneto, G. 2010a. Destruction and Solubilization of Supported Phospholipid Bilayers on Silica by the Biosurfactant Surfactin. *Langmuir* 26: 7334–7342.

Shen, H.H., Thomas, R.K., and Taylor, P. 2010b. The Location of the Biosurfactant Surfactin in Phospholipid Bilayers Supported on Silica Using Neutron Reflectometry. *Langmuir* 26: 320–327.

Sheppard, J.D. and Mulligan, C.N. 1987. The production of surfactin by *Bacillus subtilis* grown on peat hydrolysates. *Applied Microbiology and Biotechnology* 27: 110–116.

Simpson, D.R., Natraj, N.R., McInerney, M.J., and Duncan, K.E. 2011. Biosurfactant-producing *Bacillus* are present in produced brines from Oklahoma oil reservoirs with a wide range of salinities. *Applied Microbiology and Biotechnology* 91: 1083–1093.

Singh, P. and Cameotra, S.S. 2004. Potential applications of microbial surfactants in biomedical sciences. *Trends in Biotechnology* 22: 142–146.

Sivapathasekaran, C., Mukherjee, S., Ray, A., Gupta, A., and Sen, R. 2010. Artificial neural network modeling and genetic algorithm based medium optimization for the improved production of marine biosurfactant. *Bioresource Technology* 101: 2884–2887.

Sivapathasekaran, C., Mukherjee, S., Samanta, R., and Sen, R. 2009. High-performance liquid chromatography purification of biosurfactant isoforms produced by a marine bacterium. *Analytical and Bioanalytical Chemistry* 395: 845–854.

Sivapathasekaran, C., Mukherjee, S., Sen, R., Bhatacharya, B., and Samanta, R. 2011. Single step concomitant concentration, purification and characterization of two families of lipopeptides of marine origin. *Bioprocess and Biosystem Engineering* 34: 339–346.

Sorenson, D., Nielsen, T.H., Christophersen, C., Sorensen, K., and Gaijhede, M. 2001. Cyclic lipoundecapeptide amphisin from *Pseudomonas* sp. DSS73. *Acta Crystallography* 57: 1123–1124.

Sriram, M.I., Gayathiri, S., Gnanaselvi, U., Jenfer, P.S., Raj, S.M., and Gurunathan, S. 2011a. Novel lipopeptide biosurfactant produced by hydrocarbon degrading and heavy metal tolerant bacterium *Escherichia fergusonii* KLU01 as a potential tool for bioremediation. *Bioresource Technology* 102: 9291–9295.

Sriram, M.I., Kalishwaralal, K., Deepak, V., Gracerosepat, R., Srisakthi, K., and Gurunathan, S. 2011b. Biofilm inhibition and antimicrobial action of lipopeptide biosurfactant produced by heavy metal tolerant strain *Bacillus cereus* NK1. *Colloids and Surfaces B: Biointerfaces* 85: 174–181.

Syldatk, C. and Wagner, F. 1987. Production of biosurfactants. In *Biosurfactants and Biotechnology, Surfactant Science Series,* Vol. 25, Kosaric, N., Cairns, W.L., and Gray, N.C.C. (eds.). Marcel Dekker, New York, pp. 89–120.

Tanaka, T., Fujita, K.I., Takenishi, S., and Taniguchi, M. 1997. Existence of an optically heterogeneous peptide unit in poly-(γ-glutamic acid) produced by *Bacillus subtilis. Journal of Fermentation and Bioengineering* 84: 361–364.

Thavasi, V.R.M., Nambaru, S., Jayalakshmi, S., Balasubramanian, T., and Banat, I.M. 2009. Biosurfactant production by *Azotobacter chroococcum* isolated from the marine environment. *Marine Biotechnology* 11(5): 551–556.

Thimon, L., Peypoux, F., and Michel, G. 1992. Interactions of surfactin, a biosurfactant from *Bacillus subtilis* with inorganic cations. *Biotechnology Letters* 14: 713–718.

Thompson, D.N., Fox, S.L., and Bala, G.A. 2001. The effects of pretreatments on surfactin production from potato process effluent by *Bacillus subtilis. Applied Biochemistry and Biotechnology* 91–93: 487–501.

Tsuge, K., Ohata, Y., and Shoda, M. 2001. Gene *yerP*, involved in surfactin self-resistance in *Bacillus subtilis. Antimicrobial Agents and Chemotherapy* 45: 3566–3573.

Vanittanakom, N., Loeffler, W., Koch, U., and Jung, G. 1986. Fengycin—A novel antifungal lipopeptide antibiotic produced by *Bacillus subtilis* F-29-3. *Journal of Antibiotics* 39: 888–901.

Vater, J., Kablitz, B., Wilde, C., Franke, P., Metha, N., and Cameotra, S.S. 2002. Matrix-assisted laser desorption ionization-time of flight mass spectrometry of lipopeptide biosurfactants in whole cells and culture filtrates of *Bacillus subtilis* C-1 isolated from petroleum sludge. *Applied and Environmental Microbiology* 68: 6210–6219.

Vollenbroich, D., Pauli, G., Ozel, M., and Vater, J. 1997. Antimycoplasma properties and application on cell culture of surfactin, a lipopeptide antibiotic from *Bacillus subtilis. Applied and Environmental Microbiology* 63: 44–69.

Wang, J., Haddad, N.I.A., Yang, S.-Z., and Mu, B.-Z. 2010. Structural characterization of lipopeptides from *Brevibacillus brevis* HOB1. *Applied Biochemistry and Biotechnology.* 160: 812–821.

Washio, K., Lim, S.P., Roongsawang, N., and Morikawa, M. 2010. Identification and characterization of the genes responsible for the production of the cyclic lipopeptide arthrofactin by *Pseudomonas* sp. MIS38. *Bioscience, Biotechnology and Biochemistry* 74: 992–999.

Wei, Y.-H. and Chu, I.-M. 1998. Enhancement of surfactin production in iron-enriched media by *Bacillus subtilis* ATCC 21332. *Enzyme and Microbiological Technology* 22: 724–728.

Wei, Y.H. and Chu, I.M. 2002. Mn^{2+} improves surfactin production by *Bacillus subtilis. Biotechnology Letters* 24: 479–482.

Wei, Y.H., Chu, I.M., and Chang, J.S. 2007. Using taguchi experimental design methods to optimize trace element composition for enhanced surfactin production by *Bacillus subtilis* ATCC 21332. *Process Biochemistry* 42: 40–45.

Wei, Y.-H., Wang, L.-C., Chen, W.-C., and Chen, S.-Y. 2010. Production and characterization of fengycin by indigenous *Bacillus subtilis* F29-3 originating from a potato farm. *International Journal of Molecular Science* 11: 4526–4538.

Winterburn, J.B. and Martin, P.J, 2012. Foam mitigation and exploitation in biosurfactant production. *Biotechnology Letters* 34(2): 187–95.

Yakimov, M.M., Timmis, N., Wray, V., and Fredickson, H.L. 1995. Characterization of lipopeptide surfactant produced by thermotolerant and halotolerant subsurface *Bacillus licheniformis* BAS50. *Applied and Environmental Microbiology* 61: 1706–1713.

Yeh, M.S., Wei, Y.H., and Chang, J.S. 2005. Enhanced production of surfactin from *Bacillus subtilis* by addition of solid carriers. *Biotechnology Progress* 21: 1329–1334.

Yeh, M.S., Wei, Y.-H., and Chang, J.-S. 2006. Bioreactor design for enhanced carrier-assisted surfactin production with *Bacillus subtilis*. *Process Biochemistry* 41: 1799–1805.

Yoneda, T., Miyota, Y., Furuya, K., and Tsuzuki, T. 2006. Production process of surfactin. US Patent 7011969.

Youssef, N.H., Nguyen, T., Sabatini, S.A., and McInerney, M.J. 2007. Basis for formulating biosurfactant mixtures to achieve ultra low interfacial tension values against hydrocarbons. *Journal of Industrial Microbiology and Biotechnology* 34(7): 497–507.

Zhao, Z., Wang, Q., Wang, K., Brain, K., Liu, C., et al. 2010. Study of the antifungal activity of *Bacillus vallismortis* ZZ185 in vitro and identification of its antifungal components. *Bioresource Technology* 101: 292–297.

7 Biosurfactants in the Food Industry

Marcia Nitschke and Siddhartha G.V.A.O. Costa

CONTENTS

INTRODUCTION

Surfactants have been used in food industry as cleaning agents, and in food formulations as emulsifiers and fat substitutes. They comprise a significant part of food additive market, which is projected to exceed $33.9 billion by 2015 (IFT, 2012). Over the past few years, the demand of food additives follows the increasing awareness of customers who desire, more and more, for natural and environmental friendly food ingredients.

The properties demonstrated by microbial biosurfactant (BS) such as high biodegradability, low toxicity, low CMC, high surface activity, stability to extreme pH, temperature, salt concentrations, and biological activity are very useful for food industry (Bognolo, 1999). Thus, BS can replace conventional synthetic surfactants with great advantages, and additionally, their bio-based origin confers to these molecules the status of natural additives fulfilling the actual market needs.

Rhamnolipid Inc. (www.rhamnolipid.com), Saraya Co. (www.saraya.com), AGAE Technologies (www.agaetech.com), and Urumqi Unite Biotechnology Co. (www.rhamnolipid-biosurfactant.com) are some examples of BS suppliers found in international market, which offer different formulations for specific industrial needs. Lack of cost competitiveness when compared to synthetics has been overcome by the novel market trends and regulatory environmental requirements; consequently, the number of companies interested in exploring BS is increasing, and renowned surfactant vendors such BASF and Ecover have already ventured in BS market (Transparency Market Research, 2012). However, economically large-scale production of BS remains a challenge, and efforts have been done to minimize the production costs in order to facilitate commercial use. The use of cheap or waste substrates as raw material in an industrial biotransformation can be an attractive strategy for economical BS production.

Considering the potential of BSs as new promising alternatives to food industry, this chapter discusses their application as agents to form emulsions, to control microbial pathogen growth, adhesion and biofilm formation, as well as their production using wastes or by-products from food processing chain.

USE OF BIOSURFACTANTS IN FOOD PROCESSING

BIOFILM PREVENTING AND BIOFILM DISRUPTION

The perception of microbes is essentially based on the concept of pure cultures, where these organisms can be studied individually. Actually, it is known that in nature, most microorganisms are found associated to communities with different complexity degrees, and the planktonic existence seems to be only eventual (Jenkinson and Lappin-Scott, 2001). A biofilm can be defined as a sessile community of microbes that are irreversibly associated with a surface, embedded in a matrix of extracellular polymeric material (EPM) (Costerton et al., 1999).

Biofilm establishment involves a sequence of events that can be summarized in the following steps (Stoodley et al., 2002; Cloete et al., 2009):

a. Reversible adhesion: Individual cells come close and interact with a surface guided by long-range forces such as van der Waals, electrostatic, and hydrophobic (Palmer et al., 2007). At this point, cells can be detached and become planktonic and also can be easily removed by rinsing.

b. Irreversible adhesion: After initial adhesion, loosely bound organisms must maintain contact with the surface, firmly attach, and grow to form a biofilm. Irreversible attachment is mediated by short-range forces such as dipole–dipole, hydrogen, ionic, and covalent bonds (Palmer et al., 2007). The transition from weak to strong interactions with the surface is often mediated by the production of EPM, which consists of diverse polymers including polysaccharides, proteins, nucleic acids, and phospholipids. After irreversible attachment, rinsing will no longer remove the cells, and they should be removed by scraping the surface.

c. Development of biofilm architecture: Cells multiply and aggregate to form microcolonies and send out chemical signs to increase extracellular matrix production. Further development of macrocolonies and the formation of water channels take place.

d. Biofilm maturation: As biofilms mature, a tridimensional structure comprising channels and pores are developed, and the cells are redistributed away from the surface. The density and complexity increase as microorganisms begin to interact with molecules in the surrounding environment. The growth is limited by nutrient availability, oxygen, pH, temperature, osmolarity, and adequate hydrodynamic flow across the biofilm (Carpentier and Cerf, 1993).

e. Detachment and dispersion: As biofilms reach a critical mass, a dynamic equilibrium in outermost layers begins to generate planktonic organism that are free to colonize other surfaces. Exhaustion of nutrients, production of polymeric matrix degrading enzymes, fluid shear stress, and microbially generated gas bubbles are some factors important to biofilm detachment (Hunt et al., 2004).

The formation of biofilms is regulated by signaling molecules that accumulate and induce the expression of certain genes in response to population density (Annous et al., 2009). This quorum-sensing regulation is also involved in the detachment of cells from the biofilms (Hunt et al., 2004).

From an ecological point of view, it is advantageous to microorganisms to live associated on biofilms once they are more protected from adverse environmental conditions (Maukonen et al., 2003). On biofilms, microorganisms are also less susceptible to sanitizers, antibiotics, host defenses, and shear forces when compared to planktonic cells, and they also may contribute to the increasing resistance to antimicrobials (Bagge-Ravn et al., 2003; Hall-Stoodley et al., 2004; Simões et al., 2006).

Biofilms are a great concern among food processing industries; once they establish, they can result in the obstruction of pipelines, corrosion of equipments, reducing efficiency of temperature transfer systems, and in addition, to spoilage and contamination of final products with pathogens (Kumar and Anand, 1998), affecting not only the economics of the process but also the consumers' health.

Food preparation surfaces are usually in contact with organic matter such as milk and meat, which are a great source of nutrients for microorganisms, and also with inorganic materials. Inadequate cleaning procedures may lead to the accumulation of residues to equipment and surfaces promoting the initial adhesion and further biofilm development (Bagge-Ravn et al., 2003). When a fluid comes into contact with a surface, the first step that occurs is adsorption of molecules to the surface. This step is usually called conditioning. Surface conditioning is generally related to the fluid bulk composition and may have inhibiting or stimulating effects for the adhesion of microorganisms to surface (Hood and Zottola, 1995).

The aim of food industry is to provide a high-quality and safety food to consumers, and microbial control is essential to reach this goal (Hood and Zottola, 1995; Bagge-Ravn et al., 2003). Food processors have zero tolerance levels to pathogens as *Salmonella* spp. and *Listeria monocytogenes*, and the presence of a single cell may be as important as a well-developed biofilm (Hood and Zottola, 1995; Nitschke and Costa, 2007); thus, it is necessary to find economical and efficient ways to clean and disinfect food-contact surfaces. Unfortunately, even with cleaning and sanitization procedures consistent with good manufacturing practices, microorganisms could remain on equipment surfaces and survive for prolonged periods (Cloete et al., 2009), so it is imperative to search for new biofilm controlling strategies.

Some studies have shown that the conditioning of surfaces with BS can significantly reduce microbial adhesion and inhibit or reduce subsequent biofilm development.

Surfactin was tested by Mirelles et al. (2001) as an antiadhesive agent to inhibit biofilm development on catheters by *Salmonella enterica*, *Escherichia coli*, *Proteus mirabilis*, and *P. aeruginosa*. The total inhibition by preconditioning the surface was observed for all bacteria, but not for *P. aeruginosa*.

The preconditioning of stainless steel and poly(tetrafluoroethylene) PTFE surfaces with a BS obtained from *P. fluorescens* inhibits the *L. monocytogenes* L028 adhesion. A significant reduction (>90%) was attained in microbial adhesion levels on stainless steel, whereas no significant effect was observed in PTFE (Meylheuc et al., 2001). Further work demonstrated that the prior adsorption of *P. fluorescens* surfactant on stainless steel also favored the bactericidal effect of disinfectants (Meylheuc et al., 2006a). The ability of adsorbed BSs, obtained from gram-negative (*P. fluorescens*) and gram-positive (*L. helveticus*) bacteria isolated from foodstuffs, on inhibiting the *L. monocytogenes* adhesion to stainless steel was also investigated. Adhesion tests showed that both BSs were effective, strongly decreasing the surface contamination level. The antiadhesive biological coating reduced both the total adhering flora and the viable/cultivable adherent *L. monocytogenes* on stainless steel surfaces (Meylheuc et al., 2006b).

The adhesion of *E. coli* CFT073 and *Staphylococcus aureus* ATCC 29213 to polystyrene was reduced by 97% and 90% respectively using BSs obtained from *Bacillus subtilis* and *B. licheniformis*. The antiadhesive activity was observed either by coating the surface or by adding the BS to the inoculum (Rivardo et al., 2009). A BS obtained from *Lactobacillus paracasei* A20 was able to reduce adhesion of *S. aureus* to polystyrene by 76.8%, whereas it showed low activity against *P. aeruginosa* and *E. coli* (Gudiña et al., 2010).

An interesting work regarding the antiadhesive activity of BSs was conducted by Shakerifard et al. (2009). The authors evaluated the ability of lipopeptide surfactants from *B. subtilis* to modify the hydrophobicity of Teflon and stainless steel surfaces and its correlation with the adhesion of *B. cereus* spores. They concluded that there is a good correlation between surface hydrophobicity modifications promoted by BSs and attachment of spores. The best results were shown to iturin A, which reduce the adhesion of spores to Teflon by 6.5-fold at 100 mg/L.

The prior adsorption of surfactin on polypropylene and stainless steel surfaces reduced the adhesion the food pathogens *L. monocytogenes*, *S. enteritidis*, and *Enterobacter sakazakii*. The number of adhered cells of *L. monocytogenes* on stainless steel was reduced by two log units (Nitschke et al., 2009). Zeraik and Nitschke (2010) evaluated the effect of temperature on antiadhesive activity of surfactin against *S. aureus*, *L. monocytogenes*, and *M. luteus* in polystyrene surfaces. The authors demonstrated that the decrease in temperature increases the antiadhesive activity of surfactin especially at 4°C (63%–66% reduction), which is important once pathogens like *L. monocytogenes* can grow at this temperature.

Araújo et al. (2011) evaluated the antiadhesive activity of surfactin and rhamno-lipids against 15 *L. monocytogenes* strains. Polystyrene plates were preconditioned with the BSs, and further bacterial adhesion was measured during different time intervals. The adhesion of *L. monocytogenes* ATCC 7644 was reduced by 84% using surfactin, while rhamnolipids decreased the adhesion of *L. monocytogenes* ATCC15313 by 82%. The antiadhesive activity observed was not related to cell growth inhibition, and both BSs show potential to prevent and retard pathogen colonization of the surface.

A recent work describes the reduction on adhesion of individual and mixed cultures of *S. aureus*, *L. monocytogenes*, and *S. enteritidis* on polystyrene by the use of surfactin and rhamnolipids. The preconditioning with surfactin reduced the adhesion of *L. monocytogenes* and *S. enteritidis* by 42%, whereas the treatment using rhamnolipids reduced the adhesion of *L. monocytogenes* by 57.8% and the adhesion of *S. aureus* by 67.8%. BSs were less effective to avoid the adhesion of mixed cultures when compared to individual cultures (Gomes and Nitschke, 2012). Concerning their antiadhesive activity, the preconditioning using BSs is able to modify the surface of the materials altering the hydrophobicity and the acid/base character, important factors involved in bacterial adhesion. However, antiadhesive activity is dependent on the type of microorganism and the surface involved, temperature, pH, salts, nutrients, and the presence of cellular appendages as pili and flagella (Goulter et al., 2009).

Another important approach to be explored is the use of BSs to remove (disrupt) preformed biofilms. In order to be effective in removing biofilms, BSs have to penetrate into the interface between solid substrate and the biofilm so they could adsorb and reduce interfacial tension. The attractive interactions between bacterial and solid surfaces may be decreased leading to the removal of biofilm (McLansborough et al., 2006).

Kuiper et al. (2004) characterized two lipopeptide BSs produced by *Pseudomonas putida* PCL1445 and verified that both BSs were able to inhibit biofilm formation from different *Pseudomonas* species and strains on PVC surfaces. Furthermore, these BSs were also able to disrupt preexisting biofilms.

Irie et al. (2005) reported the disruption of *Bordetella bronchiseptica* biofilms using rhamnolipids, and Dusane et al. (2010) have demonstrated that *B. pumilus* biofilms were removed up to 93% using 100 mM rhamnolipids. A glycolipid BS produced by a marine *Brevibacterium casei* was effective in the disruption 24 h-old biofilms of *Vibrio parahaemolyticus*, *P. aeruginosa*, and *E. coli* formed on glass coverslips (Kiran et al., 2010).

Gomes and Nitschke (2012) evaluated the disruption of 48 h-old biofilms of individual and mixed cultures of food pathogens established on polystyrene surfaces using rhamnolipids and surfactin. After 2 h contact with 0.1% surfactin, the preformed biofilms of *S. aureus* were reduced by 63.7%, *L. monocytogenes* by 95.9% (Figure 7.1), *S. enteritidis* by 35.5%, and the mixed culture by 58.5% (Figure 7.2). The rhamnolipids at 0.25% concentration removed 58.5% of the biofilm of *S. aureus*, 26.5% of *L. monocytogenes*, 23.0% of *S. enteritidis*, and 24.0% the mixed culture. Authors found that the increase in the concentration of BS and the time of contact

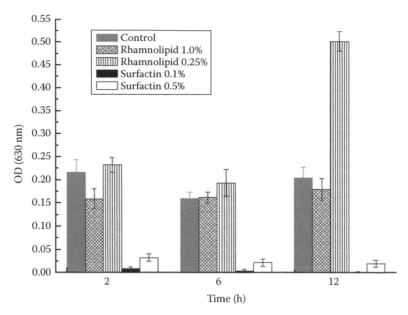

FIGURE 7.1   Removal of *Listeria monocytogenes* biofilms from polystyrene surfaces by surfactin and rhamnolipid biosurfactants after different contact times.

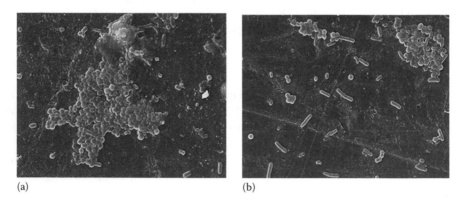

FIGURE 7.2   SEM images of 48 h biofilm of mixed culture of *L. monocytogenes*, *S. aureus*, and *S. enteritidis* established on polystyrene surface: (a) without biosurfactant addition and (b) after 2 h contact with surfactin 0.1% solution (magnification 10,000×).

decreased biofilm removal percentage. The different susceptibilities of each bacterial biofilm to the BSs can be related to the amount and chemical composition of the polymeric material produced by the strains.

Tahmourespour et al. (2011) observed that a BS produced by *L. fermentum* was able to inhibit the production of extracellular glucans by *Streptococcus mutans*, a well-known biofilm-producing bacterium involved in dental caries. It was hypothesized that two enzymes produced by *S. mutans* for polymer synthesis were inhibited by the BS. The BS showed substantial antibiofouling activity reducing the process of attachment and biofilm production. This work demonstrates that the investigation of surface-active molecules and the study of their effect on biofilm population are an unexplored area of research, which can lead to the discovery of new molecules applied to specific bacterial control.

ANTIMICROBIAL ACTIVITY

Many studies pointed out the use of BSs as agents to control microbial growth. Most studies explored their activity against human pathogenic bacteria, but BSs have been shown to inhibit the growth of yeasts, fungi, viruses, and algae. The mechanism involved in the antimicrobial activity observed to BS remains unclear; nevertheless, researchers are conscious that most BSs act by disturbing the cytoplasmic membrane, as they have an amphipathic nature that allows their interaction with phospholipids, altering the permeability with consequent cell damage (Ortiz et al., 2006; Sotirova et al., 2008). Some reviews available in the literature discuss the biological applications of BSs including their antimicrobial activity (Rodrigues et al., 2006a; Kitamoto et al., 2009; Raaijmakers et al., 2010). In this section, we will focus on the antimicrobial potential exhibited by BS on controlling food-related pathogens.

Rhamnolipid BS demonstrated antimicrobial activity against several microorganisms such as the gram-positive bacteria, *S. aureus, B. subtilis, Clostridium perfringens*, the gram-negative bacteria *S. typhimurium, E. coli, E. aerogenes*, and the fungi *Phytophthora infestans, P. capsici, Botrytis cinerea, Fusarium graminearum*, and *Mucor* spp. (Haba et al., 2003; Benincasa et al., 2004; Sha et al., 2011). A glycolipid from *Brevibacterium casei* showed bactericidal effect against *E. coli* and *Vibrio alginolyticus* showing an MBC of 38 and 41 µg/mL respectively (Kiran et al., 2010). Rivardo et al. (2010) reported that a lipopeptide from *B. licheniformis* V9T14 increases the bactericidal activity of silver ions against *E. coli* biofilms reducing the biofilm population. This synergistic interaction could be useful in disinfecting surfaces to reduce bacterial colonization and spreading of disease.

Although many studies describe the antimicrobial activity of BSs, there is a lack of experiments showing this activity directly in food matrix. An example is the recent work performed by Manoharan et al. (2012) that reported the development of surfactin nanoemulsions evaluating their antimicrobial effect against food pathogens and in the microbial population of food products. Surfactin-sunflower-based nanoemulsion showed highest antibacterial activity against *S. typhi*, followed by *L. monocytogenes* and *S. aureus*. The nanoemulsion also demonstrated fungicidal activity against *Rhizopus nigricans, Aspergillus niger*, and *Penicillium* sp. and sporicidal activity against *B. cereus* and *B. circulans*. An in situ evaluation of antimicrobial activity of surfactin nanoemulsion on raw chicken, apple juice, milk, and mixed vegetables showed a significant reduction in the native cultivable bacterial and fungal populations of these products.

Huang et al. (2011) also demonstrated that a *S. enteritidis* strain was sensitive to surfactin and polylysine, with minimum inhibitory concentrations (MICs) of 6.25 and 31.25 µg/mL, respectively. *S. enteritidis* was reduced to six log-cycles in milk when the temperature was 4.45°C, the action time was 6.91 h, and the concentration (surfactin/polylysine ratio 1:1) was 10.03 µg/mL.

The combination of nisin with rhamnolipids has extended the shelf-life and inhibited thermophilic spores in UHT soymilk. The use of compositions comprising natamycin and rhamnolipids in salad dressing prolonged its shelf-life and inhibited mold growth. Compositions comprising natamycin, nisin, and rhamnolipids in cottage cheese have extended shelf-life by inhibiting mold and bacterial growth, especially gram-positive and spore-forming bacteria (Gandhi and Skebba, 2007).

Araújo et al. (2011) demonstrated that the rhamnolipids produced by *P. aeruginosa* PA1 inhibit the growth of *L. monocytogenes* ATCC 19112 and ATCC 7644, suggesting that it could be exploited as an agent to control this important food pathogen. Based on this preliminary observation, Magalhães and Nitschke (2013) evaluated the antimicrobial activity of rhamnolipids against 32 *L. monocytogenes* isolates. Among the 32 tested cultures, 90.6% were susceptible to RL, and the MIC values varied from 78.1 to 2500 µg/mL. Rhamnolipids' activity was primarily bacteriostatic, and its interaction with nisin was also investigated. The results obtained showed a particular synergistic effect improving the efficacy of both antimicrobials. The mechanism of action for the combination of nisin and RL was not elucidated; however, as both antimicrobials acts on the same target, authors hypothesized that the interaction could occur on the cytoplasm membrane.

These findings demonstrate that the combination of BSs with antimicrobials opens a new perspective for the development of control strategies to avoid undesirable planktonic or biofilm-associated microorganisms in food industry.

BSs as Food Emulsifiers and Additives

An emulsion is a heterogeneous system, consisting of at least one immiscible liquid intimately dispersed in another in the form of droplets, whose diameter in general exceeds 0.1 µm. Two main types of emulsion are important in foods: oil-in-water (o/w) emulsions, where droplets of oil are suspended in an aqueous continuous phase, as mayonnaise, cream liqueur, whipped toppings, ice cream mixes; and water-in-oil (w/o) emulsion, which are exemplified by butter, margarines, and fat-based spreads (Stauffer, 2005; Nitschke and Costa, 2007). Emulsifiers are additives that allow normally immiscible liquids such as oil and water to form stable emulsion, preventing phase separation. Since food emulsifiers do more than simply stabilize emulsions, they are more accurately termed surfactants (Hasenhuettl, 2008). In addition to their major function of producing and stabilizing emulsions, food emulsifiers (or surfactants) contribute to numerous other functional roles as antistaling agents in bread and other baked goods; foaming stabilization in cakes and whipped toppings; crystal inhibition in salad oils; antisticking in candies and grill shortenings; viscosity modification in chocolate; controlled fat agglomeration in ice cream; freeze–thaw stabilization in whipped toppings and coffee whiteners; gloss enhancement in confectionery coatings, canned and moist pet foods; antispattering in margarines; and solubilizing agents for color and flavor systems among others (Hasenhuettl, 2008); therefore, they are indispensable additives in food formulations. Commercially available emulsifiers used in the food and drink sectors comprise two main types, lecithin, derived from soy and egg, and a range of emulsifiers produced from synthetic sources (Freire et al., 2010).

Considering their surface-active properties, BSs can be explored as emulsifiers in food formulations with advantages against their chemical counterparts especially due to their environmental-friendly nature (Mohan et al., 2006), low toxicity (Flasz et al., 1998), their unique structures and properties, and the increasing customer demand for natural or organic over synthetic ingredients. The rapid growth of functional foods requiring "green" additives represents an opportunity for bioemulsifiers because of their natural status. Emulsifiers, like lecithin, are derived from natural sources and may satisfy organic and vegetarian requirements for foodstuffs. The use of microbial-based emulsifiers represents an opening market to replace lecithin in products requiring non-GM ingredients. Besides, egg lecithin is avoided by vegan and other ethnic groups, and natural emulsifiers may have a place in products directed to such communities. For emulsifiers to be accepted by Jewish and Islamic consumers, they must be produced from Kosher-certified raw materials, precluding the use of almost all animal fats (Hasenhuettl, 2008).

Some BSs exhibit high emulsifier properties while others are better surface-active agents (Ron and Rosemberg, 2001). High-molecular mass BSs such as emulsan and liposan also referred as polymeric BSs have been shown to form stable emulsions with edible oils (Cirigliano and Carman, 1985). Emulsifier activity of BSs is generally evaluated with hydrocarbons; nevertheless, there are some examples in literature exploring the use of BS as emulsifying agent for food materials essentially edible oils. A bioemulsifier from *Candida glabrata* UCP1002 formed stable and compact emulsions with cotton seed oil showing emulsifying activity of 75% (Sarubbo et al., 2006). A mannoprotein bioemulsifier extracted from the cell wall of *Saccharomyces cerevisiae* was reported to stabilize oil-in-water emulsions with corn oil. Emulsions of 60% oil-in-water, with 8 g/L of bioemulsifier and 5–50 g/L sodium chlorine, were stable for 3 months at 4°C at pH 3–11. The authors suggested the practical application of the bioemulsifier for mayonnaise production and also other food products like meat, cakes, crackers, and ice creams (Torabizadeh et al., 1996). An extracellular compound obtained from *Candida utilis* was employed as emulsifier in salad dressing formulations (Shepherd et al., 1995), and a mannoprotein from *Kluyveromyces marxianus* was reported to form emulsions with corn oil that are stable for 3 months (Lukondeh et al., 2003). The rhamnolipids from *P. aeruginosa* LBI was able to form stable emulsions with sunflower, linseed, olive, palm, babassu, and Brazilian nut oils (Costa et al., 2006), and a BS from *B. subtilis* MTCC 441 showed high emulsification index with castor, mustard, coconut, gingelly, and sunflower oils suggesting their potential as emulsifying agent in food systems (Suresh Chander et al., 2012).

Alasan bioemulsifier obtained from *Acinetobacter radioresistens* KA53 was effective in stabilizing emulsions of food-grade vegetable and coconut oils. Alasan was claimed to stabilize o/w mixtures, emulsions, and dispersions, to reduce fat content and to increase shelf-life of foodstuffs (Rosenberg and Ron, 1998).

The formation of micro- and nanoemulsions has attracted attention in various fields of application. In food systems, they can be applied to delivering functional components (omega-3 fatty acids, vitamins, minerals, and probiotics), flavors, and antioxidants (Ré et al., 2010). Microemulsions usually require the use of surfactant mixtures with salt or alcohol, and only a few surfactants such as soybean lecithin are known to effectively form w/o microemulsion without the addition of any cosurfactants (Kitamoto et al., 2009). The mannosylerythritol lipid A (MEL-A) surfactant obtained from *Pseudozyma* spp. was reported to form stable water-in-oil microemulsion without the need of a cosurfactant showing great potential for future applications in this field (Worakikanchanakul et al., 2008). Surfactin nanoemulsions that showed antimicrobial activity (Manoharan et al., 2012) can also be explored as a carrier agent.

According to Kosaric (2001), BSs can act controlling consistency, retarding staling and solubilizing flavor oils in bakery and ice-cream formulations. The addition of rhamnolipid surfactants improved stability, texture, volume, and conservation of bakery products (Van Haesendonck and Vanzeveren, 2004). The addition of 0.10% rhamnolipid to muffins and croissants enhanced the moisture, improved the texture, and maintained freshness for a longer period of time (Gandhi and Skebba, 2007).

Dagbert et al. (2006) have evaluated the corrosion behavior of stainless steel in the presence of a BS produced by a *P. fluorescens* strain. These authors have concluded that the tested BS can delay the stainless steel surface corrosion, especially if the electrolyte to which the surface will be exposed is not too aggressive.

BSs are versatile molecules that can be applied not only in food industry as cleaning agents or to modify surface properties avoiding adhesion, corrosion, and removing biofilms but also in food formulations as antimicrobial, antiadhesive, and emulsifiers indicating their potential as multipurpose ingredients or additives.

The toxicity of BSs is already a concern to industries once their main goal is to offer safe products to their consumers. Some data are available about security levels of BS; however, food processing industry demonstrates some distrust especially due to the microbial origin of the molecules such as the rhamnolipids obtained from the opportunistic pathogen *P. aeruginosa*. Rhamnolipids are considered nontoxic, noncarcinogenic, and was approved by FDA to be used in fruit, vegetables, and legume crops as biofungicide (www.rhamnolipid.com). Nonpathogenic species of *Pseudomonas* such as *P. chlororaphis*, *P. luteola*, and *P. fluorescens* has been proposed to produce rhamnolipids (Nitschke et al., 2011), and other GRAS microbes such as the yeast *Candida glabrata* (Sarubbo et al., 2006) have been considered for BS production avoiding the use undesirable strains. Surfactin from *B. subtilis* did not show any toxicological effects at dose 2500 mg/kg after a single oral administration in rats, and no surfactin-related toxicities in survival, clinical signs, hematological parameters, and histopathological observations of hematopoietic organs were found (Hwang et al., 2009). Some studies demonstrate that BSs as rhamnolipids, sophorolipids, surfactin, and mannosylerythritol lipids can be used in cosmetics and pharmaceutical products reinforcing the idea of their low toxicity and safety (Rodrigues et al., 2006a; Kitamoto et al., 2009).

PRODUCTION OF BS FROM FOOD INDUSTRY RESIDUES

The use of renewable resource based on plant biomass, especially parts not used as foodstuff, can create a sustainable economic growth including new markets based on environmental-friendly nature. However, for the future rearrangement of a substantial economy to biological raw materials, new approaches in research and development, production, and economy are necessary. This context is linked with the concept of biorefinery, which is based on the use of carbon molecules extracted from plant in order to substitute carbons from oil and gas (Kamm and Kamm, 2004; Octave and Thomas, 2009). This industrial development of a renewable energy will contribute to the diversification of energy sources used to transport and to the development of green chemistry, which will partially substitute petrochemicals.

The 12 principles of green chemistry (Anastas and Kirchhoff, 2002) (prevention, atom economy, less hazardous chemical syntheses, designing safer chemicals, safer solvents and auxiliaries, design for energy efficiency, use of renewable feedstock, reduce derivatives, catalysis, design for degradation, real-time analysis for pollution prevention, and inherently safer chemistry for accident prevention) can help the development of new industrial process, such as the biotransformation and fermentation. Like this, the combination between biotechnological and chemical conversion of substances will play an important role.

The relatively high raw material prices and low productivities currently inhibit potential economical production of bioproducts on an industrial scale. The major challenge for the industry today is to identify alternative materials that are biodegradable, based on renewable resources, and that show reasonable cost–performance effectiveness. BSs are a very promising and interesting substance class because they are based on renewable resources, sustainable, and biologically degradable (Muller et al., 2012). Competitive production of BSs is still restricted by the high cost of production, and in the last decades, concentrated efforts have been done to minimize these production costs. The use of cheap or waste substrates to lower the initial raw material costs involved in the process together with the assessment of the substrate and product output with focus on appropriate organism is a good alternative to turn the large scale of BSs competitively comparing with petroleum-based surfactants (Makkar et al., 2011).

Water-soluble substrates, such as potato process effluent, molasses, and cassava wastewater and water-immiscible substrates, such as residues from oil refinery and waste cooking oil are examples of inexpensive agroindustrial wastes and food industry byproducts that can be used as feedstock for BS production. Their use can lead to reduce environmental problems caused by waste disposal and help establish the economical large-scale production of BSs.

According to the details provided earlier, we discuss the use of renewable agro-industrial substrates to BS production focusing on the principle that by-products can be exploited as raw material in an industrial biotransformation, aiming the optimization of the operational cultivation conditions in order to achieve high yields.

BS Production from Vegetable Oils and Fat Industries

Plant oils contain various triacylglycerols based on fatty acids with chains of 8–24 carbon length. The most represented have chains of 16 and 18 carbon molecules. The global production and consumption of vegetable oils has increased continuously in the past decades, and the two major markets for this product are food and industrial use, which includes biodiesel. The main growth has been in palm oil, soybean oil, rapeseed oil, and sunflower oil.

Great quantities of waste are generated by the oil and fat industries: residual oils, tallow, marine oils, soapstock, and frying oils. It is well known that the disposal of wastes is a growing problem, and new alternatives for the use of fatty wastes should be studied, as the treatment and disposal costs for these wastes are a vast financial burden to various industries (Nitschke et al., 2005a). The crude or unrefined oils extracted from oilseeds are generally rich in free fatty acids, mono-, di-, and triacylglycerides, phosphatides, pigments, sterols, tocopherols, glycerol, hydrocarbons, vitamins, protein fragments, trace metals, glycolipids, pesticides, and resinous and mucilaginous materials among others (Dumont and Narine, 2007).

The high content of fats, oils, and other nutrients in these waste makes them interesting and cheap raw materials for industries involved in useful secondary metabolite production (Makkar et al., 2011). Vegetable oils and residues from vegetable oil refinery are among the most used low-cost substrates for BS production (Haba et al., 2000; Nitschke et al., 2005a; Makkar et al., 2011). Soapstock is an example of vegetable oil industry waste that was used as a substrate to BS production (Benincasa et al., 2002; Nitschke et al., 2005b). Soapstock is a residue from oil neutralization process and is generated in large quantities by the vegetable oil processing industry (amounts to 2%–3% of the total oil production). Sunflower oil soapstock was assayed as the carbon source for rhamnolipid production by *P. aeruginosa* LBI strain, giving a final surfactant concentration of 12 g/L in shaker and 16 g/L in bioreactor experiments (Benincasa et al., 2002). Soapstock from soybean, cottonseed, babassu, palm, and corn oil refinery was tested to rhamnolipid production, and the soybean soapstock waste was the best substrate, generating 11.7 g/L of rhamnolipids and a production yield of 75% (Nitschke et al., 2005b).

Another example of BS production from soapstock and postrefinery fatty acids is the production of glycolipids by *Candida antarctica* and *C. apicola* in a cultivation medium supplemented with these wastes. The efficiency of glycolipids synthesis was from 7.3 to 13.4 g/L and from 6.6 to 10.5 g/L in the medium supplemented with soapstock and postrefinery fatty acids, respectively (Bednarsky et al., 2004).

Frying oil is produced in large quantities for use both in the food industry and at the domestic scale, and their use has great potential for microbial growth and transformation. Haba et al. (2000) studied a screening process for the selection of microorganism strains with a capacity to grow on frying oils (sunflower and olive oils) and accumulate surface-active compounds in the culture media. Nine *Pseudomonas* spp. strains were selected by decreasing the surface tension of the medium to 34–36 mN/m, and *P. aeruginosa* 47T2 showed a final production of rhamnolipid of 2.7 g/L as rhamnose and a production yield of 0.34 g/g. Soybean frying oil was used for rhamnolipid production in a submerged cultivation process in stirred tank reactors with the aim of optimizing the aeration rate and agitation speed (de Lima et al., 2009). *Rhodococcus erythropolis* was tested to produce glycolipids using sunflower frying oil as cheap renewable substrate, and the final BS showed high surface activity and emulsification capability with potential application to the cleanup of hydrocarbon-contaminated sites (Sadouk et al., 2008).

Palm oil (Oliveira et al., 2009; Sarachat et al., 2010), corn oil (Mata-Sandoval et al., 1999), sunflower oil (Muller et al., 2010), soybean oil (Prieto et al., 2008), olive oil, and grape seed oil (Wei et al., 2005) are examples that vegetable oils can be used directly to BS production. Ecover (Belgium N.V.) has produce sophorolipids from *C. bombicola* using vegetable oil (rapeseed oil) in combination with glucose as carbon source. The product shows cost-efficient production and excellent performance resulting in hard surface cleaning applications. This is an example of successful transformation of BS production from a laboratory development to an industrial process using renewable resource based on plant biomass.

BS Production from Sugar Industry

Molasses is one such by-product of sugar cane and sugar beets industry. The principal reasons for widespread use of molasses as substrate are its low price compared to other sources of sugar and the presence of several other compounds besides sucrose (Makkar and Cameotra, 1999). Molasses has been assayed to BS production using different microorganisms. *P. aeruginosa* GS3 was reported to produce rhamnolipids during growth on molasses and corn steep liquor as the primary carbon and nitrogen sources (Patel and Desai, 1997). *P. luteola* and *P. putida* produced rhamnolipid BSs from different sugar beet molasses concentration (1%–5% w/v). The rhamnolipid production increased with the increase in the concentration of molasses, and maximum production occurred when 5% (w/v) of molasses were used (Onbasli and Aslim, 2009). *B. subtilis* also was described to produce BS from molasses. The authors described an optimized medium containing 16% of molasses in an effort to reduce the production process costs (Abdel-Mawgoud et al., 2008).

BS from Starch-Rich Wastes

The processing of carbohydrate-rich agriculture products (cassava and potato) can generate large amount of waste. For example, the production of 1 ton of cassava flour generates 300 L of cassava wastewater; consequently, the treatment and disposal of this waste represent a great economic cost to the cassava flour industry as well as an environmental problem. Thereby there is an increased need for a better management of this waste. Nitschke and Pastore (2004) described the surfactin BS production by *B. subtilis* from cassava effluent as substrate. The BS exhibited good surface activity and produced reasonable yields of surfactin. *P. aeruginosa* also was related to produce BS from cassava wastewater, and this waste showed potential as an alternative substrate for rhamnolipid production (Costa et al., 2009).

Fox and Bala (2000) demonstrated that potato processing effluent was a suitable alternative carbon source to generate surfactant from *B. subtilis*. Das and Mukherjee (2007) evaluated the BS production using powdered potato peels as substrate, and the lipopeptide BS showed good surface activity and yield.

BS from Dairy Industry

The dairy industry involves processing raw milk into products including milk, butter, cheese, and yogurt, using processes such as chilling, pasteurization, and homogenization. Typical by-products include buttermilk, whey, and their derivatives.

Huge amounts of water are used during the process producing effluents containing dissolved sugars and proteins, fats, and possibly residues of additives. Besides, these effluents are characterized by their high BOD value, and their incorrect disposal represents an environmental problem.

Cheese whey was used as a substrate to BS production in a two-step batch cultivation process by *C. bombicola* and *Cryptococcus curvatus*. In the first step, *C. curvatus* was grown on deproteinized whey concentrates; the cultivation broth was disrupted with a glass bead mill, and it served a medium for growth and sophorolipid production by *C. bombicola* (Daniel et al., 1999). Also, whey showed as a potential substrate to BS production by *L. pentosus* (Rodrigues et al., 2006b). *Yarrowia lipolytica*, *Micrococcus luteus*, and *Burkholderia cepacia* showed capability to produce BS from whey wastewater. The BSs produced were biochemically characterized and the properties were analyzed. The BS showed good emulsification index and hemolytic and surface activities (Yilmaz et al., 2009).

BS PRODUCTION FROM UNCONVENTIONAL SUBSTRATES

Combinations of distillery waste with other industrial wastes such as whey waste, fruit processing waste, and sugar industry effluent were evaluated for BS production by four bacterial isolates from a distillery unit. The combinations of wastes improved the yields of BSs and reduced the chemical oxygen demand of the combined wastes. Total sugars, nitrogen, and phosphate levels reduced in the range of 79%–86%, 58%–71%, and 45%–59%, respectively, showing the environmental and economical benefits derived from the use of wastes as substrates to BS production (Dubey et al., 2012).

Another attractive substrate to BS production is cashew apple juice. This substrate is rich in reducing sugar, vitamins, and mineral salts and is cheap (US $ 1.00/kg), which make an interesting and inexpensive culture medium. *Acinetobacter calcoaceticus*, *B. subtilis*, and *P. aeruginosa* showed potential to growth at cashew apple juice and produce BS (Rocha et al., 2006, 2007, 2009).

Some reviews about the use of unconventional substrates and agroindustrial residues to BS production have been published (Makkar and Cameotra, 1999; Makkar and Cameotra, 2002; Nitschke and Costa, 2007; Makkar et al., 2011; Henkel et al., 2012) showing that production cost of BS still remains the limiting factor to industrial process scale, and many research have been done based on the utilization of waste conversion to bioproducts.

CONCLUSION

Increasing demand for natural ingredients and green chemistry stimulates the exploitation of BSs. These molecules show singular properties very useful in food processing even as an ingredient, a cleaning agent, or a prospective high-value product obtained from food waste substrates. The main challenges to be overcome to improve BS utilization by food industry are the generation of more data about their toxicity including in vivo tests; additionally, more research has to be done applying BS in food matrix in order to evaluate not only their properties but also their interaction with food components. Furthermore, there is a lack of information about BS contribution on sensory properties of food they are incorporated or in contact. The use of wastes or by-products can reduce costs and valorize the residues; however, simple substrates that need few preparation steps and standardization should be preferred to turn the process cost-effective. The discovery of new and noble applications to BS will contribute to the valorization of these molecules increasing even more their demand. Finally, the improvement in microbial strain productivity, safety, and stability is imperative to the success of BSs.

ACKNOWLEDGMENT

The authors thank the Fundação de Amparo a Pesquisa do Estado de São Paulo (FAPESP) for financial support.

REFERENCES

Abdel-Mawgoud, A., Aboulwafa, M., and Hassouna, N. 2008. Optimization of surfactin production by *Bacillus subtilis* isolate BS5. *Applied Biochemistry and Biotechnology,* 150:305–325.

Anastas, P. T. and Kirchhoff, M. K. 2002. Origins, current status, and future challenges of green chemistry. *Accounts of Chemical Research*, 35:686–694.

Annous, B. A., Fratamico, P. M., and Smith, J. L. 2009. Quorum sensing in biofilms: why bacteria behave the way they do. *Journal of Food Science*, 74: 24–37.

Araújo, L. V., Abreu, F., Lins, U., Anna, L. M. M. S., Nitschke, M., and Freire, D. M. G. 2011. Rhamnolipid and surfactin inhibit *Listeria monocytogenes* adhesion. *Food Research International,* 44:481–488.

Bagge-Ravn, D., Yin, N., Hjelm, M., Christiansen, J. N., Johansen, C., and Gram, L. 2003. The microbial ecology of processing equipment in different fish industries: Analysis of the microflora during processing and following cleaning and disinfection. *International Journal of Food Microbiology,* 87:239–250.

Bednarski, W., Adamczak, M., Tomasik, J., and Plaszczyk, M. 2004. Application of oil refinery waste in the biosynthesis of glycolipids by yeast. *Bioresource Technology*, 95:15–18.

Benincasa, M., Abalos, A., Oliveira, I., and Manresa, A. 2004. Chemical structure, surface properties and biological activities of the biosurfactant produced by *Pseudomonas aeruginosa* LBI from soapstock. *Antonie Van Leeuwenhoek*, 85:1–8.

Benincasa, M., Contiero, J., Manresa, M. A., and Moraes, I. O. 2002. Rhamnolipid production by *Pseudomonas aeruginosa* LBI growing on soapstock as the sole carbon source. *Journal of Food Engineering*, 54:283–288.

Bognolo, G. 1999. Biosurfactants as emulsifying agents for hydrocarbons. *Colloids and Surfaces A: Physicochemical and Engineering Aspects,* 152:41–52.

Carpentier, B. and Cerf, O. 1993. Biofilms and their consequences with particular references to hygiene in the food industry. *Journal of Applied Bacteriology*, 75:499–511.

Cirigliano, M. C. and Carman, G. M. 1985. Purification and characterization of liposan, a bioemulsifier from *Candida lipolytica*. *Applied and Environmental Microbiology,* 50:846–850.

Cloete, E., Molobela, I., Van Der Merwe, A., and Richards, M. 2009. Biofilms in the food and beverage industries: An introduction. In *Biofilms in the Food and Beverages Industries*, eds. Fratamico, P. M., Annous, B. A., and Gunther IV, N. W., pp. 3–41. Boca Raton, FL: CRC Press.

Costa, S. G. V. A. O., Lepine, F., Milot, S., Deziel, E., Nitschke, M., and Contiero, J. 2009. Cassava wastewater as substrate for the simultaneous production of rhamnolipids and polyhydroxyalkanoates by *Pseudomonas aeruginosa*. *Journal of Industrial Microbiology & Biotechnology*, 36:1063–1072.

Costa, S. G. V. A. O., Nitschke, M., Haddad, R., Eberlin, M. N., and Contiero, J. 2006. Production of *Pseudomonas aeruginosa* LBI rhamnolipids following growth on Brazilian native oils. *Process Biochemistry,* 41:483–488.

Costerton, J. W., Stewart, P. S., and Greenberg, E. P. 1999. Bacterial biofilms: A common cause of persistent infections. *Science,* 284:1318–1322.

Dagbert, C., Meylheuc, T., and Bellon-Fontaine, M. N. 2006. Corrosion behavior of AISI 304 stainless steel in presence of a biosurfactant produced by *Pseudomonas fluorescens. Electrochimica Acta,* 51:5221–5227.

Daniel, H. J., Otto, R. T., Binder, M., Reuss, M., and Syldatk, C. 1999. Production of sophorolipids of whey: Development of a two-stage process with *Cryptococcus curvatus* ATCC 20509 and *Candida bombicola* ATCC 22214 using deproteinized whey concentrates as substrates. *Applied Microbiology and Biotechnology,* 51:40–45.

Das, K. and Mukherjee, A. K. 2007. Comparison of lipopeptide biosurfactants production by *Bacillus subtilis* strains in submerged and solid state fermentation systems using a cheap carbon source: Some industrial applications of biosurfactants. *Process Biochemistry,* 42:1191–1199.

de Lima, C., Ribeiro, E., Sérvulo, E., Resende, M., and Cardoso, V. 2009. Biosurfactant production by *Pseudomonas aeruginosa* grown in residual soybean oil. *Applied Biochemistry and Biotechnology,* 152:156–168.

Dubey, K. V., Charde, P. N., Meshram, S. U., Yadav, S. K., Singh, S., and Juwarkar, A. A. 2012. Potential of new microbial isolates for biosurfactant production using combinations of distillery waste with other industrial wastes. *Journal of Petroleum & Environmental Biotechnology,* S1:002. doi:10.4172/2157–7463.S1-002.

Dumont, M. J. and Narine, S. S. 2007. Soapstock and deodorizer distillates from North American vegetable oils: Review on their characterization, extraction and utilization. *Food Research International,* 40:957–974.

Dusane, D. H., Nancharaiah, V., Zinjarde, S. S., and Venugopalan, V. P. 2010. Rhamnolipid mediated disruption of marine *Bacillus pumilus* biofilms. *Colloids and Surfaces B: Biointerfaces,* 81:242–248.

Flasz, A., Rocha, C. A., Mosquera, B., and Sajo, C. 1998. A comparative study of the toxicity of a synthetic surfactant and one produced by *Pseudomonas aeruginosa* ATCC 55925. *Medical Science Research,* 26:181–185.

Fox, S. L. and Bala, G. A. 2000. Production of surfactant from *Bacillus subtilis* ATCC 21332 using potato substrates. *Bioresource Technology,* 75:235–240.

Freire, D. M. G., Araújo, L. V., Kronemberger, F. A., and Nitschke, M. 2010. Biosurfactants as emerging additives in food processing. In *Food Engineering: New Techniques and Products,* eds. Ribeiro, C. P. and Passos, M. L., pp. 685–705. Boca Raton, FL: CRC Press.

Gandhi, N. R. and Skebba, V. L. P. 2007. Rhamnolipid compositions and related methods of use. W. O. International Application Patent (PCT) 2007/095258 A3, filed February 12, 2007, and issued August 23, 2007.

Gomes, M. Z. V. and Nitschke, M. 2012. Evaluation of rhamnolipid and surfactin to reduce the adhesion and remove biofilms of individual and mixed cultures of food pathogenic bacteria. *Food Control,* 25:441–447.

Goulter, R. M., Gentle, I. R., and Dykes, G. A. 2009. Issues in determining factors influencing bacterial attachment: A review using the attachment of *Escherichia coli* to abiotic surfaces as an example. *Letters in Applied Microbiology,* 49:1–7.

Gudiña, E. J., Rocha, V., Teixeira, J. A., and Rodrigues, L. R. 2010. Antimicrobial and antiadhesive properties of a biosurfactant isolated from *Lactobacillus paracasei* ssp. *paracasei* A20. *Letters in Applied Microbiology,* 50:419–424.

Haba, E., Espuny, M. J., Busquets, M., and Manresa, A. 2000. Screening and production of rhamnolipids by *Pseudomonas aeruginosa* 47T2 NCBI 40044 from waste frying oils. *Journal of Applied Microbiology,* 88:379–387.

Haba, E., Pinazo, A., Jauregui, O., Espuny, M. J., Infante, M. R., and Manresa, A. 2003. Physicochemical characterization and antimicrobial properties of rhamnolipids produced by *Pseudomonas aeruginosa* 47T2 NCBIM 40044. *Biotechnology and Bioengineering*, 81:316–322.

Hall-Stoodley, L., Costerton, J. W., and Stoodley, P. 2004. Bacterial biofilms: From the natural environment to infectious diseases. *Nature Reviews Microbiology*, 2:95–108.

Hasenhuettl, G. L. 2008. Synthesis and commercial preparation of food emulsifiers. In *Food Emulsifiers and Their Applications*, eds. Hasenhuettl, G. L. and Hartel, R. W., pp. 1–9. New York: Springer.

Henkel, M., Muller, M. M., Kugler, J. H., Lovaglio, R. B., Contiero, J., Syldatk, C., and Hausmann, R. 2012. Rhamnolipids as biosurfactants from renewable resources: Concepts for next-generation rhamnolipid production. *Process Biochemistry*, 47:1207–1219.

Hood, S. K. and Zottola, E. A. 1995. Biofilms in food processing. *Food Control*, 6:9–18.

Huang, X., Suo, J., and Cui, Y. 2011. Optimization of antimicrobial activity of surfactin and polylysine against *Salmonella* Enteritidis in milk evaluated by a response surface methodology. *Foodborne Pathogens and Disease*, 8:439–443.

Hunt, S. M., Werner, E. M., Huang, B., Hamilton, M. A., and Stewart, P. S. 2004. Hypothesis for the role of nutrient starvation in biofilm detachment. *Applied and Environmental Microbiology*, 70:7418–7425.

Hwang, Y. H., Kim, M. S., Song, I. B., Park, B. K., Lim, J. H., Park, S. C., and Yun, H. I. 2009. Subacute (28 day) toxicity of surfactin C, a lipopeptide produced by *Bacillus subtilis*, in rats. *Journal of Health Science*, 55:351–355.

IFT. 2012. Global food additives market to exceed 33 B by 2015. http://www.ift.org/ food-technology/daily-news/2010/june/08/global-food-additives-market-to-exceed-33-b-by-2015.aspx, accessed August 10, 2012.

Irie, Y., O'Toole, G. A., and Yuk, M. H. 2005. *Pseudomonas aeruginosa* rhamnolipids disperse *Bordetella bronchiseptica* biofilms. *FEMS Microbiology Letters*, 250:237–243.

Jenkinson, H. F. and Lappin-Scott, H. M. 2001. Biofilms adhere to stay. *Trends in Microbiology*, 9:9–10.

Kamm, B. and Kamm, M. 2004. Principles of biorefinery. *Applied Microbiology and Biotechnology*, 64:137–145.

Kiran, G. S., Sabarathnam, B., and Selvin, J. 2010. Biofilm disruption potential of a glycolipid biosurfactant from marine *Brevibacterium casei*. *FEMS Immunology and Medical Microbiology*, 59:432–438.

Kitamoto, D., Morita, T., Fukuoka, T., Konishi, M., and Imura, T. 2009. Self-assembling properties of glycolipid biosurfactants and their potential applications. *Current Opinion in Colloid & Interface Science*, 14:315–328.

Kosaric, N. 2001. Biosurfactants and their application for soil bioremediation. *Food Technology and Biotechnology*, 39:295–304.

Kuiper, I., Lagendijk, E. L., Pickford, R., Derrick, J. P., Lamers, G. E. M., Thomas-Oates, J. E., Lugtenberg, B. J. J., and Bloemberg, G. V. 2004. Characterization of two *Pseudomonas putida* lipopeptide biosurfactants, putisolvin I and II, which inhibit biofilm formation and break down existing biofilms. *Molecular Microbiology*, 51:97–113.

Kumar, C. G. and Anand, S. K. 1998. Significance of microbial biofilms in food industry: A review. *International Journal of Food Microbiology*, 42:9–27.

Lukondeh, T., Ashbolh, N. J., and Rogers, P. L. 2003. Evaluation of *Kluyveromyces marxianus* FII 510700 grown on a lactose-based medium as a source of natural bioemulsifier. *Journal of Industrial Microbiology & Biotechnology*, 30:715–720.

Magalhães, L. and Nitschke, M. 2013. Antimicrobial activity of rhamnolipids against *Listeria monocytogenes* and their synergistic interaction with nisin. *Food Control*, 29:138–142.

Makkar, R. S. and Cameotra, S. S. 1999. Biosurfactant production by microorganisms on unconventional carbon sources. *Journal of Surfactants and Detergents*, 2:237–241.

Makkar, R. S. and Cameotra, S. S. 2002. An update on the use of unconventional substrates for biosurfactant production and their new applications. *Applied Microbiology and Biotechnology*, 58:428–434.

Makkar, R. S., Cameotra, S. S., and Banat, I. M. 2011. Advances in utilization of renewable substrates for biosurfactant production. *AMB Express*, 1:5.

Manoharan, M. J., Bradeeba, K., Parthasarathi, R., Sivakumaar, P. K., Chauhan, P. S., Tipayno, S., Benson, A., and Sa, T. 2012. Development of surfactin based nanoemulsion formulation from selected cooking oils: Evaluation for antimicrobial activity against selected food associated microorganisms. *Journal of the Taiwan Institute of Chemical Engineers*, 43:172–180.

Mata-Sandoval, J. C., Karns, J., and Torrents, A. 1999. High-performance liquid chromatography method for the characterization of rhamnolipid mixtures produced by *Pseudomonas aeruginosa* UG2 on corn oil. *Journal of Chromatography A*, 864:211–220.

Maukonen, J., Mättö, J., Wirtanen, G., Raaska, T., Mattila-Sandholm, T., and Saarela, M. 2003. Methodologies for the characterization of microbes in industrial environments: A review. *Journal of Industrial Microbiology & Biotechnology*, 30:327–356.

McLandsborough, L., Rodriguez, A., PéRez-Conesa, D., and Weiss, J. 2006. Biofilms: At the interface between biophysics and microbiology. *Foods Biophysics*, 1:94–114.

Meylheuc, T., Methivier, C., Renault, M., Herry, J. M., Pradier, C. M., and Bellon-Fontaine, M. N. 2006b. Adsorption on stainless steel surfaces of biosurfactants produced by gram-negative and gram-positive bacteria: Consequence on the bioadhesive behaviour of *Listeria monocytogenes*. *Colloids and Surfaces B: Biointerfaces*, 52:128–137.

Meylheuc, T., Renault, M., and Bellon-Fontaine, M. N. 2006a. Adsorption of a biosurfactant on surfaces to enhance the disinfection of surfaces contaminated with *Listeria monocytogenes*. *International Journal of Food Microbiology*, 109:71–78.

Meylheuc, T., Van Oss, C. J., and Bellon-Fontaine, M. N. 2001. Adsorption of biosurfactant on solid surfaces and consequences regarding the bioadhesion of *Listeria monocytogenes* LO28. *Journal of Applied Microbiology*, 91:822–832.

Mirelles II, J. R., Toguchi, A., and Harshey, R. M. 2001. *Salmonella enterica* serovar Typhimurium swarming mutants with altered biofilm-forming abilities: Surfactin inhibits biofilm formation. *Journal of Bacteriology*, 183:5848–5854.

Mohan, P. K., Nakhla, G., and Yanful, E. K. 2006. Biokinetics of biodegradability of surfactants under aerobic, anoxic and anaerobic conditions. *Water Research*, 40:533–540.

Muller, M. M., Hormann, B., Syldatk, C., and Hausmann, R. 2010. *Pseudomonas aeruginosa* PAO1 as a model for rhamnolipid production in bioreactor systems. *Applied Microbiology and Biotechnology*, 87:164–174.

Muller, M. M., Kugler, J. H., Henkel, M., Gerlitzki, M., Hormann, B., Pohnlein, M., Syldatk, C., and Hausmann, R. 2012. Rhamnolipids—Next generation biosurfactants? *Journal of Biotechnology*, 162: 366–380.

Nitschke, M., Araújo, L. V., Costa, S. G. V. A. O., Pires, R. C., Zeraik, A. E., Fernandes, A. C. L. B., Freire, D. M. G., and Contiero, J. 2009. Surfactin reduces the adhesion of foodborne pathogenic bacteria to solid surface. *Letters is Applied Microbiology*, 49:241–247.

Nitschke, M. and Costa, S. G. V. A. O., 2007. Biosurfactants in food industry. *Trends in Food Science & Technology*, 18: 252–259.

Nitschke, M., Costa, S. G. V. A. O., and Contiero, J. 2005a. Rhamnolipid surfactants: An update on the general aspects of these remarkable biomolecules. *Biotechnology Progress*, 21:1593–1600.

Nitschke, M., Costa, S. G. V. A. O., and Contiero, J. 2011. Rhamnolipids and PHAs: Recent reports on Pseudomonas-derived molecules of increasing industrial interest. *Process Biochemistry*, 46:621–630.

Nitschke, M., Costa, S. G. V. A. O., Haddad, R., Gonçalves, L. A. G., Eberlin, M. N., and Contiero, J. 2005b. Oil wastes as unconventional substrates for rhamnolipid biosurfactant production by *Pseudomonas aeruginosa* LBI. *Biotechnology Progress*, 21:1562–1566.

Nitschke, M. and Pastore, G. M. 2004. Biosurfactant production by *Bacillus subtilis* using cassava processing effluent. *Applied Biochemistry and Biotechnology*, 112:163–172.

Octave, S. and Thomas, D. 2009. Biorefinery: Toward an industrial metabolism. *Biochimie*, 91:659–664.

Oliveira, F. J. S., Vazquez, L., de Campos, N. P., and de França, F. P. 2009. Production of rhamnolipids by a *Pseudomonas alcaligenes* strain. *Process Biochemistry*, 44:383–389.

Onbasli, D. and Aslim, B. 2009. Determination of rhamnolipid biosurfactant production in molasses by some *Pseudomonas* spp. *New Biotechnology*, 25:255.

Ortiz, A., Teruel, J. A., Espuny, M. J., Marqués, A., Manresa, A., and Aranda, F. J. 2006. Effects of dirhamnolipid on the structural properties of phosphatidylcholine membranes. *International Journal of Pharmaceutics*, 325:99–107.

Palmer, J., Flint, S., and Brooks, J. 2007. Bacterial cell attachment, the beginning of a biofilm. *Journal of Industrial Microbiology & Biotechnology*, 34:577–588.

Patel, R. M. and Desai, A. J. 1997. Biosurfactant production by *Pseudomonas aeruginosa* GS3 from molasses. *Letters in Applied Microbiology*, 25:91–94.

Prieto, L. M., Michelon, M., Burkert, J. F. M., Kalil, S. J., and Burkert, C. A. V. 2008. The production of rhamnolipid by a *Pseudomonas aeruginosa* strain isolated from a southern coastal zone in Brazil. *Chemosphere*, 71:1781–1785.

Raaijmakers, J. M., Bruijn, I., Nybroe, O., and Ongena, M. 2010. Natural functions of lipopeptides from *Bacillus* and *Pseudomonas*: More than surfactants and antibiotics. *FEMS Microbiology Reviews*, 34:1037–1062.

Ré, M. I., Santana, M. H. A., and d'Avila, M. A. 2010. Encapsulation technologies for modifying food performance. In *Food Engineering: New Techniques and Products,* eds. Ribeiro, C. P. and Passos, M. L., pp. 223–275. Boca Raton, FL: CRC Press.

Rivardo, F., Martinotti, M. G., Turner, R. J., and Ceri, H. 2010. The activity of silver against *Escherichia coli* biofilm is increased by a lipopeptide biosurfactant. *Canadian Journal of Microbiology*, 56:272–278.

Rivardo, F., Turner, R. J., Allegrone, G., Ceri, H., and Martinotti, M. G. 2009. Anti-adhesion activity of two biosurfactants produced by *Bacillus* spp. prevents biofilm formation of human bacterial pathogens. *Applied Microbiology and Biotechnology*, 83:541–553.

Rocha, M. V. P., Barreto, R. G., Melo, V., and Gonçalves, L. R. B. 2009. Evaluation of cashew apple juice for surfactin production by *Bacillus subtilis* LAMI008. *Applied Biochemistry and Biotechnology*, 155:63–75.

Rocha, M. V. P., Oliveira, A. H. S., Souza, M. C. M., and Gonçalves, L. R. B. 2006. Natural cashew apple juice as fermentation medium for biosurfactant production by *Acinetobacter calcoaceticus. World Journal of Microbiology and Biotechnology,* 22:1295–1299.

Rocha, M. V. P., Souza, M. C. M., Benedicto, S. C. L., Bezerra, M. S., Macedo, G. R., Saavedra, G. A. P., and Gonçalves, L. R. B. 2007. Production of biosurfactant by Pseudomonas aeruginosa grown on cashew apple juice. *Applied Biochemistry and Biotechnology*, 136/140:185–194.

Rodrigues, L., Banat, I. M., Teixeira, J., and Oliveira, R. 2006a. Biosurfactants: Potential applications in medicine. *Journal of Antimicrobial Chemotherapy*, 57:609–618.

Rodrigues, L., Moldes, A., Teixeira, J., and Oliveira, R. 2006b. Kinetic study of fermentative biosurfactant production by *Lactobacillus* strains. *Biochemical Engineering Journal*, 28:109–116.

Ron, E. Z. and Rosenberg, E. 2001. Natural roles of biosurfactants. *Environmental Microbiology*, 3:229–236.

Rosenberg, E. and Ron, E. Z. 1998. Bioemulsifiers. US Patent 5840547, filed September 30, 1996, and issued November 24, 1998.

Sadouk, Z., Hacene, H., and Tazerouti, A. 2008. Biosurfactants production from low cost substrate and degradation of diesel oil by a *Rhodococcus* strain. *Oil and Gas Science and Technology*, 63:747–753.

Sarachat, T., Pornsunthorntawee, O., Chavadej, S., and Rujivaranit, R. 2010. Purification and concentration of a rhamnolipid biosurfactant produced by *Pseudomonas aeruginosa* SP4 using foam fractionation. *Bioresource Technology*, 101:324–330.

Sarubbo, L. A., Luna, J. M., and Campos-Takaki, G. M. 2006. Production and stability studies of the bioemulsifier obtained from a new strain of *Candida glabrata* UCP 1002. *Electronic Journal of Biotechnology*, 9:400–406.

Sha, R., Jiang, L., Meng, Q., Zhang, G., and Song, Z. 2011. Producing cell-free culture broth of rhamnolipids as a cost-effective fungicide against plant pathogens. *Journal of Basic Microbiology*, 51:1–9.

Shakerifard, P., Gancel, F., Jacques, P., and Faille, C. 2009. Effect of different *Bacillus subtilis* lipopeptides on surface hydrophobicity and adhesion of *Bacillus cereus* 98/4 spores to stainless steel and teflon. *Biofouling*, 25:533–554.

Shepherd, R., Rockey, J., Sutherland, I. W., and Roller, S. 1995. Novel bioemulsifiers from microorganisms for use in foods. *Journal of Biotechnology*, 40:207–217.

Simões, M., Simões, L. C., Machado, I., Pereira, M. O., and Vieira, M. J. 2006. Control of flow-generated biofilms with surfactants—Evidence of resistance and recovery. *Food and Bioproducts Processing*, 84:338–345.

Sotirova, A. V., Spasova, D. I., Galabova, D. N., Karpenko, E., and Shulga, A. 2008. Rhamnolipid-biosurfactant permeabilizing effects on gram-positive and gram-negative bacterial strains. *Current Microbiology*, 56:639–644.

Stauffer, C. E. 2005. Emulsifiers for the food industry. In *Bailey's Industrial Oil and Fat Products*, ed. Shahidi, F., pp. 229–267. Hoboken, NJ: John Wiley & Sons.

Stoodley, P., Sauer, K., Davies, D. G., and Costeron, J. W. 2002. Biofilms as complex differentiated communities. *Annual Review of Microbiology*, 56:187–209.

Suresh Chander, C. R., Lohitnath, T., Mukesh Kumar, D. J., and Kalaichelvan, P. T. 2012. Production and characterization of biosurfactant from *Bacillus subtilis* MTCC441 and its evaluation to use as bioemulsifier for food bio-preservative. *Advances in Applied Science Research*, 3:1827–1831.

Tahmourespour, A., Salehib, R., Kermanshahic, R. K., and Eslamid, G. 2011. The anti-biofouling effect of *Lactobacillus fermentum*-derived biosurfactant against *Streptococcus mutans*. *Biofouling*, 27:385–392.

Torabizadeh, H., Shojaosadatib, S. A., and Tehrani, H. A. 1996. Preparation and characterization of bioemulsifier from *Saccharomyces cerevisiae* and its application in food products. *LWT—Food Science and Technology*, 29:734–737.

Transparency Market Research, 2012. Biosurfactants Market—Global scenario, raw material and consumption trends, industry analysis, size, share and forecasts, 2011–2018. http://www.transparencymarketresearch.com/biosurfactants-market.html accessed August 22, 2012.

Van Haesendonck, I. P. H. and Vanzeveren, E. C. A. 2004. Rhamnolipids in bakery products. W. O. International Application Patent (PCT) 2004/040984, filed November 4, 2003, and issued May, 21, 2004.

Wei, Y. H., Chou, C. L., and Chang, J. S. 2005. Rhamnolipid production by indigenous *Pseudomonas aeruginosa* J4 originating from petrochemical wastewater. *Biochemical Engineering Journal*, 27:146–154.

Worakitkanchanakul, W., Imura, T., Morita, T., Fukuoka, T., Sakai, H., Abe, M., Rujiravanit, R., Chavadej, S., and Kitamoto, D. 2008. Formation of W/O microemulsion based on natural glycolipid biosurfactant, mannosylerythritol Lipid-A. *Journal of Oleo Science*, 57:55–59.

Yilmaz, F., Ergene, A., Yalçin, E., and Tan, S. 2009. Production and characterization of biosurfactants produced by microorganisms isolated from milk factory wastewaters. *Environmental Technology*, 13:1397–1404.

Zeraik, A. E. and Nitschke, M. 2010. Biosurfactants as agents to reduce adhesion of pathogenic bacteria to polystyrene surfaces: Effect of temperature and hydrophobicity. *Current Microbiology*, 61:554–559.

8 Trehalose Biosurfactants

Nelly Christova and Ivanka Stoineva

CONTENTS

INTRODUCTION

Many prokaryotic and eukaryotic microorganisms can grow on compounds that are poorly soluble in water and often are associated with the production of surface-active compounds or biosurfactants. Biosurfactants comprise a diverse group of chemical structures with amphiphilic character. Generally, the lipophilic parts of their molecules consist of long-chain fatty acids, hydroxyl fatty acids, or α-alkyl-β-hydroxyl fatty acids, while the hydrophilic moieties can be carbohydrates, amino acids, cyclic peptides, phosphates, carboxylic acids, or alcohols. Biosurfactants can be categorized in five groups regarding their chemical composition: glycolipids, lipopeptides, phospholipids, fatty acids, and polymeric biosurfactants. Biosurfactants can also be grouped into two categories: low-molecular-mass compounds that lower surface and interfacial tension and high-molecular-mass compounds that bind tightly to surfaces (Rosenberg and Ron 1999).

The amphiphilic character of the surfactant molecules is responsible for their unique properties. They appear to act preferentially at the interface between phases with different polarity and hydrogen bonding, forming an ordered molecular film at the interface and altering significantly the interfacial energy (Georgiou et al. 1992). Recent studies show that whenever a microbe comes across an interface, biosurfactants play a significant role in several biological processes such as motility, bacterial cell signaling, biofilm formation, cellular differentiation, substrate accession, and bacterial pathogenesis (Kitamoto et al. 2002; Lang 2002; Cameotra and Makkar 2004; Van Hamme et al. 2006).

197

Potential applications of naturally occurring surface-active compounds are quite diverse and at present main applications are related to environmental concerns, such as bioremediation of hydrocarbon and heavy metal contaminated sites, enhanced oil recovery, and treatment of oil spills (Hester 2001; O'Connor 2002). Alternatively to synthetic compounds, biosurfactants can be utilized in agriculture, mining, and food-processing, with functional properties such as wetting and foaming agents and as emulsifiers in pharmaceutical and cosmetic products (Banat et al. 2000). Recently, some biosurfactants were proved to be suitable alternatives to synthetic medicines and antimicrobial agents and may be used as effective therapeutic agents (Rodrigues et al. 2006). Currently biosurfactants have not yet been commercialized extensively due to their relatively high production and recovery costs. Improvement of biosynthesis efficiency through the use of low-cost medium components coupled with efficient downstream processing techniques and development of hyperproducing strains can make biosurfactant production commercially profitable (Mukherjee et al. 2006).

Glycolipids are the most common class of biosurfactants. They are carbohydrates in combination with long-chain aliphatic or hydroxyaliphatic acids. Among the glycolipids, the best-known subclasses are the rhamnolipids from *Pseudomonas aeruginosa*, sophorolipids from yeasts, and trehalose lipids from *Mycobacteria* and related bacteria (Desai and Banat 1997). Trehalose lipids possess unique surface properties and wide biological activities, as well as other favorable characteristics like low CMC, biodegradability, and low toxicity. The present chapter focuses on their diverse chemical structures, some aspects of their production, physicochemical and biological properties, and possible applications.

TREHALOSE LIPIDS

It is of great importance to know the chemical structure of a surfactant molecule, because slight differences in structure can lead to pronounced differences in its surface properties and bioactivity. Based on the knowledge to date, the members of the trehalose glycolipid family can be divided into three general subclasses: I-subclass-6,6 substituted trehalose esters [such as fatty acid trehalose diesters (TDEs), trehalose dicorynomycolates (TDCMs), and trehalose dimycolates (TDMs)]; II-subclass-2,3-diesters of trehalose sulfates; and III-sub class-succinoyl diesters and tetraesters (Figure 8.1).

CHEMICAL STRUCTURES OF GLYCOLIPIDS PRODUCED FROM *MYCOBACTERIA*

Trehalose lipids were first discovered as an important factor in the virulence of *Mycobacteria*. The so-called cord factor is the best known and was isolated as a pure compound in 1956 from the lipids of *Mycobacterium tuberculosis* (Asselineau and Lederer 1956). The chemical structure of cord factor of *Mycobacterium tuberculosis* comprises a branched-chain mycolic acid esterified to the 6-hydroxyl group of each glucose to give 6,6'-dimycoloyl-α,α'-trehalose. Mycolic acids possess a great variety of structures, the number of their carbon atoms varies from approximately 60 to 90; they contain not only a normal long-chain saturated alkane, $C_{22}H_{45}$ or $C_{24}H_{49}$, but a different number of double bonds, methyl groups, and cyclopropane rings (Lederer 1967) (Figure 8.1, I-subclass). Later, several groups of trehalose-containing glycolipids based on the aforementioned acids were isolated mainly from *Mycobacteria*, *Nocardia*, and *Corynebacteria* as reviewed by Asselineau and Asselineau (1978). Trehalose lipids were identified as general surfactants when Suzuki et al. (1969) detected a glycolipid with strong emulsifying activity in the emulsion layer of *Arthrobacter paraffineus* KY4303 grown on *n*-paraffines. Each molecule contained trehalose and two α-branched-β-hydroxy fatty acids (corynomycolic acids) esterified to undetermined sites on the sugar. Novel types of trehalose lipid surfactants were first reported for the paraffin oxidizing bacterium *Mycobacterium paraffinicum* producing at least five trehalose lipids. Three of them were characterized as previously known fatty acyl derivatives of trehalose: 6,6′-di-*O*-mycoloyl-α,α′-D-trehalose ("cord factor"); 6-*O*-mycoloyl-α,α′-D-trehalose ("lyso cord factor"); 6,6′-di-*O*-acyl-(C_{12}—C_{16})-α,α′-D-trehalose (a low-molecular-weight analog of "cord factor"). The remaining two lipids were identified as 6-*O*-mycoloyl-6′-*O*-acyl-(C_{12}–C_{16})-α,α-D-trehalose and 2-*O*-octanoyl-3,2′-di-*O*-decanoyl-6-*O*-succinoyl-α,α′-D-trehalose (Batrakov et al. 1981).

In *Mycobacterium* were found the trehalose glycolipids termed sulfoglycolipids to account for the presence of a sulfate functionality. Although they were for the first

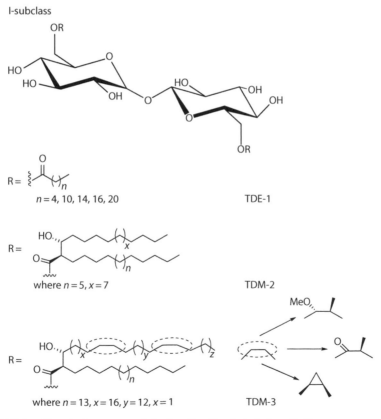

FIGURE 8.1   Chemical structure of the family of basic trehalose lipids.

(continued)

Il-subclass-sulfo trehalose lipids

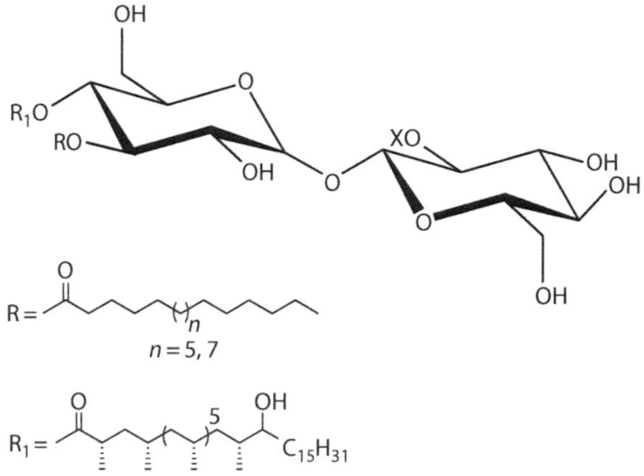

$R = $

$n = 5, 7$

$R_1 = $　　$C_{15}H_{31}$

$X = SO_3H$　　　　Diacyl trehalose sulfate

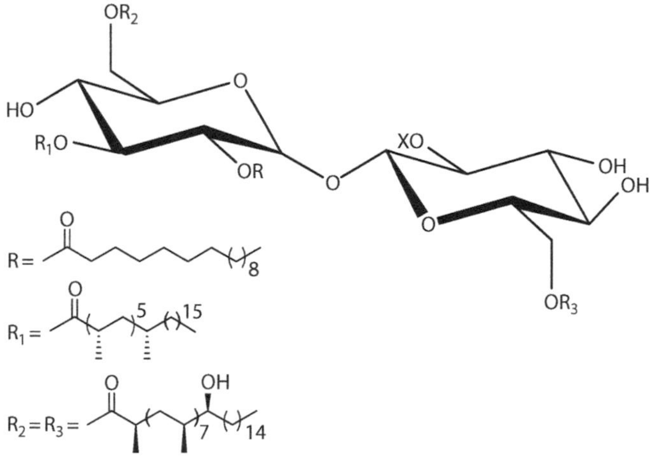

$R = $

$R_1 = $

$R_2 = R_3 = $

$X = SO_3H$　　　　Sulfolipid-1

FIGURE 8.1 (continued)　Chemical structure of the family of basic trehalose lipids.

III-subclass-succinoyl trehalose lipids

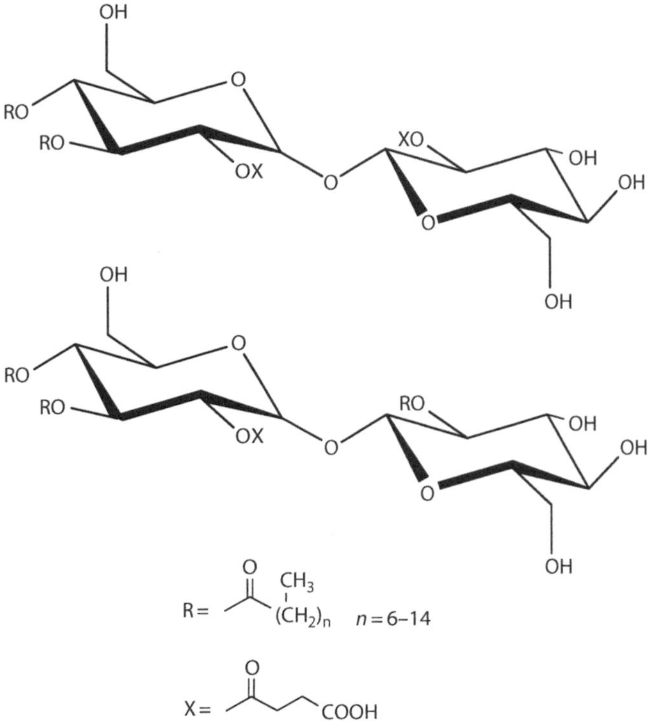

$$R = \overset{O}{\overset{\|}{C}}\ \underset{(CH_2)_n}{\overset{CH_3}{\big|}} \quad n = 6\text{--}14$$

$$X = \overset{O}{\overset{\|}{C}}\ {\sim}COOH$$

FIGURE 8.1 (continued) Chemical structure of the family of basic trehalose lipids.

time isolated in 1959 by Middlebrook et al. (1959) the chemical structure of these trehalose sulfolipids was not clarified until 1976 when Goren et al. characterized them as 2,3,6,6′-tetraesters of trehalose 2′-sulfate. After purification and detailed analytical characterization of glycolipids isolated from *Mycobacterium tuberculosis*, Domenech et al. (2004) offered another structure in contrast with the structures proposed by Goren et al. (1976) and Converse et al. (2003). The suggested chemical structure of these glycolipids is 2,3-diacyl-D-trehalose-2′-sulfates (Figure 8.1. II-subclass).

CHEMICAL STRUCTURES OF GLYCOLIPIDS PRODUCED FROM *RHODOCOCCUS*

In terms of chemical structure, a lot of research was focused particularly in the *Rhodococcus* genus and trehalose-containing glycolipids, which often occur as a complex mixture during the growth of rhodococci on *n*-alkanes, have been extensively studied. Thus, Kretschmer et al. (1982) detected trehalose-6-monocorynemycolates, trehalose-6,6′-diacylates (3-oxo-2-alkyl alkanoyc acid), and trehalose-6-acylates (3-oxo-2-alkyl alkanoic acid) in the organic extract of *Rhodococcus erythropolis* DSM 43215 grown on *n*-alkanes. The major glycolipid species isolated from *Rhodococcus* strain H13–A was identified as 2.3.4.6,2′,3′,4′,6′–octa-acyltrehalose and was acylated with C10–C22 saturated and unsaturated fatty acids, C35–C40 mycolic acids, hexanedioic, dodecanedioic acids, and 10-methyl hexadecanoic and 10–methyl octadecanoic acids (Singer et al. 1990). After alkaline hydrolysis of the glycolipids of *R. erythropolis* S-1 Kurane et al. (1995) detected both glucose and trehalose, and a mixture of mycolic acids ranging from C_{32} to C_{40}. Further research revealed that trehalose lipids produced by rhodococci and related genera possess high structural diversity that can be summarized as follows: monocorynomycolates (Batrakov et al. 1981; Kretschmer and Wagner 1983), dicorynomycolates (Rapp et al. 1979; Philp et al. 2002), tricorynomycolates (Tomiyasu et al. 1986), as well as mono-, di-, tetra-, hexa-, and octa-acylated derivatives of trehalose (Kretschmer and Wagner 1983; Singer et al. 1990; Philp et al. 2002) that exemplify the nonionic type of trehalose lipid biosurfactants, and trehalose tetraesters (Ristau and Wagner 1983; Kim et al. 1990; Espuny et al.1995; Rapp and Gabriel-Jurgens 2003; Tuleva et al. 2008; Marques et al. 2009; Tuleva et al. 2009) and succinoyl trehalose lipids (Uchida et al. 1989a; Tokumoto et al. 2009) constituting the anionic type of biosurfactants (Figure 8.1. III subclass).

PRODUCTION

BIOSYNTHESIS

The proposed biosynthetic pathways for trehalose lipids are reviewed by Lang and Philp (1998). According to them, the carbohydrate and fatty acid moieties are synthesized independently and are subsequently esterified. The formation of the mycolate is considered to be a Claisen-type condensation of two fatty acids, while the synthesis of the final sugar residue, trehalose–6–phosphate (TPS), is catalyzed by a trehalose–6–phosphate synthetase which links two D–glycopyranosyl units at C1 and C1'. UDP–glucose and glucose–6–phosphate act as the precursors (Lang and Philp 1998). Labeling studies with homologous lipid moieties established a biosynthetic pathway where the complete lipid moiety (corynomycolic acid) was formed and subsequently linked by a stepwise esterification to α,α'-trehalose, forming α,α'-trehalose 6,6'-dicorynomycolates (Figure 8.2) (Kretschmer and Wagner 1983).

A detailed pathway to the biosynthesis of all mycolic acids in *M. tuberculosis* and TDMs was proposed by Takayama et al. (2005). The final step in the synthesis of the cell wall of *M. tuberculosis* is the attachment of mycolic acids to the peptidoglycan–arabinogalactan complex of the cell wall. The authors proposed that the synthesis of TDM begins by transfer of mycolic acids from mycolyl–S–Pks13 (polyketide synthase) to D–mannopyranosyl–1–phosphoheptaprenol to yield 6–O–mycolyl–β–D–mannopyranosyl–1–phosphoheptaprenol and then to trehalose 6–phosphate to yield phosphorylated trehalose monomycolate (TMM–P). Phosphatase releases the phosphate group to yield trehalose monomycolate (TMM) which is immediately transported outside the cell by an ABC transporter. Antigen 85 (mycolyltransferase) then catalyzes the transfer of a mycolyl group from TMM to the cell wall arabinogalactan and to other TMMs to produce arabinogalactan–mycolate and TDM, respectively (Takayama et al. 2005).

However, it should be noted that there is still a lack of fundamental research on the metabolic pathways involved in trehalose lipid synthesis from different carbon sources and on the regulation mechanisms allowing optimum biosurfactant production.

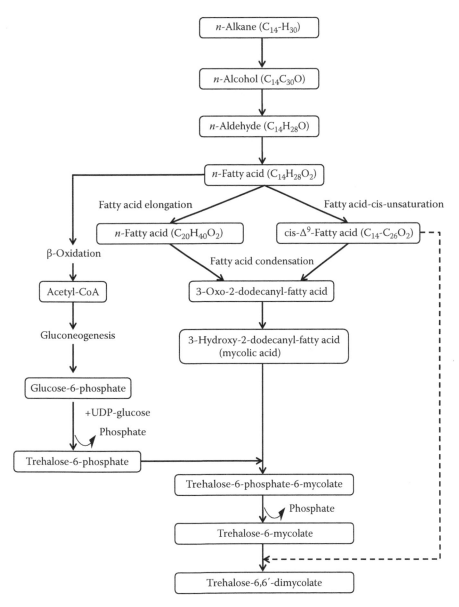

FIGURE 8.2 Scheme of biosynthesis of treholose mono- and dimycolate by *Rhodococcus erythropolis* DSM 4315 from *n*-alkanes. (Modified by Kretschmer, A. and Wagner, F., *Biochim. Biophys. Acta*, 753, 306, 1983.)

CARBON SOURCE AND GROWTH CONDITIONS

It was assumed that the alkanothrophic *Rhodococcus* strains respond specifically to the presence of hydrocarbons in the medium by producing cell-bound trehalose lipid surfactants (Lang and Philp 1998). Several studies were conducted confirming this capability. For example, we tested various hydrocarbons as substrates for *Rhodococcus wratislaviensis* biosurfactant production and found that strain *BN38* was able to utilize *n*-alkanes ranging from *n*-octane to *n*-heptadecane (Table 8.1). Short-chain *n*-alkanes (C_5–C_7) were not able to support growth and only after *n*-undecane the organism produced significant amounts of biomass and biosurfactant with the surface tension lowered to 29–32 mN m^{-1}. Surfactant concentration reached the maximal level of 3.1 g L^{-1} on hexadecane. Similar results were obtained by Philp et al. (2002) with the alkanotrophic strain *Rhodococcus ruber* IEGM 231. The surfactant production of *R. ruber* increased with the increase in *n*-alkanes chain length. Thus, during growth on undecane and dodecane, the surfactant concentrations were minimal, but increased from 1.8–2.2 g L^{-1} on tridecane and tetradecane, to 4.0 g L^{-1} on pentadecane, reaching its maximal level of 9.9 g L^{-1} on hexadecane. Haddadin et al. (2009) tested four carbon sources (diesel, naphtalene, crude oil, and benzene) and found that biosurfactant production of *R. erythropolis* and *R. ruber* strains was optimal on naphthalene and diesel oil. More recently, some authors reported the production of extracellular trehalose lipids from *n*-alkanes. Thus, *Rhodococcus* sp.

TABLE 8.1
Growth and Surfactant Production of *R. wratislaviensis BN38* on *n*-Alkanes

Growth Substrate	Biomass (g L⁻¹)	Surface Tension (mN m⁻¹)	Surfactant Concentration (g L⁻¹)
Pentane	—	59.0	—
Hexane	—	62.0	—
Heptane	—	62.0	—
Octane	0.54	40.3	0.05
Nonane	0.73	39.0	0.15
Decane	0.86	35.0	1.50
Undecane	0.90	32.0	0.75
Dodecane	1.65	23.3	1.20
Tridecane	1.53	29.0	1.20
Tetradecane	1.80	30.2	1.50
Pentadecane	2.30	29.0	1.85
Hexadecane	3.35	28.4	3.10
Heptadecane	2.54	29.0	2.75

Source: Tuleva, B., Christova, N., Cohen, R., Stoev, G., and Stoineva, I.: Production and structural elucidation of trehalose tetraester (biosurfactant) from a novelalkanothrophic alkanothrophic *Rhodococcus wratislaviensis* strain. *J. Appl. Microbiol.* 2008. 104. 1703–1710. Copyright Wiley-VCH Verlag GmbH & Co. KGaA. Reproduced with permission.

SD-74 produced extracellular succinoyl trehalose lipids on *n*-hexadecane as the sole carbon source (Tokumoto et al. 2009). On the other hand, when cultivated on the same substrate, two types of biosurfactants (free fatty acids and trehalose lipids) were detected in the supernatant of the bacterial culture of *R. erythropolis* strain 3C–9 (Peng et al. 2007). Furthermore, the ability of some rhodococci to produce extracellular biosurfactants on soluble substrates was also studied. *Rhodococcus erythropolis* ATTCC 4277, when grown on glycerol as the sole carbon source, released all the trehalose lipids into the medium, whereas the production was partially cell-bound when the strain was cultivated on *n*-hexadecane. *Rhodococcus erythropolis* strain EK–1 produced mainly emulsifiers during growth on hydrophilic substrates and surface active substances on hydrophobic ones (Pirog et al. 2004).

Growth conditions for trehalose lipid production on hydrocarbons have been studied in several batch and continuous culture experiments. Cell-growth-associated production, production under growth-limiting conditions and by resting cells determine the accumulation of different types and quantities of trehalose lipids (Lang and Philp 1998). Thus, when *R. wratislaviensis* BN38 was grown on *n*-hexadecane, trehalose lipid accumulation was growth-associated and biomass increased with substrate depletion (Figure 8.3) (Tuleva et al. 2008). Similar correlation between glycolipid production, biomass formation, and *n*-hexadecane consumption was observed in *R. erythropolis*

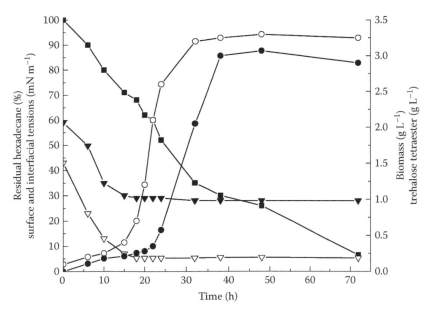

FIGURE 8.3   Growth and trehalose lipid production by *R. wratislaviensis BN38* during batch cultivation on 20 g L^{-1} *n*-hexadecane. Incubation was done at 28°C with shaking at 130 rpm. θ, Surface and σ, interfacial tensions of the culture; ■, residual hexadecane; o, biomass; ●, trehalose tetraester. (From Tuleva, B., Christova, N., Cohen, R., Stoev, G., and Stoineva, I.: Production and structural elucidation of trehalose tetraester (biosurfactant) from a novelalkanothrophic alkanothrophic *Rhodococcus wratislaviensis* strain. *J. Appl. Microbiol.* 2008. 104. 1703–1710. Copyright Wiley-VCH Verlag GmbH & Co. KGaA. Reproduced with permission.)

and *R. ruber* strains producing nonionic glycolipids (Rapp et al. 1979; Philp et al. 2002; Haddadin et al. 2009). The formation of trehalose–2,2′,3,4–tetraesters was favored both by nitrogen limitation and temperature shift from 30°C to 22°C during growth of *R. erythropolis* DSM432215 on hydrocarbons (Ristau and Wagner 1983). Trehalose–2,2′,3,4–tetraester production was growth-linked in a biosurfactant-producing *Rhodococcus* sp. 51T7 (Espuny et al. 1996), while the extracellular accumulation of mono- and disuccinoyl trehalose lipids by a strain *R. erythropolis* SD-74 was evidently not growth-associated (Uchida et al. 1989b). Kim et al. (1990) used both nitrogen limitation and resting cells for enhanced synthesis of an anionic tetraester during cultivation on *n*-alkanes and the successful conversion of the technical grade C10 *n*-alkane into more than 20 g L^{-1} of trehalolipids.

OPTIMIZATION OF PRODUCTION

According to Mukherjee et al. (2006), a large number of biosurfactant producers belonging to the genera *Rhodococcus*, *Gordonia*, and *Acinetobacter* have not yet been exploited properly for the economic production of naturally occurring surface-active compounds. This refers particularly to trehalose lipids that are often cell-bound; therefore, product recovery costs increase and production yields are low. Some practical approaches have been adopted with the aim of reducing biosurfactant production costs: (1) use of cheap and waste substrates, (2) bioprocess development including optimal production and recovery, (3) development of overproducing strains (Mukherjee et al. 2006).

As pointed earlier, there are still limited attempts for reducing production costs by using cheap substrates for trehalose lipid production. However, recent publications proved that rhodococci could grow well on vegetable oils and successfully produced biosurfactants. Thus, Sadouk et al. (2008) reported that *R. erythropolis* 16 LM.USTHB, grown on residual sunflower frying oil, produced extracellular glycolipids, lowering in their crude form the surface tension of water until 31.9 mN m^{-1}. Ruggeri et al. (2009) isolated *Rhodococcus* sp. BS32 capable of producing extracellular biosurfactants on rapeseed oil.

Several studies are focused on optimization of the cultural factors affecting biosurfactant production. For example, Uchida et al. (1989) reported that the production of succinoyl trehalose lipids in *R. erythropolis* SD-74 was optimized at high concentration of phosphate buffer and neutral pH and a yield of 40 g L^{-1} was achieved. Improvement of the cultivation conditions (oxygen supply, pH) and usage of 100 g L^{-1} technical grade C10 *n*-alkane resulted in a yield of 32 g L^{-1} trehaloselipid in 20 L bioreactor (Kim et al. 1990). Espuny et al. (1996) optimized sodium nitrate, potassium phosphate, and iron concentration, and the production of trehalose tetraester by *Rhodococcus* strain 51T7 increased from 0.5 to 3.0 g L^{-1}. More recently, the production of biosurfactant from *Rhodococcus* spp. MTCC 2574 was effectively enhanced by response surface methodology (Mutalik et al. 2008). The yield of biosurfactant before and after optimization was 3.2 and 10.9 g L^{-1}, respectively. Franzetti et al. (2009) applied a steepest ascent procedure and a Central Composite Design to obtain a second-order polynomial function fitting the experimental data near the optimum. In the optimized cultural condition, they obtained a fivefold increase in the biosurfactant concentration compared to the unoptimized medium.

The optimization procedure did not change the number and type of the glycolipid biosurfactants produced by *Gordonia* sp. BS29.

There is only one report for the development of a recombinant strain *Gordonia amarae* through electroporating the gene *vgb* encoding *Vitreoscilla* (bacterial) hemoglobin (VHb) (Dogan et al. 2006). The result was increased biosurfactant production in the engineered strain by 1.4-fold and 2.4-fold for limited and normal aeration, respectively.

PHYSICOCHEMICAL PROPERTIES

The surfactant properties of trehalose lipids have been studied and documented predominantly in the *Rhodococcus* genus. Most of them were found to be powerful biosurfactants, lowering in their pure form the surface tension of water from 72 mN m^{-1} to values between 19 and 43 mN m^{-1}, and the interfacial tension between water and *n*-hexadecane (decane or kerosene) to less than 1 mN m^{-1} at CMC values between 0.7 and 37 mg L^{-1} (Niescher et al. 2006; Peng et al. 2007; Tuleva et al. 2008, 2009; Marques et al. 2009; Tokumoto et al. 2009).

For example, the trehalose corynomycolates from *R. erythropolis* DSM43215 reduced interfacial tension from 44 to 18 mN m^{-1}. Their interfacial properties were stable in solutions with a wide range of pH and ionic strength, and the estimated critical micelle concentration was extremely low in high-salinity solutions (Kretschmer et al. 1982). Tokumoto et al. (2009) demonstrated that the succinoyl trehalose lipid (STL-1) from *R. erythropolis* SD-74 and its sodium salt (NaSTL-1) exhibited excellent surface activity at quite low concentrations. The estimated CMC and γ_{cmc} values for STL-1 were 5.6 × 10^{-6} M and 19 mN m^{-1}, and those for NaSTL-1 were 7.7 × 10^{-6} and 32.7 mN m^{-1}, respectively. In the case of the purified from *R. wratislaviensis* trehalose tetraester, the surface tension of water was reduced to 24.4 mN m^{-1} at a concentration of 5 mg L^{-1} (the CMC), while the interfacial tension at the water/hexadecane interface decreased to 1.3 mN m^{-1}.

Trehalolipids also show the ability to emulsify various hydrocarbons and oils. Thus, Bicca et al. (1999) registered emulsion indices (E_{24}) from 20% to 60% for the *R. ruber* biosurfactant with *n*-alkanes, aromatic hydrocarbons, and petroleum fractions. Similar data were obtained by Ivshina et al. (1998) for the biosurfactants produced by *R. erythropolis*, *R. longus*, *R. opacus*, and *R. ruber* with a maximum E_{24} of 62.5%. Tuleva et al. (2008) studied the emulsifying properties of the purified trehalose tetraester from *R. wratislaviensis* with several hydrophobic substrates in comparison with synthetic surfactants. The pure product successfully emulsified aliphatic and aromatic hydrocarbons, carbon mixtures, and several oils, and showed for almost all tested hydrocarbons higher activity than the chemical surfactants (Table 8.2).

The hydrophilic–lipophilic balance (HLB) indicates whether a surfactant will promote water-in-oil or oil-in-water emulsions with a view to possible fields of industrial use. For example, trehalolipids from *R. erythropolis* 517T with the HBL value of 11 produced stable emulsions with water and paraffin or isoporpylmyristate (Marques et al. 2009). This property makes them useful as detergents or as stabilizers for oil-in-water emulsions. On the other hand, Kuyukina et al. (2006) determined an HLB value of 6.4–6.7 for the crude biosurfacatant from *R. ruber* IEGM 231, indicating

TABLE 8.2

Emulsifying Activity of the Purified Trehalose Tetraester from *R. wratislaviensis BN38*

Hydrocarbon Substrate	Trehalose Tetraester	Triton X-100	Tween 20	Tween 80
Toluene	54 ± 1	35 ± 1	28 ± 1	32 ± 1
Benzene	68 ± 1	42 ± 1	31 ± 1	34 ± 1
Xylene	62 ± 1	47 ± 2	34 ± 1	39 ± 1
Hexadecane	65 ± 2	49 ± 1	47 ± 1	49 ± 3
n-Alkanes (C_{12-22})	47 ± 1	50 ± 1	42 ± 2	47 ± 2
Kerosene	50 ± 1	49 ± 2	35 ± 1	39 ± 1
Crude oil	23 ± 2	17 ± 3	16 ± 2	18 ± 1
Mineral oils	52 ± 2	50 ± 1	47 ± 2	43 ± 1
Olive oil	69 ± 1	54 ± 1	43 ± 2	46 ± 1
Sunflower oil	63 ± 1	59 ± 1	40 ± 2	42 ± 1
Almond oil	70 ± 2	52 ± 2	44 ± 1	45 ± 3

Source: Tuleva, B., Christova, N., Cohen, R., Stoev, G., and Stoineva, I.: Production and structural elucidation of trehalose tetraester (biosurfactant) from a novelalkanothrophic alkanothrophic *Rhodococcus wratislaviensis* strain. *J. Appl. Microbiol.* 2008. 104. 1703–1710. Copyright Wiley-VCH Verlag GmbH & Co. KGaA. Reproduced with permission.

[a] Results are expressed as percentages of the total height occupied by the emulsion; values are means of at least three determinations.

its prevailing hydrophobic properties and confirm its potential as a water-in-oil emulsifier. However, it should be pointed that available information is still limited on the physicochemical properties of trehalose lipids and that is one of the reasons why their industrial application is quite insufficient.

PHYSIOLOGICAL ROLE AND BIOLOGICAL ACTIVITIES

The synthesis of biosurfactants is often associated with the assimilation of hydrophobic substrates with low water solubility. The results presented in several studies (Bouchez-Naitali et al. 1999, 2001; Bouchez-Naitali and Vandecasteele 2008) on the role of biosurfactants in the mode of alkane uptake show that it is strongly related to the surface properties of the microbial degraders. Rapp et al. (1979) noted that only 10% of the total TDMs were released in the culture medium, thus conferring high hydrophobicity of the cell surface. Cell-bound biosurfactants promote high cell surface hydrophobicity, allowing direct contact of the cells with oil droplets (Lang and Philp 1998). This is consistent with our observations of a spontaneous attachment of the highly hydrophobic cells of *R. wratislaviensis* and *Micrococcus luteus* to the oil surface in the early exponential growth phase (Tuleva et al. 2008, 2009). It should also be mentioned that at that time, growth was restricted to the hydrocarbon–aqueous interface, confirming the high affinity of the cells to the hydrophobic substrate. Furthermore, microorganisms can use their biosurfactants to regulate their surface properties, in order to attach or detach from surfaces, according to needs (Ron and Rosenberg 2001). Franzetti et al. (2008) described that *Gordonia* cells presented high values of cell surface hydrophobicity in the early exponential growth, and remained attached to large hydrocarbon drops, so access to the substrate was by direct contact. During the late exponential phase, cells became hydrophilic, adhesion to hydrocarbons decreased, and the authors hypothesized that this led to a change in the mode by which *Gordonia* cells access the substrate during growth on hydrocarbons. At the same time, cells excreted extracellular bioemulsifier, allowing the hydrophilic cells to attach to the hydrophilic outer layer of the emulsified oil droplets.

The present interest in the physiological role of cell-wall trehalose lipids from a medical point of view has been focused exclusively on mycobacterial cord factors because of their toxic properties and key role in the pathogenecity of *Mycobacteria*. Trehalose 6,6′-dimycolate (TDM) is the most abundant, most granulomagenic, and most toxic lipid extractable from the surface of virulent *Mycobacterium tuberculosis* (Hunter et al. 2006). TDM is the main virulence factor for *M. tuberculosis* that makes it resistant to anti-tuberculoses medications. According to Barry et al. (1998), TDM, together with arabinogalactan mycolate and TMM, forms an integral part of the cell-wall cytoskeleton, conferring high cell surface hydrophobicity and acid fastness. Intensive research revealed that mycolic-acid-containing glycolipids, namely TDM as their best-studied representative, exert a number of immunomodifying effects (Ryll et al. 2001). They are able to stimulate innate, early adaptive, and both humoral and cellular adaptive immunity. Most functions can be associated with their ability to induce a wide range of chemokines (MCP-1, MIP-1alpha, IL-8) and cytokines (e.g., IL-12, IFN-gamma, TNF-alpha, IL-4, IL-6, IL-10).

Trehalose lipids from *Rhodococcus* and related genera showed no growth inhibition against Gram-negative bacteria and yeasts (Kitamoto et al. 2002). However, trehalolipids from *R. erythropolis* DSMZ 43215 inhibited the conidia germination of the fungus *Glomerella cingulata* at a concentration of 300 mg L^{-1} (Kitamoto et al. 2002). It was also reported that mice inoculated with TDM emulsion acquired high resistance to intranasal infection by influenza virus (Azuma et al. 1987; Hoq et al. 1997).

Several publications have recently appeared on the biological activities of trehalose lipid biosurfactants on phospholipid membranes aiming to gain insight into the molecular mechanisms of these interactions. Aranda et al. (2007) described molecular interactions between trehaloselipid and phosphatidylcholine resulting in an alteration of bilayer stability, and in this respect indicating that trehalose lipid is able to permeabilize phospholipid membranes. Ortiz et al. (2008) studied the effect of a purified from *Rhodococcus* sp. trehalose lipid on the thermotropic and structural properties of phosphatidylethanolamine membranes of different chain length and saturation. They found that the biosurfactant affected the gel-to-liquid crystalline phase transition of phosphatidylethanolamines, broadening and shifting the transition to lower temperatures. The trehalose lipid did not modify the macroscopic bilayer organization of saturated phosphatidylethanolamines and presented good miscibility both in the gel and the liquid crystalline phases. The conclusion is that the trehalose lipid incorporates into the phosphatidylethanolamine bilayers and produces structural perturbations that might affect the function of the membrane. Similar results were also obtained with phosphatidylserine membranes (Ortiz et al. 2009). Zaragoza et al. (2009) proposed a mechanism of membrane permeabilization in which trehalose lipid incorporates into phosphatidylcholine membranes and segregates within lateral domains that may constitute membrane defects or "pores." A recent study on the hemolytic activity of a succinoyl trehalose lipid, produced by *Rhodococcus* sp., revealed that the biosurfactant caused the hemolysis of human erythrocytes by a colloid–osmotic mechanism, most likely by the formation of enhanced permeability domains, or "pores" enriched in the biosurfactant within the erythrocyte membrane (Zaragosa et al. 2010).

POTENTIAL APPLICATION OF TREHALOSE LIPIDS

Biosurfactant application in environmental and industrial biotechnologies is very promising due to their low toxicity, biodegradability, functional stability, and environmental compatability. At present, biosurfactants can be used as emulsifiers, deemulsifiers, wetting and foaming agents, detergents, and food ingredients (Banat et al. 2000). The main commercial use of biosurfactants is in hydrocarbon bioremediation for enhancement the solubility of hydrophobic compounds and their biodegradation in contaminated soils. Another field of application is the oil industry for microbial enhanced oil recovery (MEOR), and oil storage tank cleaning (Singh et al. 2007). Biosurfactants were also found to possess several properties of therapeutic and biomedical importance (Rodrigues et al. 2006). Furthermore, they have the potential to be used as anti-adhesive biological coatings for biomaterials, thus reducing hospital infections and the use of synthetic drugs and chemicals.

Interest in the use of trehalose biosurfactants, like other biosurfactants, in various fields has been increasing in the past years as a result of their unique multifunctional features. Their present potential applications have been comprehensively reviewed by Franzetti et al. (2010) and Kuyukina and Ivshina (2010). The wide diversity of trehalose biosurfactants makes them an attractive group of compounds for potential use as green alternatives and eco-friendly methods in a great variety of industrial, biotechnological, and biomedical applications (Figure 8.4). For example, in remediation technologies, where product purity is of less concern, trehalose lipids attracted attention as solubilizing agents for different hydrophobic compounds.

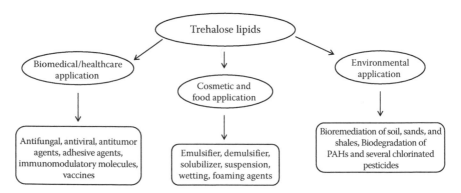

FIGURE 8.4   Scheme of potential application of trehalose lipids in environmental and biomedical fields.

Thus, Page et al. (1999) found that a biosurfactant from *Rhodococcus* strain H13–A was 35-fold more effective than the synthetic Tween 80 in increasing transfer of polycyclic aromatic hydrocarbons (PAHs) into the aqueous phase. It was also shown that some *Rhodococcus* strains produce biosurfactant complexes that remove almost 100% of oil sorbed onto sands and oil shales (Ivshina et al. 1998), and improved the displacement of crude oil from rock cores (Singer et al. 1990). From a medical point of view, mycobacterial sulfoglycolipids are of particular interest, because they are known to induce an immune response and are therefore potential candidates to be tested in novel vaccines against tuberculosis (Gilleron et al. 2004). It was found that naturally occurring diacyl trehalose sulfates are most potent T-cell activators, while other derivatives containing saturated or monounsaturated polymethylated fatty acids at the three-position showed modest activity. The presence of T cells that have the ability to recognize 2,3-diacyl-D-trehalose-2'-sulfates in patients with active or latent tuberculosis demonstrate the potential use of sulfoglycolipids in the development of new tuberculosis vaccines (Guiard et al. 2009).

Other beneficial effect of trehalose lipids is their ability to trigger cell differentiation instead of cell proliferation. Isoda et al. (1997) investigated the biological activities of seven glycolipids, including succinoyl-trehalose lipids STL-1 and STL-3. It was found that STLs possess differentiation-inducing activity in HL-60 human promyelocytic leukemia cell line, which was attributed to a specific interaction with the plasma membrane instead of a simple detergent-like effect. It was also found that the biological effects of STL-3 and its analogs depended on the structure of the hydrophobic moiety of STL-3 (Sudo et al. 2000).

Mycobacterial trehalose-6,6'-dimycolate (TDM) also exhibited strong antitumor activity (Orbach-Arbouys et al. 1983; Natsuhara et al. 1990), but its biomedical application is limited because of its high toxicity and the pathogenicity of the producing strain. In this respect, it is particularly important to search for trehalolipid surfactants with pronounced biological activities among the nonpathogenic producers. Thus, in our recent experiments, the trehalose tetraester from *M. luteus* BN56 displayed strong cytotoxic effect on human leukemic cell lines HL-60, BV-173 and SKW-3, and provides the backbone for the use of this glycolipid as novel reagent for the treatment of tumor cells (unpublished data).

CONCLUSION

Trehalose lipids have been extensively investigated over the past several decades. Now we possess a good amount of research regarding their production, chemical structures, and properties. Despite their favorable physicochemical and biological activities, which determine their potential applications in environmental and industrial biotechnologies, these affordable biosurfactants have not been commercialized successfully. This is mainly due to the fact that trehalose lipids are mostly cell-bound and are produced from nonrenewable carbon sources. Unfortunately, there is still lack of research on the development of efficient bioprocess, using low-cost and renewable resources, and optimization of the culture conditions, including cost-effective recovery processes. Although that different attractive aspects of trehalose lipids have been recently documented, such as their biomedical and therapeutic properties, future fundamental research should be focused toward the development of novel genetically engineered hyperproducing strains coupled with economization of the biosurfactant production process. The efficient combination of these options could open up perspectives for higher yields and successful large-scale economically profitable production of these unique biomolecules.

REFERENCES

Aranda, F.J., Teruel, J.A., Espuny, M.J., Marques, A., Manresa, A., Palacios-Lidon, E., and Ortiz, A. 2007. Domain formation by a *Rhodococcus* sp. biosurfactant trehalose lipid incorporated into phosphatidylcholine membranes. *Biochimica et Biophysica Acta* 1768:2596–2604.

Asselineau, C. and Asselineau, J. 1978. Trehalose-containing glycolipids. *Progress in the Chemistry of Fats and Other Lipids* 16:59–99.

Asselineau, J. and Lederer, E. 1956. The chemical structure of the cord factor of *Mycobacterium tuberculosis*. *Biochimica et Biophysica Acta* 20:299–309.

Azuma, M., Suzutani, T., Sazaki, K., Yoshida, I., Sakuma, T., and Yoshida, T. 1987. Role of interferon in the augmented resistance of trehalose-6,6′-dimycolate-treated mice to influenza virus infection. *Journal of General Virology* 68:835–843.

Banat, I.M., Makkar, R.S., and Cameotra, S.S. 2000. Potential commercial applications of microbial surfactants. *Applied Microbiology and Biotechnology*, 53:495–508.

Barry, C.E. 3rd, Lee, R.E., Mdluli, K., Sampson, A.E., Schroeder, B.G., Slayden, R.A., and Yuan, Y. 1998. Mycolic acids: Structure, biosynthesis and physiological functions. *Progress in Lipid Research* 37:143–179.

Batrakov, S.G., Rozynov, B.V., Koronelli, T.V., and Bergelson, L.D. 1981. Two novel types of glycolipids. *Chemistry and Physics of Lipids* 29:241–266.

Bicca, F.C., Fleck, L.C., and Ayub, M.A.Z. 1999. Production of biosurfactant by hydrocarbon degrading *Rhodococcus ruber* and *Rhodococcus erythropolis*. *RevistadeMicrobiologia* 30:231–236.

Bouchez-Naitali, M., Blanchet, D., Bardin, V., and Vandecasteele, J.P. 2001. Evidence for interfacial uptake in hexadecane degradation by *Rhodococcus equi*: The importance of cell flocculation. *Microbiology* 147:2537–2543.

Bouchez-Naitali, M., Rakatozafy, H., Marchal, R., Leveau, J.Y., and Vandecasteele, J.P. 1999. Diversity of bacterial strains degrading hexadecane in relation to the mode of substrate uptake. *Journal of Applied Microbiology* 8:421–428.

Bouchez-Naitali, M. and Vandecasteele, J.P. 2008. Biosurfactants, an help in the biodegradation of hexadecane? The case of *Rhodococcus* and *Pseudomonas*. *World Journal of Microbiology and Biotechnology* 24:1901–1907.

Cameotra, S.S. and Makkar, R.S. 2004. Recent applications of biosurfactants as biological and immunological molecules. *Current Opinion of Microbiology* 7:262–266.

Converse, S.E., Mougous, J.D., Leavell, M.D., Leary, J.A., Bertozzi, C.R., and Cox, J.S. 2003. MmpL8 is required for sulfolipid-1 biosynthesis and Mycobacterium tuberculosis virulence. *Proceedings of the National Academy of Sciences USA* 100:6121–6126.

Desai, J.D. and Banat, I.M., 1997. Microbial production of surfactants and their commercial potential. *Microbiology and Molecular Biology Reviews* 61:47–64.

Dogan, I., Pagilla, K.R., Webster, D.A., and Stark, B.C. 2006. Expression of *Vitreoscilla haemoglobin* in *Gordonia amarae* enhances biosurfactant production. *Journal of Industrial Microbiology and Biotechnology* 33:693–700.

Domenech, P., Reed, M.B., Dowd, C.S., Manca, C., Kaplan, G., and Barry, C.E. 2004. The Role of MmpL8 in sulfatide biogenesis and Virulence of *Mycobacterium tuberculosis*. *Journal of Biological Chemistry* 279:21257–21265.

Espuny, M.J., Egido, S., Mercade, M.E., and Manresa, A. 1995. Characterization of trehalose tetraester produced by a waste lubricant oil degrader *Rhodococcus* sp. *Toxicological and Environmental Chemistry* 48:83–88.

Espuny, M.J., Egido, S., Rodón, I., Manresa, A., Mercadé, M.E. 1996. Nutritional requeriments of a biosurfactant producing strain *Rodococcus* sp. 51T7. *Biotechnology Letters* 18:521–526.

Franzetti, A., Bestetti, G., Caredda, R., La Colla, P., and Tamburini, E. 2008. Surface active substances and their role in bacterial access to hydrocarbons in *Gordonia* strains. *FEMS Microbiology Ecology* 63:238–248.

Franzetti, A., Caredda, P., La Colla, P., Pintus, M., Tamburini, E., Papacchini, M., and Bestetti, G. 2009. Cultural factors affecting biosurfactant production by *Gordonia* sp. BS29. *International Biodeterioration and Biodegradation* 63:943–947.

Franzetti, A., Isabella Gandolfi, I., Bestetti, G., Smyth, T., and Banat, I.M. 2010. Production and applications of trehalose lipid biosurfactants. *European Journal of Lipid Science and Technology*, 112:617–627.

Georgiou, G., Lin, S.C., and Sharma, M.M. 1992. Surface-active compounds from microorganisms. *Biotechnology* 10:60–65.

Gilleron, M., Stenger, S., Mazorra, Z. et al. 2004. Diacylated sulfoglycolipids are novel mycobacterial antigens stimulating CD1-restricted T cells during infection with *Mycobacterium tuberculosis*. *Journal of Experimental Medicine* 199:649–659.

Goren, M.B., Brokl, O., Roller, P., Fales, H.M., and Das, B.C. 1976. Sulfatides of Mycobacterium tuberculosis: The structure of the principal sulfatide (SL-I). *Biochemistry* 15:2728–2735.

Guiard, J., Collmann, A., Garcia-Alles, L.F., Mourey, L., Brando, T., Mori, L., Gilleron, M., Prandi, J., De Libero, G., and Puzo, G. 2009. Fatty acyl structures of *Mycobacterium tuberculosis* sulfoglycolipid govern T cell response. *Journal of Immunology* 182: 7030–7037.

Haddadin, M.S.Y., Arqoub, A.A.A., Reesh, I.A., and Haddadin, J. 2009. Kinetics of hydrocarbon extraction from oilshale using biosurfactant producing bacteria. *Energy Conversion and Management* 50:983–990.

Hester, A. 2001. Market forecast. *Industrial Bioprocess* 23:3–4.

Hoq, M.M., Suzutani, T., Toyoda, T., Horiike, G., Yoshida, I., and Azuma, M. 1997. Role of gamma delta TCR+ lymphocytes in the augmented resistance of trehalose 6,6'-dimycolate-treated mice to influenza virus infection. *Journal of General Virology* 78:1597–1603.

Hunter, R.L., Olsen, M., Jagannath, C., and Actor, J.K. 2006. Trehalose 6,6'-dimycolate and lipid in the pathogenesis of caseating granulomas of tuberculosis in mice. *American Journal of Patholology* 168:1249–1261.

Isoda, H., Kitamoto, D., Shinmoto, H., Matsumura, M., and Nakahara, T. 1997. Microbial extracellular glycolipid induction of differentiation and inhibition of protein kinase C activity of human promyelocytic leukaemia cell line HL60. *Bioscience, Biotechnology and Biochemistry* 61:609–614.

Ivshina, I.B., Kuyukina, M.S., Philp, J.C., and Christo, N. 1998. Oil desorption from mineral and organic materials using biosurfactant complexes from *Rhodococcus* species. *World Journal of Microbiology and Biotechnology* 14:711–717.

Kim, J.-S., Powalla, M., Lang, S., Wagner, F., Lunsdorf, H., and Wray, V. 1990. Microbial glycolipid production under nitrogen limitation and resting cell conditions. *Journal of Biotechnology* 13:257–266.

Kitamoto, D., Isoda, H., and Nakahara, T. 2002. Functions and potential applications of glycolipid biosurfactants—From energy saving materials to gene delivery carriers. *Journal of Bioscience Bioengineering* 94:187–201.

Kretschmer, A., Bock, H., and Wagner, F. 1982. Chemical and physical characterization of interfacial-active lipids from rhodococcus erythropolis grown on n-alkanes. *Applied and Environmental Microbiology* 44:864–870.

Kretschmer, A. and Wagner, F. 1983. Characterization of biosynthetic intermmediates of trehalose dicorynomycolates from *Rhodococcus eythropolis* grown on n-alkanes. *Biochimica
   et Biophysica Acta* 753:306–313.
Kurane, R., Hatamochi, K., Kakuno, T., Kiyohara, M., Tajima, T., Hirano, M., and Taniguchi, Y.
   1995. Chemical structure of lipid bioflocculant produced by *Rhodococcus eythropolis.
   Bioscience Biotechnology and Biochemistry* 59:1652–1656.
Kuyukina, M.S. and Ivshina, I.B. 2010. Application of *Rhodococcus* in bioremediation of
   contaminated environments, *Biology of Rhodococcus, Microbiology Monographs,
   Springer*, 16:231–262.
Kuyukina, M.S., Ivshina, I.B., Gavrin, Yu. A., Podorozhko, E.A., Lozinsky, V.I., Jeffree, C.E.,
   and Philp, J.C. 2006. Immobilization of hydrocarbon-oxydizing bacteria in poly(vinyl
   alcohol) cryogels hydrophobized using a biosurfactant. *Journal of Microbiological
   Methods* 65:596–603.
Lang, S. 2002. Biological amphiphiles microbial biosurfactants. *Current Opinion in Colloid
   and Interface Science* 7:12–20.
Lang, S. and Philp, J.C. 1998. Surface-active lipids in rhodococci. *Antonie Van Leeuwenhoek*
   74:59–70.
Lederer, E. 1967. Glycolipids of mycobacteria and related microorganisms. *Chemistry and
   Physics of Lipids* 1:294–315.
Marques, A.M., Pinazo Farfan, A.M., Aranda, F.J., Teruel, J.A., Ortiz, A., Manresa, A., and
   Espuny, M.J. 2009. The physicochemical properties and chemical composition of tre-
   halose lipids produced by *Rhodococcus erythropolis* 51T7. *Chemistry and Physics of
   Lipids* 158:110–117.
Middlebrook, G., Coleman, C.M., and Schaefer, W.B. 1959. Sulfolipid from virulent tubercle
   bacilli. *Proceedings of the National Academy of Sciences USA* 45:1801–1804.
Mukherjee, S., Das, P., and Sen, R. 2006. Towards commercial production of microbial surfactants. *Trends in Biotecnolology* 24:509–515.
Mutalik, S.R., Vaidya, B.K., Joshi, R.M., Desai, K.M., and Nene, S.N. 2008. Use of response
   surface optimization for the production of biosurfactant from *Rhodococcus* spp. MTCC
   2574. *Bioresource Technology* 99:7875–7880.
Natsuhara, Y., Oka, S., Kaneda, K., Kato, Y., and Yano, I. 1990. Parallel antitumor, granu-
   loma-forming and tumornecrosis-factor-priming activities of mycoloyl glycolipids
   from Nocardia rubra that differ in carbohydrate moiety: Structure-activity relationships.
   Cancer Immunology Immunotherapy 31: 99–106.
Niescher, S., Wray, V., Lang, S., Kashabek, S.R., and Schlomann, M. 2006. Identification and
   structural characterization of novel trehalose dinocardiomycolates from n-alkane grown
   Rhodococcus opacus 1CP. *Applied Microbiology and Biotechnology* 70:605–611.
O'Connor, I. 2002. Market forecast: Microbial biosurfactants. *Industrial Bioprocess* 24:10–11.
Orbach-Arbouys, S., Tenu, J.P., and Petit, J.F. 1983. Enhancement of in vitro and in vivo anti-
   tumor activity by cord factor (6–60-dimycolate of trehalose) administered suspended in
   saline. *International Archives of Allergy and Applied Immunology* 71:67–73.
Ortiz, A., Teruel, J.A., Espuny, M.J., Marques, A., Manresa, A., and Aranda, F.J. 2008.
   Interaction of *Rhodococcus sp.* biosurfactant trehalose lipid with phosphatidylethanol-
   amine membranes. *Biochimica et Biophysica Acta* 1778:2806–2813.
Ortiz, A., Teruel, J.A., Espuny, M.J., Marques, A., Manresa, A., and Aranda, F.J. 2009.
   Interactions of a bacterial biosurfactant trehalose lipid with phosphatidylserine mem-
   branes. *Chemistry and Physics of Lipids* 158:46–53.
Page, C.A., Bonner, J.S., Kanga, S.A., Mills, M.A., and Autenrieth, R.L. 1999. Biosurfactant
   solubilization of PAHs. *Environmental Engineering Science* 16:465–474.
Peng, F., Liu, Z., Wang, L., and Shao, Z. 2007. An oil-degrading bacterium: *Rhodococcus
   erythropolis* strain 3C-9 and its biosurfactants. *Journal of Applied Microbiology*
   102:1603–1611.

Philp, J.C., Kuyukina, M.S., Ivshina, I.B., Dunbar, S.A., Christofi, N., Lang, S., and Wray, V. 2002. Alkanotrophic *Rhodococcus ruber* as a biosurfactant producer. *Applied Microbiology and Biotechnology* 59:318–324.

Pirog, T.P., Shevchuck, T.A., Voloshina, I.N., and Karpenko, E.V. 2004. Production of surfactants by *Rhodococcus erythropolis* strain EK-1, grown on hydrophilic and hydrophobic substrates. *Applied Biochemistry and Microbiology* 40:470–475.

Rapp, P., Bock, H., Wray, V., and Wagner, F. 1979. Formation, isolation and characterization of trehalose dimycolates from *Rhodococcus erythropolis* grown on n-alkanes. *Journal of General Microbiology* 115:491–503.

Rapp, P. and Gabriel-Jurgens, L.H.E. 2003. Degradation of alkanes and highly chlorinated benzenes, and production of biosurfactants, by a psychrophilic *Rhodococcus sp.* and genetic characterization of its chlorobenzene dioxygenase. *Microbiology* 149:2879–2890.

Ristau, E. and Wagner, F. 1983. Formation of novel anionic trehalose tetraesters from Rhodococcus erythropolis under growth limiting conditions. *Biotechnology Letters* 5:95–100.

Rodrigues, L., Banat, I.M., Texeira, J., and Oliviera, R. 2006. Biosurfactants: Potential applications in Medicine. *Journal of Antimicrobial Chemotherapy* 57:609–618.

Ron, E.Z. and Rosenberg E. 2001. Natural roles of biosurfactants. *Environmental Microbiology* 3:229–236.

Rosenberg, E. and Ron, E.Z. 1999. High- and low-molecular-mass microbial surfactants. *Applied Microbiology and Biotechnology* 52:154–162.

Ruggeri, C., Franzetti, A., Bestetti, G., Caredda, P., La Colla, P., Pintus, M., Sergi, S., and Tamburini, E. 2009. Isolation and characterization of surface active compound-producing bacteria from hydrocarbon-contaminated environments. *International Biodeterioration and Biodegradation* 63:936–942.

Ryll, R., Kumazawa, Y., and Yano, I. 2001. Immunological properties of trehalose dimycolate (cord factor) and other mycolic acid-containing glycolipids—A review. *Microbiology and Immunology* 45:801–811.

Sadouk, Z., Hacene, H., and Tazerouti, A. 2008. Biosurfactants production from low cost substrate and degradation of diesel oil by a *Rhodococcus* strain. *Oil and Gas Science and Technology* 63:747–753.

Singer, M.E.V., Finnerty, W.R., and Tunelid, A. 1990. Physical and chemical properties of a biosurfactant synthesized by *Rhodococcus* species H13-A. *Canadian Journal of Microbiology* 36:746–750.

Singh, A., van Hamme, J.D., and Ward, O.P. 2007. Surfactants in microbiology and biotechnology: Part 2. Application aspects. *Biotechnology Advances* 25:99–121.

Sudo, T., Zhao, X., Wakamatsu, Y. et al. 2000. Induction of the differentiation of human HL-60 promyelocytic leukemia cell line by succinoyl trehalose lipids. *Cytotechnology* 33: 259–264.

Suzuki, T., Tanaka, K., Matsubara, I., and Kinoshita, S. 1969. Trehalose lipid and α-branched-ß-hydroxy fatty acid formed by bacteria grown on n-alkanes. *Agricultural and Biological Chemistry* 33:1619–1627.

Takayama, K., CindyWang, C., and Besra, G.S. 2005. Pathway to synthesis and processing of mycolic acids in *Mycobacterium tuberculosis*. *Clinical Microbiology Reviews* 18:81–101.

Tokumoto, Y., Nomura, N., Uchiama, H., Imura, T., Morita, T., Fukuoka, T., and Kitamoto, D. 2009. Structural characterization and surface-active properties of succionyltrehalose lipid produced by *Rhodococcus* sp. SD-74. *Journal of Oleo Science* 58:97–102.

Tomiyasu, I., Yoshinaga, J., Kurano, F., Kato, Y., Kaneda, K., Imaizumi, S., and Yano, I. 1986. Occurrence of a novel glycolipid, 'trehalose 2,3,6′-trimycolate' in a psychrophylic, acid—fast bacterium, *Rhodococcus aurantiacus* (Gordona aurantiaca). *FEBS Letters* 203:239–242.

Tuleva, B., Christova, N., Cohen, R., Antonova, D., Todorov, T., and Stoineva, I. 2009. Isolation and characterization of trehalose tetraester biosurfactant from a soil strain *Micrococcus luteus* BN56. *Process Biochemistry* 44:135–141.

Tuleva, B., Christova, N., Cohen, R., Stoev, G., and Stoineva, I. 2008. Production and structural elucidation of trehalose tetraester (biosurfactant) from a novelalkanothrophic alkanothrophic *Rhodococcus wratislaviensis* strain. *Journal of Applied Microbiology* 104:1703–1710.

Uchida, Y., Misawa, S., Nakahara, T., and Tabuchi, T., 1989b. Factors affecting the production of succinoyl trehalose lipids By *Rhodococcus Erythropolis* SD-74 Grown On *N-Alkanes, Agricultural and Biological Chemistry* 53:765–769.

Uchida, Y., Tsuchiaya, R., Chino, M., Hirano, J., and Tabichi, T. 1989a. Extracellular accumulation of mono- and di-succinoyl trehalose lipids by a strain of *Rhodococcus erythropolis* grown on n-alkanes. *Agricultural and Biological Chemistry* 53:757–763.

Van Hamme, J.D., Singh, A., and Ward, O.P. 2006. Physiological aspects. Part I in a series of papers devoted to surfactants in microbiology and biotechnology. *Biotechnology Advances* 24:604–620.

Zaragoza, A., Aranda, F.J., Espuny, M.J., Teruel, J.A., Marques, A., Manresa, A., and Ortiz, A. 2009. Mechanism of membrane permeabilization by a bacterial trehalose lipid biosurfactant produced by *Rhodococcus* sp. *Langmuir* 26:8567–8572.

Zaragoza, A., Aranda, F.J., Espuny, M.J., Teruel, J.A., Marques, A., Manresa, A., and Ortiz, A. 2010. Hemolytic activity of a bacterial trehalose lipid biosurfactant produced by *Rhodococcus* sp.: Evidence for a colloid-osmotic mechanism. *Langmuir* 26:8567–8572.

9 Biosurfactant-Mediated Nanoparticle Synthesis

A Green and Sustainable Approach

Vivek Rangarajan, Snigdha Majumder, and Ramkrishna Sen

CONTENTS

INTRODUCTION

Nanotechnology deals with particles of nanoscale dimensions and holds the promise to bring incredible revolution in health science and biotechnology. This technology has the potential to create new materials and devices with a wide range of applications. On the other hand, nanotechnology raises many issues concerning human health safety, environmental and other public concerns about social and ethical issues. In recent years, nanotechnology has witnessed major advances and expanded its capabilities in areas such as drug delivery, surgery, and clinical research. Manipulation of materials at the molecular scale will result in the conception of many new functional nanomaterials with unusual but useful features that differ from bulk materials. The most striking features are changes in electrical, mechanical, and magnetic properties, and chemical properties such as solubility, reactivity, and catalytic activity (Schmid, 1992; Daniel and Astruc, 2004). Currently in the nanotechnology field, the development of efficient and cost-effective methods of nanoparticle synthesis has gained importance in attaining monodispersity and uniform dimension in an eco-friendly manner. It is due to the emergence of green chemistry approaches that almost all research endeavors have been primarily focused on cutting down the use of hazardous chemicals and replacing them with environment-friendly and sustainable reagents for nanoparticle synthesis. Particularly, the use of greener templates and nonhazardous reducing agents can significantly contribute to the formulation of an eco-friendly synthesis process.

Considering the expanding horizon of biosurfactant research, mainly due to its versatile applications in food, cosmetic, pharmaceutical, oil recovery, and environmental industries (Cameotra and Makkar, 2004; Mulligan, 2005; Sen, 2008; Kanlayavattanakul and Lourith, 2010), the current review focuses on the possibilities and potential of biosurfactants in the green synthesis of nanoparticles of both inorganic and organic types, with special emphasis on nanometals, which occupy a significant space in the research of nanomaterials.

PROPERTIES OF NANOPARTICLES

In recent years, efforts for finding or synthesizing new materials with properties that differ from those of bulk materials have increased to meet the increasing demand in various applications. Nanoparticles with unique properties hold the potential to cater to the future demands in various fields of rapidly advancing science and technology. The properties that merit the consideration of nanoparticles as potential candidates (Goesmann and Feldmann, 2010) over the existing bulk materials are as follows: (1) surface-dependent properties, in which the nanoparticles supersede the behavior of bulk materials, owing to their high surface area-to-volume ratio. Various applications that arise due to surface effects include catalysis, surface conductivity, thin film applications, etc. (Baxter and Schmuttenmaer, 2006; Lee et al., 2006; Moshfegh, 2009); (2) size-dependent properties in which the remarkably smaller size of nanoparticles makes them suitable for applications in optics, molecular biology, and medicine (Salata, 2004; Xu et al., 2006); and (3) size-dependent quantum effects, which meet the requirements for the applications of nanoparticles in surface-enhanced Raman spectroscopy (SERS) (Li et al., 2010; Kumar et al., 2011). So, on the whole, nanoparticles are categorized as those, on the one hand, with larger surface area that can act as functional-binding and adsorbing agents for drugs, probes, and proteins, and, on the other hand, with a distinct surface property that makes them chemically more reactive as compared to their fine analogs. In addition to these unique characteristics, another important criterion is the green aspect of nanomaterial synthesis that forms the essence of this review in the use of nontoxic chemicals, environmental benign solvents, and renewable materials during their synthesis.

METHODS OF SYNTHESIS

A number of methods have so far been adopted for the synthesis of nanoparticles, which are broadly categorized as gas-phase methods, liquid-phase methods, and grinding methods. Among them, liquid-phase method of nanoparticle synthesis such as forced hydrolysis, hydrothermal synthesis, sol–gel process, reverse-microemulsion method, etc., have gained considerable attention from the commercial point of view, as they can generate many functional nanoparticulate materials with improved properties (Eastoe and Warne, 1996; Hanh et al., 2003; Liao et al., 2006; Moncada et al., 2007; Goesmann and Feldmann, 2010). However, the most widely adopted is the chemical reduction method, in which the metal salts are converted to metal atoms using suitable reducing agents. The most common reducing agents are citrate (Polte et al., 2010), hydrides (Schwartzberg et al., 2006), ethylene glycol (Komarneni et al., 2002), and hydrazine (Maillard et al., 2003), which pose potential environmental and health hazards (Raveendran et al., 2003; Zhang et al., 2007). Though a vast number of literature demonstrate the use of well-established protocols for the chemical synthesis of nanoparticles, there has been a growing interest in recent years for the use of biotechnological approaches as a cost-effective and scalable synthesis options (Kiran et al., 2010; Narayanan and Sakthivel, 2010). Biotechnological-based approaches, in contrast to chemical synthesis method, do not require extreme processing conditions such as high temperature and pressure (Rao et al., 2003), and generally leave no toxic residues that may affect the environment.

GREEN METHODS FOR NANOPARTICLE SYNTHESIS

In a mission toward greener chemistry, it is important to eliminate or minimize the generation of waste. To address these concerns, Anastas and Warner (1998) put forward 12 fundamental principles for green chemistry. Of them, the most critical principles requiring immediate attention are adoption of less hazardous chemical synthesis routes, design of safer chemicals, use of renewable feed stocks, and design of degradable substrates and products (Raveendran et al., 2003).

With these critical principles as prime goals, two routes for green synthesis of nanoparticles have so far been identified as the most promising options that can emerge as imminent substitutes for the conventional routes of synthesis. One option is the use of microbial systems for facilitating a conducive environment for the synthesis of nanoparticles, and the other option is to employ the metabolites from microorganisms and plants, which turn out to be sources of stabilizers or substrates in nanoparticles synthesis. As the microbial system and its associated metabolisms are considered to be intricate by nature, the former route of synthesis where nanoparticles are formed, either intracellularly or extracellularly during growth (Mandal et al., 2006; Ghorbani et al., 2011), with the aid of biological macromolecules may not result in controlled synthesis of nanoparticles. In addition, the separation of the synthesized nanoparticles from the microbial system or the media environment would be a challenging task.

In the latter case, the starting materials that assist the nanoparticle synthesis are the products of microorganisms, such as proteins, peptides, DNA, oligonucleotides, lipids, and biosurfactants (Sweeney et al., 2004; Sotiropoulou et al., 2008). The self-assembly of these biological macromolecules gives rise to a variety of nanostructural templates that include shapes like tubes/wires, helixes, spheres (e.g., micelles and reverse micelles), fibrils, spindles, and rings (Chimentão et al., 2004; Elazzouzi-Hafraoui et al., 2008; Sotiropoulou et al., 2008). The embedded space within these soft macromolecular templates acts as a mold for casting of nanoparticles and further for enhancing their structural, functional, and physicochemical properties. The various functional groups and topological features of the biological macromolecules can be exploited for the self-assembly of in situ synthesized nanostructures without subjecting them to any harsh chemical treatments. It is believed that the biotemplates-based nanoparticle synthesis with its tunable size, shape, and monodispersibility will turn out to be a competent alternative that can surpass the hitherto achieved efficiencies by other conventional approaches.

METAL NANOPARTICLE SYNTHESIS

BIOSURFACTANTS AS CAPPING AGENTS

Properties of Biosurfactants

Biosurfactants are surface-active molecules that are stable at extremes of pH, temperature, and salinity. When compared to their synthetic counterparts, they have unique properties such as bulky and complicated structures, multiple chiral centers, ability to form supramolecular assemblies and liquid crystals, and other various biological activities (Xie et al., 2005). Biosurfactants are biodegradable in nature and are also reported to be nontoxic or less toxic as compared to synthetic surfactants. The aforementioned superior properties of biosurfactants elevate their status to "green molecules" for eco-friendly applications. One of the most important properties exploited in the nanoparticles synthesis is their micelle-forming ability. The noncovalent interactions that arise due to solvophobic effects of hydrophobic tails form the basis for self-aggregation into structures like micelles and vesicles. Micelles exist in different morphologies, such as spherical, ellipsoidal, and cylindrical structures, while vesicles are hollow spheres enveloped by bilayers of ampiphilic surfactants (Engberts and Kevelam, 1996; Davies et al., 2006). Biosurfactants above the critical micelle concentration (CMC) form micelles. Being amphiphiles, they can partition at air–water or water–oil (hydrocarbon) interfaces. Their use as emulsifiers in the food industry is attributable to this property.

The major concern in nanoparticle synthesis is tuning up these structures to obtain aggregates with desired morphology and properties. The morphology of these supramolecular structures can be significantly varied by changing the pH, temperature, surfactant concentration, and the ionic strength of solution.

Metal nanoparticles, owing to their size/shape-dependent, unique, and tunable properties (e.g., quantum confinement Yanhong et al., 2004), plasmon resonance (Hutter et al., 2001), and light scattering (Derkacs et al., 2008) etc., find applications in wide areas such as electronics, optics, catalysis, and biotechnology. By controlling the synthesis, it is possible to alter properties of nanosystems including surface area, optical and electrical properties, and the accessibility of the guest species.

Experimental conditions, such as pH, temperature, viscosity of solution, and processing conditions such as rate of reduction and adsorption mechanisms of stabilizing/capping agents with metal ions are some of the factors that can be controlled in the design of nanoparticles (Pradeep and Anshup, 2009; Ghorbani, 2011). Particularly, capping agents play a very important role in determining the final quality of the particles (Raveendran et al., 2003). It essentially reduces the tendency of nanoparticles to agglomerate, by protecting the surface by either causing steric or electrostatic stabilization (Cushing et al., 2004). Besides, it serves as a diffusion barrier for further growth of nanoparticles (Gutiérrez-Wing et al., 2012). Compounds like SDS, CTAB, Triton X100, AOT, and polyvinyl pyrrolidone cap the nanoparticles by steric stabilization, while trisodium citrate cap by electrostatic stabilization forming a double layer over the metals (Yonezawa et al., 2000; Chen and Hsieh, 2002; Gui et al., 2003; Ghorbani et al., 2011). Biosurfactant, as capping agents in nanoparticle synthesis, also facilitates the uniform dispersion of the nanoparticles in the liquid medium. The studies involving the use of biosurfactants as dispersion agents were proved to be very effective for the uniform distribution of zirconia (Biswas and Raichur, 2008) and colloidal alumina (Raichur, 2007) nanoparticles by rhamnolipids, and Cobalt nanoparticles (Kasture et al., 2007) by modified sophorolipids.

Chemically synthesized amphiphilic molecules were initially used as templates for nanoparticle synthesis. Zhu et al. (2005) synthesized glycolipid nanotubes functionalized with amino groups for chelation with Cu^{2+} ions, and glucose moieties for reducing the chelated Cu^{2+} ions to form copper nanoparticles. SEM and TEM analyses revealed that Cu^{2+} ions were properly seeded onto the glycolipid nanotubes, which further facilitated proper alignment of dense copper nanoparticles. The host organic nanotube templates could be easily removed by annealing the Cu-nanotube composite at 500°C in argon atmosphere.

Contemporarily, capping agents such as oleic acid, linoleic acid, and their derivatives were also used. However, the inherent limitations with these compounds, such as poor solubility of fatty acid and the exposure of methyl group to the solvent side, caused poor dispersion of nanoparticles (Kasture et al., 2007). Lately, biosurfactants such as rhamnolipids, sophorolipids, and lipopeptides have emerged as feasible alternatives for higher fatty acids.

Temperature influenced the kinetics of nanoparticle synthesis to a greater extent. Kasture et al. (2008) demonstrated the synthesis of silver nanoparticles at different temperatures between 30°C and 90°C using oleic acid and linoloeic-acid-derived sophorolipids as capping and reducing agents. These sophorolipids acted as a better capping agent than oleic acid. TEM image and DLS analysis revealed that nanoparticles were polydispersed due to slow reduction rate at 40°C. On the other hand, faster reduction rate at 90°C yielded monodispersed silver nanoparticles with an average particle diameter of 5.5 nm. Besides temperature another factor that influenced the particle size by causing slow reduction rate was the stronger association of Ag^+ ions with two cis double bonds of linoleic acid. A modified form of sophorolipid with improved properties was synthesized by hydrolyzing the sophorolipids with KOH (Kasture et al., 2007). The as-synthesized acid sophorolipids not only served as a better capping agent but also facilitated uniform dispersion of cobalt nanoparticles in aqueous solution. These sophorolipid-capped cobalt nanoparticles would be beneficial for applications in site-selective delivery systems, magnetic imaging, and magnetic separation techniques. Singh et al. (2010) evaluated cyto- and genotoxicities of sophorolipid-conjugated gold and silver nanoparticles. Of the two metal systems, gold nanoparticles were found to be more cyto- and genocompatible than silver nanoparticles.

Surfactin, a well-studied lipopeptide biosurfactant, synthesized by *Bacillus* sp., can form micelles at a very low concentration of 0.005% (Sen, 1997). Liao et al. (2010) prepared Surfactin-stabilized supermagnetic ironoxide nanoparticles (SPION) that can serve as a contrast agent for magnetic resonance imaging. The as-synthesized SPION particles in organic solvent dispersed easily in aqueous phase, to form a stable suspension of spherical nanoparticles. Singh et al. (2011) used Surfactin in the synthesis of cadmium sulfide nanoparticles. Results showed no evidence of color change or agglomeration over a period of 120 days. The improved stability of nanoparticles was due to the interaction of cadmium sulfide nanoparticles with the free amine groups of Surfactin. The presence of Surfactin had not only stabilized the particles but also acted as the capping agents during synthesis, which confined the phase-transitioned cubic nanoparticles size to 3–4 nm. Concentration of biosurfactants has always played a significant role in determining the structure of nanoparticles. The thickness of the petals in the rose-like ZnO nanostructure was greatly affected by the concentration of initial Surfactin in the precursor solution (Reddy et al., 2011). Increase in Surfactin concentration decreased the thickness of petals. The as-synthesized ZnO nanoparticles were further tested for their efficacy in the photocatalytic degradation of methylene blue.

Reddy et al. (2009) studied the effect of pH on size and stability of gold nanoparticle obtained through Surfactin-mediated synthesis. Nanoparticles were found to be more stable at pH 7 and 9, while they were aggregated within 24 h at pH 5. Furthermore, particles were of uniform size and shape at pH 7, but polydispersed at pH 5 and 9.

Other structures like porous microtubules were also reported for the synthesis of metal nanoparticles using biosurfactants. Rehman et al. generated rhamnolipid microtubules by refluxing native rhamnolipids with gold salts. These microtubules served as templates for the synthesis and self-assembly of gold nanoparticles to obtain rhamnolipid gold nanoparticle composite microtubules. Further heat treatment of these composite microtubules yielded porous gold microwire–like structures that find applications in electronics, optics, catalysis, and sensing. Similarly, Narayanan et al. (2010) used rhamnolipid for capping ZnS nanoparticles in an aqueous medium. SAXS analysis revealed uniformly dispersed nanospheres with the size range of 4.5 nm.

MICROEMULSION-BASED NANOPARTICLE SYNTHESIS

Biomolecules as building blocks in nanoparticle synthesis can maintain and integrate the structural and functional diversity of biosystems with the inherent properties of nanomaterials (Rehman et al., 2010). One of the most important and challenging steps that influences the overall properties of final product is the generation of a suitable self-assembled microstructure. Various customized protocols have been developed for the self-assembly of biosurfactants to obtain microstructures of required sizes and shapes. Elucidating the underlying mechanisms in the formation of microemulsion is important in designing a controlled process for nanoparticle synthesis. Microemulsion systems are pseudoternary systems that require the formation of a clear homogeneous system comprising water, oil, surfactant, and an alcohol-based cosurfactant (Eastoe et al., 2006). In general, the phase behavior of the pseudoternary systems can be studied from the phase diagrams, in which the weight ratio of two components is generally held constant (Xie et al., 2005). The formation of a clear single phase indicates the presence of monophasic system. The various effects such as solvent, surfactants, and cosurfactants, added electrolyte, reagent concentration, water content that influence the particle size and other effects such as nature of templates, influence of ion/molecular adsorption that influences particle shapes were comprehensively reviewed (Eastoe et al., 2006). The use of long-chain alcohol such as butanol in the pseudoternary system, with rhamnolipid as biosurfactant, widened the phase existence of the monophasic microemulsion system (Xie et al., 2006).

As surfactants are amphiphilic molecules, they are soluble both in water and hydrocarbon. In a ternary system containing water, oil, and cosurfactant, the surfactant will form spherical aggregates called reverse micelles by making ion–dipole interactions with polar cosurfactant. The polar head group orients toward the center, while the hydrocarbon tails face outward from the core. Here the cosurfactant acts as a spacer molecule, which minimizes the electrostatic repulsion between the polar head groups (Cushing et al., 2004). When water is added to these suspensions, they are attracted toward the center as a result of ion–dipole and dipole–dipole interactions.

The spherical micelles are characterized by the molar water-to-surfactant ratio:

$$\Omega_0 = \frac{[\text{Water}]}{[\text{S}]}$$

The relationship between Ω_0 and micellar radius R_m is

$$R_m = \frac{3V_s}{\Sigma_s} + \frac{3V_w\Omega_0}{\Sigma_s}$$

where

V_s and V_w are the molar volumes of surfactant and water, respectively (Cushing et al., 2004)

Σ_s is the molar interfacial area at the surfactant–oil boundary

The size of the micelle can be varied by varying the water-to-surfactant ratio. The higher the ratio, the larger will be the size of reverse micelles. Nevertheless, the challenge is to obtain a microemulsion system containing reverse micelles with uniform size and shape. The frequent collisions between the small-sized reverse micelles due to Brownian movements exclude some of the surfactant molecules into the bulk oil phase. This results in the formation of short-lived dimers that can mutually exchange the contents between them. The continuous collision of reverse micelles and the subsequent formation of short-lived dimers in a homogeneous microemulsion will facilitate the uniform distribution of contents. The fact that contents entrapped in two reverse-micellar microemulsion solutions can be exchanged, in turn, can necessitate a chemical reaction. So, a typical procedure for reverse-miceller-based nanoparticle synthesis

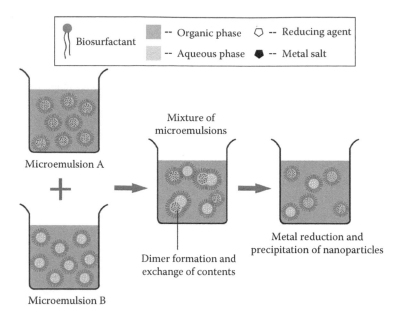

FIGURE 9.1 **(See color insert.)** Mechanism for microemulsion-based nanoparticles synthesis. (Adapted from Capek, I., *Adv. Colloid Interface Sci.*, 110, 49, 2004.)

involves the mixing of two suspensions of reverse-micellar solutions, one containing the salt of a metal and the other having a reducing agent as given in Figure 9.1.

Further, the presence of surfactants in the reverse micelles can act as capping agents that prevent aggregation of products (Soma and Papadopoulos, 1996).

Xie et al. (2006) successfully generated a single-phase microemulsion system involving pseudoternary mixtures of ionized-rhamnolipid/n-butanol/n-heptane/ water. Butanol as cosurfactant formed a stable single-phase microemulsion when compared to other lower alcohols, as revealed by the phase diagrams. Here the rhamnolipids inside the micelles prevented the aggregation of silver nanoparticles by inducing both electrostatic repulsion as well as steric hindrance around the nanoparticles. TEM studies revealed that the particles were of the size range of 2–8 nm and, further, they were found to be consistent in size over a period of 60 days when suspended in a solution. Using a similar approach, Kiran et al. (2010) investigated the synthesis of silver nanoparticles using glycolipid biosurfactant of *Brevebacterium Casei* MSA19. The particles were reported to be homogeneous in size and were stable over a period of 2 months.

Maity et al. (2011) synthesized crystalline calcium pyrophosphate nanoparticles of different morphologies by varying the water-to-surfactant (W/S) ratio in water/hexane/Surfactin reverse-microemulsion system. Nanoparticles with size distribution of 50–500 nm were obtained when the as-synthesized noncalcinated calcium phosphate (synthesized at room temperature) particles were subjected to calcination at a temperature of 800°C. The water-to-surfactant ratio influenced the shape and size of the micelle to a greater extent, which in turn changed the morphology of the calcinated nanoparticles. TEM image revealed transformation from regular sphere-like structure to irregular plate-like structures as the W/S ratio was increased from 250 to 40,000.

pH plays a vital role in determining the morphology of the biosurfactant structure (micelles or vesicles) and in turn the nanoparticles' morphology. The morphology of biosurfactant was transformed from lamellar to micellar when the pH of the solution was increased from pH = 6 (Palanisamy, 2008). The as-synthesized $Ni(OH)_2$ nanoparticles in rhamnolipid/n-heptane/n-butanol/water pseudoternary system were flaky at pH 8.0. However, at higher pH values, the particles agglomerated to form a mixture of flaky and spherical-shaped $Ni(OH)_2$ particles. At higher pH values, the supersaturation of hydroxyl ions increased the rate of nucleation, thereby forming smaller size particles (Palanisamy and Raichur, 2009). Calcination of these nickel hydroxide precipitates at 600°C, resulted in the formation of NiO nanorods in the diameter range from 8 to 35 nm.

Though the greener synthesis of nanoparticles has made significant advances in inorganic nanometals research, its utilization is yet to be realized in organic nanoparticle research. The use of biosurfactants in the organic nanoparticles synthesis has just started its expedition, with Worakitsiria et al. (2011) reporting the use of rhamnolipids as soft templates in the synthesis of polyaniline (PANI) nanoparticles. Oxidative polymerization of aniline as a starting monomer, in the presence of hydrochloroacid as a dopant and ammoniumpersulfate as oxidant, yielded PANI nanorods and nanotubes. The aspect ratio of the final product was manipulated by changing the weight ratios of ANI-to-rhamnolipid in the initial reaction mixture.

CONCLUSION

The research in the field of nanotechnology is advancing rapidly and it encompasses fields as diverse as electronics, medicine, surface science, organic chemistry, molecular biology, etc. Thus, there is a remarkable increase in the applications of nanoparticles and hence the need for its increased production. But, traditional chemical synthesis methods raise concerns about toxicity and environmental impact. This prompted us to switch over to "Green" synthesis options. The promising examples cited in this review on the use of biosurfactants as an alternative to synthetic surfactants give us a positive hope to make greener processes. Biosurfactants that are produced from environmentally safe renewable resources are preferred for nanoparticle synthesis as they use low temperature, cost-effective methods, retain sustainability, and generate minimum waste. It can minimize the problem of environmental hazards and the high cost of organic solvents and fatty acids used as capping agents. Cheaper raw materials, optimized and efficient bioprocesses, recombinant strains achieves maximum productivity of biosurfactant (Mukherjee et al., 2006; Rangarajan et al., 2012), which results in synthesizing nanoparticles more efficiently and economically. The research on biosurfactants for extended applications has been continuing. However, there is still a considerable amount of basic works pertaining to characterizing micelles and vesicles for the demanding reaction conditions that need to be carried out before they are applied to nanoparticles synthesis. Also, as the data on the phase behavior of different biosurfactants capable of forming microemulsion/reverse-microemulsion systems are scanty, a blueprint manifesting reliable behavior of such systems and that can further necessitate making of standard experimental protocols has to be established. It is strongly believed that from the growing evidence of specialized researches elucidating the morphologies of biosurfactants in oil-in-water and water-in-oil microemulsion systems and also estimating their stabilities under various physicochemical conditions will certainly pave the way for the successful realization of biosurfactants in greener syntheses of nanoparticles.

ACKNOWLEDGMENT

The authors acknowledge the help extended by B. Chandrakanth, B.Tech (4th year) student of the Department of Biotechnology, IIT Kharagpur.

REFERENCES

Anastas, P. T. and Warner, J. C. 1998. *Green Chemistry: Theory and Practice*. Oxford University Press: New York, p. 30.

Baxter, J. B. and Schmuttenmaer, C. A. 2006. Conductivity of ZnO nanowires, nanoparticles, and thin films using time-resolved. Terahertz spectroscopy. *Journal of Physical Chemistry B*, 110: 25229–25239.

Biswas, M. and Raichur, A. M. 2008. Electrokinetic and rheological properties of nano zirconia in the presence of rhamnolipid biosurfactant. *Journal of American Ceramic Society*, 91: 197–3201.

Cameotra, S. S. and Makkar, R. S. 2004. Recent applications of biosurfactants as biological and immunological molecules. *Current Opinion in Microbiology*, 7: 262–266.

Capek, I. 2004. Preparation of metal nanoparticles in water-in-oil (w/o) microemulsions. *Advances in Colloid and Interface Science*, 110: 49–74.

Chen, D. H. and Hsieh, C. H. 2002. Synthesis of nickel nanoparticles in aqueous cationic surfactant solutions. *Journal of Materials Chemistry*, 12: 2412–2415.

Chimentão, R. J., Kirm, I., Medina, F., Rodríguez, X., Cesteros, Y., Salagre, P., and Sueiras, J. E. 2004. Different morphologies of silver nanoparticles as catalysts for the selective oxidation of styrene in the gas phase. *Chemical Communications*, 4: 846–847.

Cushing, B. L., Kolesnichenko, V. L., and O'Connor, C. J. 2004. Recent advances in the liquid-phase syntheses of inorganic nanoparticles. *Chemical Reviews*, 104: 3893–3946.

Daniel, M. C. and Astruc, D. 2004. Gold nanoparticles: Assembly, supramolecular chemistry, quantum-size- related properties, and applications toward biology, catalysis, and nanotechnology. *Chemical Reviews*, 104: 293.

Davies, T. S., Ketner, A. M., and Raghavan, S. R. 2006. Self-assembly of surfactant vesicles that transform into viscoelastic wormlike micelles upon heating. *Journal of American Chemical Society*, 128, 6669–6675.

Derkacs, D., Chen, W. V., Matheu, P. M., Lim, S. H., Yu, P. K. L., and. Yub, E. T. 2008. Nanoparticle-induced light scattering for improved performance of quantum-well solar cells. *Applied Physics Letters*, 93: 91–107.

Eastoe, J., Hollamby, M. J., and Hudson, L. 2006. Recent advances in nanoparticle synthesis with reversed micelles. *Advances in Colloid and Interface Science*, 128–130: 5–15.

Eastoe, J. and Warne, B. 1996. Nanoparticle and polymer synthesis in microemulsions. *Current Opinion in Colloid and Interface Science*, 1: 800–805.

Elazzouzi-Hafraoui, S., Nishiyama, Y., Putaux, J. L., Heux, L., Dubreuil, F., and Rochas, C. 2008. The shape and size distribution of crystalline nanoparticles prepared by acid hydrolysis of native cellulose. *Biomacromolecules*, 9: 57–65.

Engberts, J. B. F. N. and Kevelam, J. 1996. Formation and stability of micelles and vesicles. *Current Opinion in Colloid and Interface Science*, 1: 779–789.

Ghorbani, H. R., Safekordi, A. A., Attar, H., and Sorkhabadi, S. M. R. 2011. Biological and non-biological methods for silver nanoparticles synthesis. *Chemical and Biochemical Engineering Quarterly*, 25: 317–326.

Goesmann, H. and Feldmann, C. 2010. Nanoparticulate functional materials. *Angewandte Chemie International Edition*, 49: 1362–1395.

Gui, Z., Fan, R., Mo, W., Chen, X., Yang, L., and Hu, Y. 2003. Synthesis and characterization of reduced transition metal oxides and nanophase metals with hydrazine in aqueous solution. *Materials Research Bulletin*, 38: 169–176.

Gutiérrez-Wing, C., Velázquez-Salazar, J. J., and José-Yacamán, M. 2012. Procedures for the synthesis and capping of metal nanoparticles. *Methods Mol. Biol.*, 906: 3–19.

Hanh, N., Quy, O. K., Thuy, N. P., Tung, L. D., and Spinu, L. 2003. Synthesis of cobalt ferrite nanocrystallites by the forced hydrolysis method and investigation of their magnetic properties. *Physica B*, 327: 382–384.

Hutter, E., Fendler, J. H., and Roy, D. 2001. Surface plasmon resonance studies of gold and silver nanoparticles linked to gold and silver substrates by 2-aminoethanethiol and 1,6-hexanedithiol. *Journal of Physical Chemistry B*, 105: 11159–11168.

Kanlayavattanakul, M. and Lourith, N. 2010. Lipopeptides in cosmetics. *International Journal of Cosmetic Science*, 32: 1–8.

Kasture, M., Singh, S., Patel, P., Joy, P. A., Prabhune, A. A., Ramana, C. V., and Prasad, B. L. V. 2007. Multiutility sophorolipids as nanoparticle capping agents: Synthesis of stable and water dispersible Co nanoparticles. *Langmuir*, 23: 11409–11412.

Kasture, M. B., Patel, P., Prabhune, A. A., Ramana, C. V., Kulkarni, A. A., and Prasad, B. L. V. 2008. Synthesis of silver nanoparticles by sophorolipids: Effect of temperature and sophorolipid structure on the size of particles. *Journal of Chemical Sciences*, 120: 515–520.

Kiran, G. S., Sabua, A., and Selvin, J. 2010. Synthesis of silver nanoparticles by glyco-lipid biosurfactant produced from marine *Brevibacterium casei* MSA19. *Journal of Biotechnology*, 148: 221–225.

Komarneni, S., Li, D., Newalkar, B., Katsuki, H., and Bhalla, A. S. 2002. Microwave-polyol process for Pt and Ag nanoparticles. *Langmuir*, 18: 5959–5962.

Kumar, G. V. P., Rangarajan, N., Sonia, B., Deepika, P., Rohman, N., and Narayana, C. 2011. Metal-coated magnetic nanoparticles for surface enhanced Raman scattering studies. *Bulletin of Materials Science*, 34: 207–216.

Lee, D., Rubner, M. F., and Cohen, R. E. 2006. All-nanoparticle thin-film coatings. *Nano Letters*, 6: 2305–2312.

Li, W. W., Guo, Y., and Zhang, P. 2010. SERS-active silver nanoparticles prepared by a simple and green method. *Journal of Physical Chemistry C*, 114: 6413–6417.

Liao, Q., Tannenbaum, R., and Wang, Z. L. 2006. Synthesis of FeNi3 alloyed nanoparticles by hydrothermal reduction. *Journal of Physical Chemistry B*, 110: 14262–14265.

Liao, Z., Wang, H., Wang, X., Wang, C., Hu, X., Cao, X., and Chang, J. 2010. Biocompatible surfactin-stabilized superparamagnetic iron oxide nanoparticles as contrast agents for magnetic resonance imaging. *Colloids and Surfaces A: Physicochemical and Engineering Aspects*, 370: 1–5.

Maillard, M., Monchicourt, P., and Pileni, M. P. 2003. Multiphoton photoemission of self-assembled silver nanocrystals. *Chemical Physics Letters*, 380: 704–709.

Maity, J. P., Lin, T., Cheng, H. P., Chen, C. Y., Reddy, A. S., Atla, S. B., Chang, Y., Chen, H., and Chen, C. 2011. Synthesis of brushite particles in reverse microemulsions of the biosurfactant surfactin. *International Journal of Molecular Sciences*, 12: 3821–3830.

Mandal, D., Bolander, M. E., Mukhopadhyay, D., Sarkar, G., and Mukherjee, P. 2006. The use of microorganisms for the formation of metal nanoparticles and their application. *Applied Microbiology and Biotechnology*, 69: 485–492.

Moncada, E., Quijada, R., and Retuert, J. 2007. Nanoparticles prepared by the sol–gel method and their use in the formation of nanocomposites with polypropylene. *Nanotechnology* 18: 335606 (7pp).

Moshfegh, A. Z. 2009. Nanoparticlecatalysts. *Journal of Physics D: Applied Physics*, 42: 233001.

Mukherjee, S., Das, P., and Sen, R. 2006. Towards commercial production of microbial surfactants. *Trends in Biotechnology*, 24: 509–515.

Mulligan, C. N. 2005. Environmental applications for biosurfactants. *Environmental Pollution*, 133: 183–198.

Narayanan, J., Ramji, R., Sahu, H., and Gautam, P. 2010. Synthesis, stabilisation and characterization of rhamnolipid-capped ZnS nanoparticles in aqueous medium. *IET Nanobiotechnology*, 4: 29–34.

Narayanan, K. B. and Sakthivel, N. 2010. Biological synthesis of metal nanoparticles by microbes. *Advances in Colloid and Interface Science*, 156: 1–13.

Palanisamy, P. 2008. Biosurfactant mediated synthesis of NiO nanorods. *Materials Letters*, 62: 743–746.

Palanisamy, P. and Raichur, A. M. 2009. Synthesis of spherical NiO nanoparticles through a novel biosurfactant mediated emulsion technique. *Materials Science and Engineering C*, 29: 199–204.

Polte, J., Ahner, T. T., Delissen, F., Sokolov, S., Emmerling, F., Thünemann, A. F., and Kraehnert, R. 2010. Mechanism of gold nanoparticle formation in the classical citrate synthesis method derived from coupled in situ XANES and SAXS evaluation. *Journal of American Chemical Society*, 132(4): 1296–1301.

Pradeep, T. and Anshup. 2009. Noble metal nanoparticles for water purification: A critical review. *Thin Solid Films*, 517: 6441–6478.

Raichur, A. M. 2007. Dispersion of colloidal alumina using a rhamnolipid -biosurfactant. *Journal of Dispersion Science and Technology*, 28: 1272–1277.

Rangarajan, V., Dhanarajan, G., Kumar, R., Sen, R., and Mandal, M. 2012. Time-dependent dosing of Fe^{2+} for improved lipopeptide production by marine *Bacillus megaterium*. *Journal of Chemical Technology Biotechnology*, DOI 10.1002/jctb.3814.

Rao, C. N. R., Kulkarni, G. U., John Thomas, P., Agrawal, V. V., Gautam, U. K., and Ghosh, M. 2003. Nanocrystals of metals, semiconductors and oxides: Novel synthesis and applications. *Current Science*, 85: 1041–1045.

Raveendran, P., Fu, J., and Wallen, S. L. 2003. Completely "green" synthesis and stabilization of metal nanoparticles. *Journal of American Chemical Society*, 125: 13940–13941.

Reddy, A, S., Chen, C. Y., Chen, C. C., Jean, J. S., Fan, C. W., Chen, H. R., Wang, J. C., and Vanita, R. 2009. Synthesis of gold nanoparticles via an environmentally benign route using a biosurfactant. *Journal of Nanoscience and Nanotechnology*, 9: 6693–6699.

Reddy, A. S., Hao, Y. K., Atla, S. B., Chen, C. Y., Chen, C. C., Shih, R. C., Chang, Y. F., Maity, J. P., and Chen, H. J. 2011. Low-temperature synthesis of rose-like ZnO nanostructures using surfactin and their photocatalytic activity. *Journal of Nanoscience and Nanotechnology*, 11: 5034–5041.

Rehman, A., Raza, Z. A., Saif-ur-Rehman, Khalid, Z. M., Subramani, C., Rotello, V. M., and Hussain, I. 2010. Synthesis and use of self-assembled rhamnolipid microtubules as templates for gold nanoparticles assembly to form gold microstructures. *Journal of Colloid and Interface Science*, 347: 332–335.

Salata, O. V. 2004. Applications of nanoparticles in biology and medicine. *Journal of Nanobiotechnology*, 2: 3.

Schmid, G. 1992. Large clusters and colloids. Metals in the embryonic state. *Chemical Reviews*, 92: 1709.

Schwartzberg, A. M., Olson, T. Y., Talley, C. E., and Zhang, J. Z. 2006. Synthesis, characterization, and tunable optical properties of hollow gold nanospheres. *Journal of Physical Chemistry B*, 110: 19935–19944.

Sen, R. 1997. Response surface optimization of the critical media components for the production of surfactin. *Journal of Chemical Technology and Biotechnology*, 68: 263–270.

Sen, R. 2008. Biotechnology in petroleum recovery: The microbial EOR. *Progress in Energy and Combustion Science*, 34: 714–724.

Singh, B. R., Dwivedi, S., Al-Khedhairy, A. A., and Musarrat, J. 2011. Synthesis of stable cadmium sulfide nanoparticles using surfactin produced by *Bacillus amyloliquifaciens* strain KSU-109. *Colloids and Surfaces B: Biointerfaces*, 85: 207–213.

Singh, S., D'Britto, V., Prabhune, A. A., Ramana, C. V., Dhawan, A., and Prasad, B. L. V. 2010. Cytotoxic and genotoxic assessment of glycolipid-reduced and -capped gold and silver nanoparticles. *New Journal of Chemistry*, 34: 294–301.

Soma, J. and Papadopoulos, K. D. 1996. Ostwald ripening in sodium dodecyl sulfate-stabilized decane-in-water emulsions. *Journal of Colloid Interface Science*, 181: 225–237.

Sotiropoulou, S., Sierra-Sastre, Y., Mark, S. S., and Batt, C. A. 2008. Biotemplated nanostructured materials. *Chemistry of Materials*, 20: 821–834.

Sweeney, R. Y., Mao, C., Gao, X., Burt, J. L., Belcher, A. M., Georgiou, G., and Iverson, B. L. 2004. Bacterial biosynthesis of cadmium sulfide nanocrystals. *Chemistry and Biology*, 11: 1553–1559.

Worakitsiria, P., Pornsunthorntawee, O., Thanpitchaa, T., Chavadeja, S., Weder, C., and Rujiravanita, R. 2011. Synthesis of polyaniline nanofibers and nanotubes via rhamnolipid biosurfactant templating. *Synthetic Metals*, 161: 298–306.

Xie, Y., Ye, R., and Liu, H. 2006. Synthesis of silver nanoparticles in reverse micelles stabilized by natural biosurfactant. *Colloids and Surfaces A: Physicochemical and Engineering Aspects*, 279: 175–178.

Xie, Y. W., Li, Y., and Ye, R. 2005. Effect of alcohols on the phase behavior of microemulsions formed by a biosurfactant—Rhamnolipid. *Journal of Dispersion Science and Technology*, 26: 455–461.

Xu, J., Han, X., Liu, H., and Hu, Y. 2006. Synthesis and optical properties of silver nanoparticles stabilized by gemini surfactant. *Colloids and Surfaces A Physicochemical and Engineering Aspects*, 273: 179–183.

Yanhong, L., Dejun, W., Qidong, Z., Min, Z., and Qinglin, Z. 2004. A study of quantum confinement properties of photogenerated charges in ZnO nanoparticles by surface photovoltage spectroscopy. *Journal of Physical Chemistry B*, 108(10): 3202–3206.

Yonezawa, T., Onoue, S. Y., and Kimizuka, N. 2000. Preparation of highly positively charged silver nanoballs and their stability. *Langmuir*, 16(12): 5218–5220.

Zhang, W., Qiao, X., and Chena, J. 2007. Synthesis of silver nanoparticles—Effects of concerned parameters in water/oil microemulsion. *Materials Science and Engineering B*, 142: 1–15.

Zhu, H., John, G., and Wei, B. 2005. Synthesis of assembled copper nanoparticles from copper-chelating glycolipid nanotubes. *Chemical Physics Letters*, 405: 49–52.

10 Enhancement of Remediation Technologies with Biosurfactants

Catherine N. Mulligan

CONTENTS

INTRODUCTION

The indiscriminate dumping of materials, bankrupt and abandoned manufacturing plants, and insufficient methods for waste storage, treatment, and disposal facilities have contributed to the contamination of many sites. Chemical waste categories include organic liquids such as solvents from dry cleaning; oils including lubricating oils, automotive oils, hydraulic oils, fuel oils, and organic sludges/solids; and organic aqueous wastes and wastewaters. Manufacturing plants produce chemical wastes that usually correlate with the amount of chemicals produced. Most soil contamination is the result of accidental spills and leaks, generation of chemical waste leachates and sludges from cleaning of equipment, residues left in used containers, and outdated materials. Smaller generators of chemical contaminants include improperly managed landfills, automobile service establishments, maintenance shops, and photographic film processors. Household wastes including pesticides, paints, and cleaning and automotive products may also contribute significantly as sources of organic chemicals (LaGrega et al., 2001). The more common heavy metals include lead (Pb), cadmium (Cd), copper (Cu), chromium (Cr), nickel (Ni), iron (Fe), mercury (Hg), and zinc (Zn).

A variety of in situ and ex situ remediation techniques exist to manage the contaminated sites. For evaluation of the most appropriate technique, the procedure in Figure 10.1 should be followed. Ex situ techniques include excavation, contaminant fixation or isolation, incineration or vitrification, washing, and biological treatment processes. In situ processes include (1) bioremediation, air or steam stripping or thermal treatment for volatile compounds; (2) extraction methods for

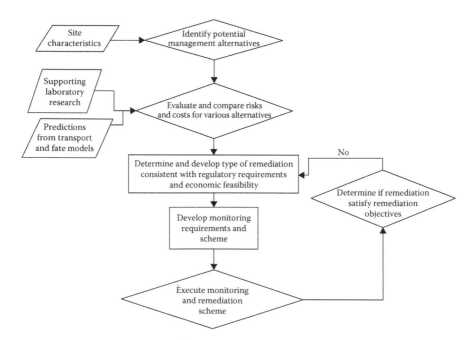

FIGURE 10.1 Process for evaluating soil remediation processes.

soluble components; (3) chemical treatments for oxidation or detoxification; and (4) stabilization/solidification with cements, limes, and resins for heavy metal contaminants. Phytoremediation, although less developed, has also been used. The most suitable types of plants must be selected based on pollutant type and recovery techniques for disposal of the contaminated plants.

Most in situ remediation techniques are potentially less expensive and disruptive than ex situ ones, particularly for large contaminated areas. Natural or synthetic additives can be utilized to enhance precipitation, ion exchange, sorption, and redox reactions (Mench et al., 2000). The sustainability of reducing and maintaining reduced solubility conditions is key to the long-term success of the treatment. Ex situ techniques are expensive and can disrupt the ecosystem and the landscape. For shallow contamination, remediation costs, worker exposure, and environmental disruption can be reduced by using in situ remediation techniques. In this chapter, the addition and effectiveness of various biological surface agents used to enhance bioremediation, in situ flushing, and soil-washing processes for soil and sediment remediation will be examined. The main focus will be on biologically produced surfactants due to their biodegradability, low toxicity, and effectiveness.

Soil remediation can be performed with or without excavation via soil washing or in situ flushing (Mulligan et al., 2001a). Solubilization of the contaminants can be performed with water alone or with additives. The solubility of the contaminant is thus a key factor. Contaminants such as trichloroethylene (TCE), polycyclic aromatic hydrocarbons (PAHs), and polychlorinated biphenyls (PCBs) are of very low solubility. Effective bioremediation of organic contaminants leads to complete mineralization of the contaminants. Often the process is enhanced by the addition of nutrients, electron acceptors, or bioaugmentation (addition of bacteria) (Yong and Mulligan, 2004). Inorganic contaminants such as metals and radionuclides are commonly found with the organic contaminants, further complicating the bioremediation process. The inorganic contaminants may be toxic to the bacteria, cannot be biodegraded, but may be converted from one form to another.

ENVIRONMENTAL TECHNOLOGIES

SOIL FLUSHING

To remove nonaqueous phase liquids (NAPLs) from the groundwater, extraction of the groundwater can be performed by pumping to remove the contaminants in the dissolved and/or free-phase NAPL zone. However, substantial periods of time can be required and effectiveness can be limited. Drinking water standards of the extracted water can be achieved after treatment with activated carbon, ion exchange, membranes, and other methods. Extraction solutions can be introduced into the soil using surface flooding, sprinklers, leach fields, and horizontal or vertical drains to enhance the removal rates of the contaminants. Water with surfactants or solvents or without additives is employed to solubilize and extract the contaminants as shown in Figure 10.2 in soil flushing. These additives include organic or inorganic acids or bases, water-soluble solvents, complexing or chelating agents, and surfactants.

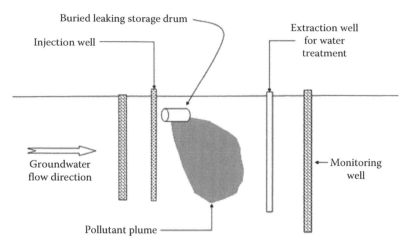

FIGURE 10.2 Schematic of soil flushing process.

Due to the sorption of residuals during flushing, the additives must be nontoxic and biodegradable. Contaminant removal efficiencies are affected by various factors including soil pH, type, porosity and moisture content, cation exchange capacity, particle size distribution, organic matter content, permeability, and the type of contaminants. Soil flushing is appropriate for highly permeable soils as the washing solution must be pumped through the soil by injection wells or surface sprinklers or other means of infiltration. The washing solution should be treated to remove the contaminants and the water reused.

High soil permeabilities (greater than 1×10^{-3} cm/s) are considered to be beneficial for such procedures. Depth to groundwater can increase costs. However, the spreading of contaminants and the fluids must be contained and recaptured. Control of these infiltrating agents may be difficult, particularly if the site hydraulic characteristics are not well-understood. Emissions of volatile organic compounds (VOCs) should be monitored and treated if required. Recycling of additives is desirable to improve process economics and reduce material use. Metals, VOCs, PCBs, fuels, and pesticides can be removed through soil flushing.

In choosing the most appropriate remediation technology, factors to be considered must include exposure routes, future land use, acceptable risks, regulatory guidelines, level and type of contaminants, site characteristics, and resultant emissions. Laboratory and field treatability tests should be performed to obtain site-specific information. Soil flushing has been demonstrated at numerous superfund sites with costs in the range of $18–$50/m^3 for large easy to small difficult sites. A schematic illustration of the criteria and tools for evaluating technologies and protocols for environmental management of contaminated soils and sediments is shown in Figure 10.3.

Soil Washing

Soil washing has been suggested for a variety of soils contaminated with metals, mixed contaminants, and organic contaminants (El-Shafey and Canepa, 2003).

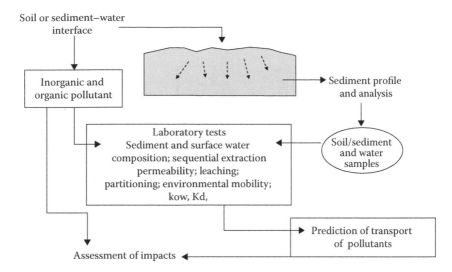

FIGURE 10.3 Schematic showing procedures required to evaluate potential for pollutant transport.

Soil washing is a process that uses water to remove contaminants from soil and sediments by physical and/or chemical techniques. Contaminated sediments are problematic as they can potentially release contaminants severely impacting water quality. Soil washing involves the addition of a solution with the contaminated soil to transfer the contaminants to the wash solution. It is most appropriate for weaker bound metals in the form of hydroxides, oxides, and carbonates. Mercury, lead, cadmium, copper, nickel, zinc, and chromium can be recovered by electrochemical processes if the levels of organic compounds are not significant. Metals can also be removed from precipitation or ion exchange. Precipitation is not applicable for metal sulfides. Pretreatment to remove uncontaminated coarser fractions can be used. Various additives can be employed such as bases, surfactants, acids, or chelating agents. Nitric, hydrochloric, and sulfuric acids can be used. However, if sulfuric acid is used, 50% of the amount is required compared to hydrochloric acid (Papadopoulos et al., 1997).

Figure 10.4 illustrates a typical soil-washing process where the separation consists of size separation, washing, rinsing, and other technologies similar to those used in the mineral processing industry. Larger particles are separated from the smaller ones as they have lower contamination levels. The smaller volumes of soil can be treated less expensively. Surfactants may be added in the washing water. The more contaminated size range is 0.24–2 mm due to the surface charges of the soil clay particles that attract anionic metal contaminants and the organic fraction that binds organic contaminants. Wash water and additives should be recycled, or treated prior to disposal. The mechanical dewatering of particles is performed via a filter press, conveyer filtration, centrifugal separation, etc. Froth flotation by the introduction of air bubbles into a slurry may also

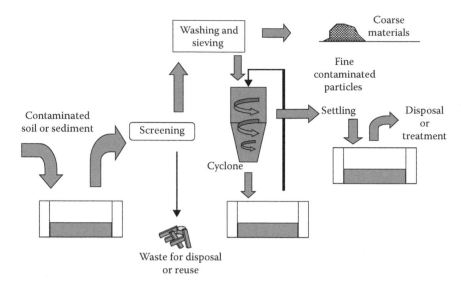

FIGURE 10.4 Schematic of soil or sediment washing of contaminated material.

be used (Venghuis and Werther, 1998). The disposal of the treated fine particles varies depending on the type and levels of the contaminants.

Mixtures of metals and organic contaminants may require sequential washing with different additives to target the various contaminants. Soil-washing processes generally use hot water to reduce the viscosity of hydrocarbons. The increased temperature also increases the solubilities of metal salts. The treated soil can then be washed to remove any residual wash solution prior to disposal. Ideally, the wash solution should be recycled. Costs of soil washing are usually in the order of USD \$70–\$190/m^3 depending on site size and complexity (Racer software, Remedial action plan 2006). Although extensively used in Europe, full-scale processes are less common in the United States. Feasibility tests should be conducted to determine optimal conditions (chemical type and dosage, contact time, agitation, temperature, and extraction steps to meet regulatory requirements). As spent washwater can be a mixture of soluble contaminants and fine particles, treatment is thus required to meet reuse or disposal requirements. Full-scale demonstrations may be required to demonstrate the feasibility of newly developed treatment processes. Presently, wastewater management systems act as a foundation for modern public health and environment protection. The idea of most suitable wastewater management systems is to use less energy, allow for elimination, or beneficial reuse of biosolids, restore natural nutrient cycles, have much smaller footprints, be more energy-efficient, and design to eliminate exposed wastewater surfaces, odors, and hazardous by-products (Daigger, 2005). Some of the incentives for the industry to incorporate sustainability into their wastewater solutions are as follows (Mosley, 2006). In addition to the technical aspects of a wastewater treatment technology, selection of a particular technology should be based on all aspects that determine its sustainability such as the human and environmental activities that surround it.

BACKGROUND ON SURFACTANTS AND BIOSURFACTANTS

SURFACTANTS

Surfactants are amphiphilic compounds with a hydrophobic portion with little affinity for the bulk medium and a hydrophilic group that is attracted to the bulk medium. The free energy of the system is reduced by replacing the bulk molecules of higher energy at an interface. Surfactants concentrate at interfaces (solid–liquid, liquid–liquid, or vapor–liquid). An interfacial boundary exists between two immiscible phases. The hydrophobic portion concentrates at the surface, while the hydrophilic portion is oriented toward the solution.

Applications are based on their abilities to lower surface tensions, increase solubility, detergency power, wetting ability, and foaming capacity (Mulligan and Gibbs, 1993). For example, due to the properties of solubilizing and mobilizing contaminants, they are used for soil washing or flushing. The petroleum industry has traditionally been the major user, as in enhanced oil removal applications to increase the solubility of petroleum components (Falatko, 1991). They have also been used for mineral flotation and in the pharmaceutical industries (Mulligan and Gibbs, 1993).

The choice of surfactant is primarily based on product cost (Mulligan and Gibbs, 1993). In general, surfactants are used to save energy and, consequently, energy costs (such as the energy required for pumping in pump-and-treat techniques). Surfactants have been used to enhance metal removal (Holden et al., 1989). Charge-type, physicochemical behavior, solubility, and adsorption behavior are some of the most important selection criteria for surfactants, in addition to their ability to enhance bioremediation of contaminated land sites (Oberbremer et al., 1990; Samson et al., 1990).

BIOSURFACTANTS

Biosurfactants have been effective for the reduction of the interfacial tension (IFT) of oil and water in situ, the viscosity of the oil, the removal of water from the emulsions prior to processing, and in the release of bitumen from tar sands. The high molecular weight Emulsan has been commercialized for this purpose (Anonymous, 1984). It contains a polysaccharide with fatty acids and proteins. Other high molecular weight biosurfactants have been reviewed by Ron and Rosenberg (2002).

Most biosurfactants are produced from hydrocarbon substrates (Syldatk and Wagner, 1987) and can be growth-associated. In this case, they can either use the emulsification of the substrate (extracellular) or facilitate the passage of a hydrophobic substrate through the membrane (cell-membrane-associated). Biosurfactants, however, can also be produced from carbohydrates, which are very soluble.

In this chapter, most of the emphasis will be placed on three well-studied low molecular weight biosurfactants including rhamnolipids, sophorolipids, and surfactin. In each case, environmental applications will be examined including enhancing solubilization and biodegradation, soil treatment (in situ and ex situ), and water and waste treatment. A summary of biosurfactants is shown in Table 10.1.

TABLE 10.1
Type and Microbial Origin of Biosurfactants Used
in Environmental Applications

Type of Surfactant	Microorganism
Glycolipids	*Arthrobacter paraffineus*
Trehalose lipids	*Corynebacterium* spp.
	Mycobacterium spp.
	Rhodococus erythropolis
	Nocardia sp.
Rhamnolipids	*Pseudomonas aeruginosa*
	Pseudomomas sp.
	Serratia rubidea
Sophorose lipids	*Candida apicola*
	Candida bombicola
	Candida lipolytica
	Candida bogoriensis
Cellobiose lipids	*Ustilago maydis*
Polyol lipids	*Rhodotorula glutinus*
	Rhodotorula graminus
Diglycosyl diglycerides	*Lactobacillus fermentii*
Lipopeptides and proteins	
Arthrofactin	*Arthrobacter* sp.
Fengycin	*Bacillus* sp.
Lichenysin A	*Bacillus licheniformis*
Lichenysin B	
Surfactin	*Bacillus subtilis*
	Bacillus pumilus
Sulfonylipids	*T. thiooxidans*
	Corynebacterium alkanolyticum
Viscosin	*Pseudomonas fluorescens*
Ornithine, lysine peptides	*Thiobacillus thiooxidans*
	Streptomyces sioyaensis
	Gluconobacter cerinus
Streptofactin	*Streptomyces tendae*
Phospholipids and fatty acids	
Phospholipids	*Acinetobacter* sp.
Fatty acids (corynomycolic acids, spiculisporic acids, etc.)	*Capnocytophaga* sp.
	Penicillium spiculisporum
	Corynebacterium lepus
	Arthrobacter paraffineus
	Talaramyces trachyspermus
	Nocardia erythropolis
Polysaccharides and polymeric surfactants	
Lipopolysaccharides	*Acinetobacter calcoaceticus (RAG1)*
	Pseudomonas sp.

TABLE 10.1 (continued)
Type and Microbial Origin of Biosurfactants Used
in Environmental Applications

Type of Surfactant	Microorganism
	Candida lipolytica
Bioemulsan BS 29	*Gardonia* sp.
Glycoprotein	
Alasan	*Acinetobacter radioresistens*
Particulate surfactant (PM)	*Pseudomonas marginalis*
Biosur PM	*Pseudomonas maltophilla*

Source: Adapted from Mulligan, C.N. and Gibbs, B.F., Factors influencing the economics of biosurfactants, in *Biosurfactants, Production, Properties, Applications*, Kosaric, N. (ed.), Marcel Dekker, New York, 1993, pp. 329–371.

ENVIRONMENTAL APPLICATIONS OF BIOSURFACTANTS

RHAMNOLIPIDS

Effect of Rhamnolipids on Contaminant Biodegradation

Petroleum Hydrocarbons

The various components of petroleum hydrocarbons are alkanes (also called aliphatic compounds), cycloalkanes, aromatics, PAHs, asphaltenes, and resins. Alkanes, represented by the formula C_2H_{2n+2}, (where n is the number of carbons and 2n + 2 is the number of hydrogens), increase in the number of isomers as the number of carbons increases; low molecular weight alkanes are easily degraded by microorganisms.

Various studies have examined the effect of rhamnolipids on biodegradation of organic contaminants with mixed results. There has been a particular focus on various hydrocarbons of low solubility. A review by Maier and Soberon-Chavez (2000) indicated that rhamnolipid addition can enhance biodegradation of hexadecane, tetradecane, pristine, creosote, and hydrocarbon mixtures in soils, in addition to hexadecane, octadecane, n-paraffin, and phenanthrene in liquid systems. Two mechanisms for enhanced biodegradation are possible: enhanced interaction with the cell surface, which increases the hydrophobicity of the surface allowing hydrophobic substrates to permeate more easily, and increased solubility of the substrate for the microbial cells as shown in Figure 10.5 (Shreve et al., 1995, Zhang and Miller, 1992). Zhang and Miller (1992) demonstrated that a concentration of 300 mg/L of rhamnolipids increased the mineralization of octadecane to 20% compared to 5% for the controls. Beal and Betts (2000) showed that the cell surface hydrophobicity during growth on hexadecane increased by the biosurfactant strain more than a nonbiosurfactant-producing one. The rhamnolipids also increased the hexadecane solubility to 22.8 from 1.8 µg/L. Other studies by Churchill et al. (1995) showed that rhamnolipid addition with a fertilizer (Inipol EAp-22) enhanced biodegradation of aromatic and aliphatic compounds in aqueous phase and soil reactors.

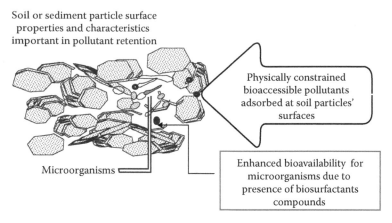

Soil or sediment particle surface
properties and characteristics
important in pollutant retention

Physically constrained
bioaccessible pollutants
adsorbed at soil particles'
surfaces

Microorganisms

Enhanced bioavailability for
microorganisms due to
presence of biosurfactants
compounds

FIGURE 10.5 Enhancement of the bioavailability of pollutants by biosurfactants.

Noordman et al. (2002) determined that the biosurfactant from *Pseudomonas aeruginosa* could enhance biodegradation if the process is rate-limited, such as in the case of small soil pore sizes (6 nm), the hexadecane is entrapped and of limited availability. The rhamnolipid can enhance release of entrapped substrates and uptake by cells (if the substrate is available). This could then stimulate bacterial degradation under in situ conditions. Al-Awadhi et al. (1994) showed that during in situ studies on oil-contaminated desert sands, up to 82.5% reduction of total petroleum hydrocarbons and 90.5% reduction of total alkanes could be achieved in 12 months.

Rahman et al. (2003) examined the bioremediation of n-alkanes in a petroleum sludge. The sludge contained an oil and grease content of 87.4%. The addition of a bacterial consortium, nutrients, and the rhamnolipids to 10% sludge led to the degradation of 100% of the C8–C11 alkanes, 83%–98% of the C12–C21 fraction, 80%–85% of the C22–C31 and 57%–73% of the C32–C40. Lower rates of biodegradation occurred as the chain length increased but were still significant even for C32–C40 compounds that are of low solubility.

The feasibility of biosurfactant production to bioremediate a soil contaminated with a mixture of petroleum hydrocarbons and heavy metals was evaluated using batch experiments (Jalali and Mulligan, 2007). After 50 days, the activity of the indigenous soil microorganisms was enhanced, enabling reduction of total petroleum hydrocarbon (TPH) by 36%. The biosurfactant concentration during this period reached three times the critical micelle concentration, which increases the average concentration of TPH and metals in the filtrate from 2.1% and 2.2% to 8.3% and 4.4%, respectively. The results indicate that biosurfactant production can enhance bioremediation of cocontaminated soils.

Li et al. (2006) isolated six diesel oil degrading bacteria from an oil-contaminated soil. One of the isolates produced a rhamnolipid biosurfactant with a CMC of the biosurfactant of 65 mg/L in water and 185 mg/L in soil. The difference in the soil was due to sorption on to the soil. This biosurfactant at concentrations of 0.01% and 0.02% was then evaluated for enhancement of diesel oil biodegradation in water and soil. Rhamnolipid improved both the extent and rate of biodegradation when the concentrations were higher than the CMC.

Another study on diesel was performed by Whang et al. (2008) to evaluate two biosurfactants, rhamnolipid and surfactin. A quantity of 80 mg/L of rhamnolipid showed a greater than twofold enhancement. Enhancement of diesel-contaminated soil remediation was also shown. Surfactin (40 mg/L) was able to enhance biomass production and diesel biodegradation by more than two times compared to the control. Due to its antibiotic characteristics, higher surfactin concentrations (up to 400 mg/L) showed substantial growth and biodegradation inhibition.

As oil spills can pose significant threats and damage to the marine and coastal ecosystems, the objective of the study by Vasefy and Mulligan (2008) was to evaluate the effectiveness of a rhamnolipid biosurfactant (JBR 425™) in combination with two commercial biological products, ASAP™ and Degreaser™ on the biodegradation of weathered light crude oil, heavy crude oil, and diesel fuel spilled on saline water. The two products were used as supplementary additives to enhance the biodegradation rate as they contain bacterial consortia and nutrients. Chemical and microbiological analyses were performed. At 20°C and 35 g/L salinity, a rhamnolipid solution: oil ratio (1:1, v/v) showed 65% removal of diesel fuel, 70% of light crude oil, and 59% of heavy crude oil after 28 days of biodegradation. The biodegradation was in the order of diesel fuel > light crude oil > heavy crude oil in terms of removal percentage and microbial densities.

Mehdi and Giti (2008) evaluated 25 strains of bacteria isolated from the marine environment and determined that *Pseudomonas* strain could reduce the quantity of crude oil by 83%. A correlation was found between the emulsification ability, cell adherence to hydrocarbon and growth rate in the crude oil.

Although there have been numerous evaluation tests regarding biosurfactants, there have been few efforts to develop models. An attempt was thus made to simulate biosurfactant enhanced bioremediation by Yu et al. (2012). A pilot scale model for BTEX contamination was designed to verify the developed model which indicated a high accuracy. However the mechanisms of desorption and enhanced biodegradation are not clear. Further investigation at full scale is necessary.

Rhamnolipid (JBR 425) was evaluated for the ability to enhance bioremediation of refinery oil sludge (Zhang et al., 2011). Optimal biodegradation occurred at a 100:50:10 of C:N:P ratio nutrient and 400 mg/kg of rhamnolipid addition. Up to 50.8% TPH reduction occurred by the strain *Luteibacter sp.*

Cameotra and Singh (2008) also evaluated the bioremediation of a soil contaminated with an oily sludge in field tests in India. They used a consortium of two isolates of *P. aeruginosa* and a rhodococcal strain. When nutrients and a crude biosurfactant were added together with inocula, more than 98% of the hydrocarbons could be removed compared to a negligible amount for the control without any additives or inoculum. The biosurfactant mixture consisted of 11 rhamnolipid congeners and was produced by the consortium. It stimulated biodegradation and was able to be used in crude form without expensive purification procedures.

Yan et al. (2011) investigated the washing of oil-based drill cuttings by a rhamnolipid biosurfactant prior to biodegradation by a mixed culture. The cuttings are wastes generated during petroleum exploration and production. The organic content decreased by 83% from 85,000 mg/kg. A rhamnolipid concentration above 360 mg/L was not beneficial. The polar fraction was removed the most (88.4%), followed by the saturates (81.6%), and then the aromatics (63.8%). A 120 day biodegradation period further reduced the contaminants to 5470 mg/kg, a 67.3% removal. Therefore, this ex situ two-stage process was an effective treatment.

The effect of rhamnolipid on phenol degradation by laccase was investigated (Liu et al. 2012). It was determined that 98% removal of phenol (up to 400 mg/L) could be achieved in 24 h in the presence of 318 μM dirhamnolipid. Removal was optimal at pH 6 and 50°C. This was up to 6.4-fold better than the control at the same conditions, which indicates potential for enhanced treatment of phenol in water.

Southam et al. (2001) studied the effect of biosurfactants on waste hydrocarbon degradation. The bacteria must adsorb first onto the surfactant–oil interface of 25–50 nm in thickness. Approximately 1% of the biosurfactant is needed to emulsify the oil. Growth on the oil occurred by microbial uptake of the nanometer-sized oil droplets. This study was enabled by using a transmission electron microscope. More of this type of research is required to determine the mechanism of hydrocarbon metabolism and biosurfactant applications.

PAHs

Polycyclic or polynuclear aromatic hydrocarbons (PAHs) are components of creosote, produced during petroleum refining, coke production, and wood preservation (Park et al., 1990). Many are suspected to be carcinogens. A general form for PAHs is $C_{4n+2}H_{2n+4}$, where n is the number of rings. A four-ring PAH would be $C_{18}H_{12}$. As the ring number increases, the degradation of the compounds becomes more difficult due to decreasing volatility and solubility and increased sorption to soil. PAHs are degraded one ring at a time like single-ring aromatics.

Researchers (Vipulanandan and Ren, 2000) compared the solubilization of the PAH, naphthalene, by a rhamnolipid, sodium dodecyl sulfate (SDS), an anionic surfactant, and Triton X-100, a nonionic surfactant. Although the biosurfactant increased the solubility of naphthalene by 30 times, the biodegradation of naphthalene (30 mg/L) took 40 days in the presence of biosurfactant (10 g/L) compared to 100 h for Triton X-100 (10 g/L). The biosurfactant was used as a carbon source instead of the naphthalene decreasing its efficiency, which did not occur in the case of Triton X-100. However, naphthalene was not biodegraded in the presence of SDS.

Deschênes et al. (1994) also showed that the rhamnolipids from the UG2 strain were more effective than SDS (up to five times) in a bioslurry. The biosurfactant enhanced the solubilization of four-ring PAHs more significantly than three-ring PAHs. SDS showed higher levels of toxicity compared to the biosurfactant as the surfactant concentration increased above 100 mg/kg. Higher molecular weight PAHs were not biodegraded even in the presence of the surfactants.

Providenti et al. (1995) showed that the effect of UG2 biosurfactants on phenanthrene mineralization in soil slurries led to decreased lag times and increased degradation. Further experiments were carried out by Dean et al. (2001) who investigated either rhamnolipid or a biosurfactant-producing strain of *P. aeruginosa* ATCC 9027 addition. Results were mixed and difficult to interpret as one strain (strain R) showed enhanced biodegradation when the surfactant was added but the other (strain P5-2) did not. Co-addition of the two strains showed enhanced mineralization of phenanthrene by strain R only. However, there seemed to be some interaction between the two strains, which will need further investigation. Also, although addition of the rhamnolipid enhanced release of the phenanthrene, it did not necessarily enhance biodegradation.

Rahman et al. (2002) showed that addition of rhamnolipid produced by *Pseudomonas sp.* DS10-129 along with poultry litter and coir pith enhanced ex situ bioremediation of a gasoline-contaminated soil. Another strain, *P. marginalis*, produced biosurfactants that solubilized PAHs such as phenanthrene and enhanced biodegradation (Burd and Ward, 1996). The addition of rhamnolipids led to attachment to the phenanthrene that enhanced bioavailability and hence degradation by *P. aeruginosa* (Garcia-Junco et al., 2001).

Straube et al. (2003) evaluated the addition of *P. aeruginosa* strain 64, a biosurfactant-producer with slow-release nitrogen and a bulking agent (ground rice hulls) to enhance bioremediation of PAH (13,000 mg/kg) and pentachlorophenol (PCP) (1,500 mg/kg) contaminated soil during landfarming. However, the added strain does not degrade PAHs. This biostimulation/bioaugmentation approach led to an 87% decrease in total PAHs and a 67% decrease in total benzo(a)pyrene (BaP) toxic equivalents compared to the control, 23% and 48% respectively, in microcosm studies. Larger-scale pan experiments showed decreases of 86% in the PAHs and 87% in total BaP toxicity by biostimulation/bioaugmentation after 16 months compared to a 12% decrease in PAHs for the control. Overall, the biosurfactants produced in the soil by the bacterial strain 64 enabled PAH biodegradation.

Yin et al. (2008) evaluated the characteristics of a rhamnolipid biosurfactant produced by an isolate of *P. aeruginosa* from oil-contaminated wastewater. The surface tension was 33.9 mN/m of the biosurfactant and the CMC was 50 mg/L. The solubility of phenanthrene increased by 23 times in the presence of the biosurfactant. The stability under a range of pH and salinities was good. Thus, it has potential for bioremediation of crude oil. The weight solubilization ratio (WSR) and micelle-phase/aqueous-phase partition coefficient (k_m) were determined to be 0.2022 and 4.82 (log). These are much higher than the synthetic surfactants, Tween and Triton.

Electrokinetic treatment of contaminated soil is not very efficient for low solubility organic contaminants. Chang et al. (2009) compared Triton X-100 with rhamnolipid addition during electrokinetic treatment of phenanthrene-contaminated soil. Rhamnolipid enhanced electroosmotic flow more than the synthetic flow. Microbial growth may also be improved. Gonzini et al. (2010) evaluated the addition of rhamnolipid to enhance the treatment process. Gazoil removal of 86% could be achieved from a 20,000 ppm contaminated soil with addition of 2 g/kg of rhamnolipid. In addition, the condition of the soil after treatment was evaluated. The contents of nitrogen, carbonates, organic matter, and salts did not decrease, and thus the microbial activity of the soil should be maintained for future use.

Chlorinated Hydrocarbons

Halogenated aromatic compounds include pesticides, DDT, 2,4-D and 2,4,5-T, plasticizers, pentachlorophenol, PCBs, among others. Their high stability and toxicity are causes of great concern for the environment and public health. The position and number of halogens are important in determining the rate and mechanism of biodegradation. The mineralization of PCBs was followed after the addition of rhamnolipid

R1 (molecular mass 650 Da with a CMC of 54 mg/L and surface tension of 36 mN/m) from *P. aeruginosa* to an acclimated culture of *Alcaligenes eutrophus* (Robinson et al., 1996). Using 4 g/L of biosurfactant, 4,4′ chlorobiphenyl was mineralized by 213 times more than the control. This work seemed promising for the application of biosurfactants for the enhanced biodegradation of PCBs and NAPLs that are soil-bound. Other work by Fiebig et al. (1997) has shown that glycolipids (GL-K12) from *P. cepacia* enhanced the degradation of Arochlor 1242 by mixed cultures, particularly for congeners with up to three Cl atoms where degradation was 100%. Batch experiments have also shown that free-phase and soil-phase PCB degradation were enhanced by rhamnolipid R1 addition (Robinson et al., 1996).

Pesticides are another group of contaminants that have been studied. Mata-Sandoval et al. (2000) compared the ability of the rhamnolipid mixture to solubilize the pesticides, trifluralin, coumaphos, and atrazine, with the synthetic surfactant Triton X-100. The synthetic surfactant was able to solubilize approximately twice as much of all pesticides as the rhamnolipid. The biosurfactant seems to bind trifluralin tightly in the micelle and releases the pesticide slowly to the aqueous phase, which could have implications for microbial uptake. This approach of utilizing micellar solubilization capacities and aqueous–micelle solubilization rate coefficients and micellar–aqueous transfer rate coefficients could be useful for future studies on microbial uptake. Addition of rhamnolipid in the presence of cadmium enabled biodegradation of the hydrocarbon naphthalene to occur as if no cadmium was present (Maslin and Maier, 2000).

Further work by Mata-Sandoval et al. (2001) was performed on the biodegradation of the three pesticides in liquid cultures in the presence of rhamnolipid or Triton X-100. Trifluralin biodegradation was enhanced in the presence of both surfactants, while atrazine decreased. Coumaphos biodegradation increased at rhamnolipid concentrations above 3 mM but declined when Triton concentrations were above that of the CMC. In soil slurries, trifluralin degradation decreased as both surfactant concentrations increased. As the concentration of rhamnolipid increased, biodegradation rates of coumaphos decreased but removal increased. The concentration of rhamnolipid also decreased, indicating biodegradation of the rhamnolipid.

The effect on the biodegradation of chlorinated hydrocarbons has also been determined (Uysal and Turkman, 2007). Unacclimated sludge was used to biodegrade 4-chlorophenol (4-CP) in the presence of a biosurfactant. Chemical oxygen demand (COD) and 4-CP, biomass and substrate removal were followed. The biosurfactant lowered transient times to steady state and sludge age requirements (<15 days). However, biosurfactant-enhanced biodegradation was not proven.

Harenda and Vipulanandan (2008) studied continuously stirred reactors to evaluate the degradation of PCE by a biosurfactant (UH) and an SDS. Biosurfactant, up to 500 mg/L, solubilized per gram more PCE. The PCE degradation in the presence of Ni/Fe particles occurred in 3 h, was first order, and led to the production of no other residual by-products other than chlorine ions. The C-O double bonds were the active groups in the biosurfactants.

A study by Hamidi and Mulligan (2010) assessed the effectiveness of rhamnolipid and two environmentally friendly surfactant-based commercial products (ASAP™ (A) and Degreaser™ (B)) for enhancing the bioremediation of PCE.

Following the USEPA's biological effectiveness test methods, and gas chromatographic (GC) analysis, 21 day experiments were conducted. Biological agents with the addition of the rhamnolipid enhanced significantly the remediation for PCE removal. PCE removal occurred in the following order for the various agents: biological agent + rhamnolipid > biological agent > rhamnolipid > control. Microbial analysis showed a direct correlation between microbial density and PCE removal, indicating that PCE removal occurred through biodegradation.

Ex Situ Washing Studies by Rhamnolipids

Hydrocarbons   Besides studies on biodegradation, rhamnolipid surfactants have been tested to enhance the release of low solubility compounds from soil and other solids. They have been found to release three times as much oil as water alone from the beaches in Alaska after the Exxon Valdez tanker spill (Harvey et al., 1990). Removal efficiency varied according to contact time and biosurfactant concentration.

Scheibenbogen et al. (1994) found that the rhamnolipids from *P. aeruginosa* UG2 were able to effectively remove a hydrocarbon mixture from a sandy loam soil and that the degree of removal (from 23% to 59%) was dependent on the type of hydrocarbon removed and the concentration of the surfactant used. Van Dyke et al. (1993) had previously found that the same strain could remove at a concentration of 5 g/L, approximately 10% more hydrocarbons from a sandy loam soil than a silt loam soil and that SDS was less effective than the biosurfactants in removing hydrocarbons. Lafrance and Lapointe (1998) also showed that injection of low concentrations of UG2 rhamnolipid (0.25%) enhanced transport of pyrene more than SDS with less impact on the soil.

Bai et al. (1998) showed that after only two pore volumes, 60% of the hexadecane was removed by a 500 mg/L concentration of rhamnolipid at pH 6 with 320 mM sodium. Various biological surfactants were compared by Urum et al. (2003) for their ability to wash a crude-oil-contaminated soil, including aescin, lecithin, rhamnolipid, saponin, tannin, and SDS. The following conditions were evaluated: surfactant concentration (0.004%, 0.02%, 0.1%, and 0.5%), surfactant volume (5, 10, 15, and 20 mL), temperature (5°C, 20°C, 35°C, and 50°C), shaker speed (80, 120, 160, and 200 strokes/min), and wash time (5, 10, 15, and 20 min). A temperature of 50°C and 10 min were optimal for most of the surfactants. More than 79% of the oil was removed by SDS, rhamnolipid, and saponin.

Rhamnolipid was also evaluated for the removal of oil for contaminated sandy soil (Santa Anna et al., 2007). The removal of oil was monitored for 101 days and the biosurfactant was shown to be effective. The composition of the aromatic and paraffinic oil did not change and could be recycled.

A completely different approach for oil cleanup was performed by Shulga et al. (2000) who examined the use of the biosurfactants for oil removal from coastal sand, and the feathers and furs of marine birds and animals. *Pseudomonas* PS-17 produces a biosurfactant and biopolymer that reduced the surface tension of water to 29.0 mN/m and the IFT against heptane to between 0.01 and 0.07 mN/m. The molecular weight of the biopolymer is from 3 to 4×10^5. The biosurfactant/biopolymer was able to remove oil from marine birds and animals contaminated by oil.

Desorption of phenanthrene from a marine sediment (Zhu et al., 2011) by a rhamnolipid was found to be more effective than SDS. Sorption of the rhamnolipid was also occurring as the CMC increased to 111.6 mg/L in the presence of the sediment.

Four different biosurfactants were compared for washing of a low level and a high TPH-contaminated soil (Lai et al., 2009). The biosurfactants were more effective than synthetic ones. Highest TPH removal was by rhamnolipids in the range of 23% and 63% for the two soils, respectively. Surfactin was almost as effective and was superior to serrawettin and a novel bioemulsifier.

Pei et al. (2009) studied the effect of biosurfactant on sorption of phenanthrene on soil. Organic matter content was highly influential. Phenanthrene sorption decreased in the presence of biosurfactant for the black loamy soil, whereas for the red sandy soil, the biosurfactant increased sorption. The biosurfactant was shown to be very biodegradable, indicating that multiple additions would be required.

Nguyen et al. (2008) compared mixtures of biological and synthetic surfactants. A mixture of rhamnolipids and synthetic surfactants reduced the IFT, thus enabling mobilization of the hydrocarbons. The hydrophobicity of the mixture was increased to close to that of the hydrocarbon and reduced oil–solution IFT by a factor or two to less than 0.1 mN/m compared to the individual surfactants. This means they could be used for remediation purposes or enhanced oil recovery and could improve the economic viability of the system.

The feasibility of removing styrene from contaminated soil by rhamnolipid was evaluated by Guo and Mulligan (2006). More than 70% removal of styrene could be achieved for an initial content in soil of 32,750 mg/kg after one day and 88.7% removal after 5 days while a 90% removal was obtained for an initial 16,340 mg/kg of styrene after 1 day. Longer contact times (5 days) enhanced removal efficiencies (up to 90%). A weight solubilization ratio of 0.29 g styrene was solubilized per gram of rhamnolipid added ($R^2 = 0.9738$). After removal from the soil, the leachate containing styrene and rhamnolipid must be treated. It was determined that more than 70% of the styrene in the leachate could be biodegraded by an anaerobic biomass (Guo and Mulligan, 2006).

Nguyen et al. (2008) mixed hydrophilic rhamnolipids with synthetic hydrophobic biosurfactants to reduce IFT for environmental remediation applications. The IFT values were reduced to less than 0.1 mN/m for all hydrocarbons, including toluene, hexane, decane, and hexadecane. For example, a ratio of biosurfactant to C12, 13-8PO sulfate of 4 to 1 (wt/wt basis) reduced IFT of toluene to 0.032 mN/m. More hydrophobic biosurfactants will need to be investigated.

Further studies were performed by Guo et al. (2011) to determine the aggregation behavior of rhamnolipid mixed with styrene by small angle neutron scattering (SANS). Styrene was used as a representative of hydrophobic molecules commonly found in contaminated soils. Deuterated and hydrogenated styrene was used to resolve the morphologies of aggregates. A structural transformation from cylindrical micelles to a binary mixture of cylindrical micelles and vesicles was induced by both elevated rhamnolipid and styrene concentrations. The resultant structure of the aggregates, vesicle, is different from the "oil droplets" commonly reported in the microemulsions of water–oil–surfactant mixtures. It was also determined that styrene solubility may be constrained by various factors that are not known at this time.

Chemical and biosurfactants were intercalated with layered double hydroxides (LDHs) to remove organic pollutants from water (Chuang et al., 2010). A concentration of 1000 mg/L of rhamnolipid intercalated in a 2:1 ratio with uncalcined LDH (K_d of 2160 mg/kg) was more effective than the chemical surfactant (K_d of 1770 mg/kg) for naphthalene sorption and is nontoxic.

Chlorinated Hydrocarbons

A soya lecithin biosurfactant followed by a second step of photocatalytic treatment of the effluents was evaluated for the remediation of a PCB-contaminated site (Occulti et al., 2007). Compared to Triton X-100, the lecithin removed less soil organic content, was of lower ecotoxicity, and was more effective in removing PCBs. Although both surfactants decreased the efficiency of the photocatalytic treatment for PCBs, lecithin performed better. In a later study, Occulti et al. (2008) further tested a soya lecithin biosurfactant-based washing for a PCB-contaminated soil. Photocatalytic treatment of the leachate followed the washing process. They postulated that the system was a sustainable remediation as it was an integrated chemical, microbiological, and ecotoxicological monitoring procedure. Scale-up and field application, however, would be required as a future step.

Hexachlorobenzene (HCB)-contaminated soil can be treated by surfactant washing. However, to recover the surfactant for reuse, methods are needed. Granular activated carbon was evaluated by Wan et al. (2011) to recover the contaminant from the rhamnolipid solution after washing spiked kaolin or an actual contaminated soil to enable reuse of the surfactant. With 10 g/L of powdered activated carbon (PAC), 99% of the HCB could be removed. In addition, it was found that if the surfactant solution (25 g/L of rhamnolipid) was combined with the PAC (10 g/L) as a washing solution, two cycles could remove more than 86%–88% of the PAC from the soil. Therefore, this coupling was very effective.

Clifford et al. (2007) evaluated the removal of perchloroethylene (PCE) by a rhamnolipid biosurfactant. The PCE–biosurfactant solution IFT was 10 mN/m, which is quite high, indicating that mobilization is not likely. This is beneficial as it minimizes vertical mobilization. However, partitioning of PCE did occur with a WSR of 1.2 g of PCE/g of rhamnolipid (mainly by the monorhamnolipid). Thus, this biosurfactant is a good candidate for surfactant-enhanced recovery of PCE. Albino and Nambi (2009) evaluated solubilization of PCE and TCE by surfactin and rhamnolipid. WSR of 3.83 and 12.5 for these compounds, respectively, were found for surfactin and 2.06 and 8.36 for rhamnolipid (mainly a dirhamnolipid), respectively. Both were superior to synthetic surfactants (SDS, Tween 80, and Triton X-100) but surfactin was superior to rhamnolipid. Solubilities of TCE increased from pH 7 to 5, but the rhamnolipid and surfactin started to precipitate at the lower pH.

Heavy Metals

The anionic nature and the complexation ability of rhamnolipids enable them to remove metals from soil and ions such as cadmium, copper, lanthanum, lead, and zinc (Herman et al., 1995; Ochoa-Loza, 1998; Tan et al., 1994). The nature and mechanism of the biosurfactant–metal complexes are being studied. Ochoa-Loza et al. (2001) determined stability constants by an ion–exchange resin technique.

Cations of lowest to highest affinity for rhamnolipid were $K^+ < Mg^{2+} < Mn^{2+} < Ni^{2+} < Co^{2+} < Ca^{2+} < Hg^{2+} < Fe^{3+} < Zn^{2+} < Cd^{2+} < Pb^{2+} < Cu^{2+} < Al^{3+}$. The affinities were in the same order of magnitude or higher than those of organic acids (acetic, citric, fulvic, and oxalic acids) with metals, thus indicating the potential of the rhamnolipid for metal remediation. Molar ratios (MRs) of the rhamnolipid to individual metals were 2.31 for copper, 2.37 for lead, 1.91 for cadmium, 1.58 for zinc, and 0.93 for nickel. Common soil cations, magnesium and potassium, had lower MRs of 0.84 and 0.57, respectively.

In the presence of oil contamination, rhamnolipids were added to soil (Mulligan et al., 1999a,b) and sediments to remove heavy metals (Mulligan et al., 2001b). Although 80%–100% of cadmium and lead can be removed from artificially contaminated soil, from field samples the results were more in the range of 20%–80% due to increased bonding of the contaminants over time (Fraser, 2000). Biosurfactant could also be added as a soil-washing process for excavated soil. Due to the foaming property of the biosurfactant, metal–biosurfactant complexes can be removed by addition of air to cause foaming and then the biosurfactant can be recycled through precipitation by reducing the pH to 2.

Neilson et al. (2003) studied lead removal by rhamnolipids. A 10 mM solution of rhamnolipid removed about 15% of the lead after 10 washes. High levels of Zn and Cu did not have any influence on lead removal. Mulligan et al. (1999a, 2001b) showed that lead could be removed from the iron oxide, exchangeable and carbonate fractions. These removal levels are very low and the process could be improved if the biosurfactants could be added in multiple cycles (Neilson et al., 2003).

Rhamnolipids have also been added to another metal-contaminated media, mining residues, to enhance metal extraction (Dahrazma and Mulligan, 2004). Batch tests using a 2% rhamnolipid concentration showed that 28% of the copper was extracted. Although concentrations higher than 2% extracted more copper, the rhamnolipid solution was very viscous and became difficult to work with. The addition of 1% NaOH with the rhamnolipid enhanced the removal up to 42% at a concentration of 2% rhamnolipid but decreased removal at higher surfactant concentrations. Sequential extraction studies were also being performed to characterize the mining residue and to determine the types of metals being extracted by the biosurfactants. Approximately 70% of the copper was associated with the oxide fraction, 10% with the carbonate, 5% with the organic matter, and 10% with the residual fraction. After washing for 6 days with 2% biosurfactant (pH 6), 50% of the carbonate fraction and 40% of the oxide fraction were removed. In summary, rhamnolipids are effective for hydrocarbon and heavy-metal removal and could also be effective for the removal of mixed (hydrocarbon and metal) contaminants. However, studies have not been performed at large scale.

Other studies have also been performed to evaluate the feasibility of metal removal by biosurfactants. Juwarkar et al. (2007) investigated the removal of cadmium and lead by a biosurfactant produced by *P. aeruginosa* BS2 in column tests. Cadmium removal was more than Pb. Within 36 h, more than 92% of Cd and 88% of Pb was removed by the rhamnolipid (0.1%) The rhamnolipid was also able to decrease toxicity, allow microbial activity (*Azotobacter* and *Rhizobium*) to take place and to not degrade soil quality. Cost-effectiveness, though, needs to be evaluated.

Asci et al. (2007) evaluated the ability of rhamnolipid to remove Cd(II) from kaolinite. Of the various sorption models evaluated for Cd(II), the Kolbe-Corrigan model fitted best. The effects on desorption by pH, rhamnolipid concentration, and sorbed Cd(II) concentration were determined. The optimal conditions were pH 6.8 for an initial Cd concentration of 0.87 mM, and a rhamnolipid concentration of 80 mM. A removal of 71.9% of Cd(II) was achieved.

Sepiolite and feldspar, two soil components were compared for their ability to sorb cadmium by Asci et al. (2008a). Sepiolite was found to be a superior accumulator of cadmium than feldspar. The desorption of the rhamnolipid from feldspar (96%) was much higher than from sepiolite (10%). Asci et al. (2008b) then examined the removal of zinc from Na-feldspar (a soil component) by a rhamnolipid biosurfactant. Significant sorption of zinc was shown onto the feldspar. Optimal pH for removal was found to be 6.8. This was postulated to be due to the small vesicles and micelles at a pH > 6.0. Low IFTs in this range would facilitate sorption of the biosurfactant and subsequent metal contact. A concentration of 25 mM was optimal for 98.8% removal of 2.2 mM of zinc (a 12.2:1 ratio).

Sorption of biosurfactants can reduce their potential for removal of contaminants from various components of the soil. Some preliminary tests by Guo and Mulligan 2007 indicated that organic and clay contents increased sorption and that the monorhamnolipid (R1) adsorbed more than R2. Therefore, further rhamnolipid sorption tests were performed (Ochoa-Loza et al., 2007). The R1 sorption was dependent on concentration and followed the order of hematite > kaolinite > MnO_2 ~ illite ~ Ca-montmorillonite > gibbsite > humic acid coated silica for low R1 concentrations for organic degradation. For higher concentrations, illite $\gg$ humic acid coated silica > Ca-montmorillonite > hematite > MnO_2 > gibbsite ~ kaolinite. R1 was also found to sorb more strongly than dirhamnolipid (R2) but is more efficient for metal removal as sorption reduces the efficiency of rhamnolipid removal. This information will enable predictions to occur regarding the feasibility of rhamnolipid treatment and the quantity of rhamnolipid for rhamnolipid treatment.

Kim and Vipulanandan (2006) determined that the removal by a biosurfactant from lead-contaminated soil (kaolinite) could be represented by a linear isotherm. The biosurfactant was produced from vegetable oil. Over 75% of the lead also could be removed from 100 mg/L contaminated water at 10 times the CMC. The biosurfactant-to-lead ratio for optimal removal was 100:1. FTIR spectroscopy indicated that the carboxyl group of the biosurfactant was implicated in the removal. Micelle partitioning could also be represented by Langmuir and Freundlich models. The biosurfactant micelle partitioning was more favorable than the synthetic surfactants, SDS and Triton X-100.

Batch washes of biosurfactants (rhamnolipids, saponin, and mannosyl–erytritol lipids [MEL]) were used to remove heavy metals from soil from a construction site and a lake sediment (Mulligan et al., 2007). The soil contained 890 mg/kg of zinc, 260 mg/kg copper, 170 mg/kg nickel, and 230 mg/kg total petroleum hydrocarbons, and the sediment contained 4440 mg/kg zinc, 94 mg/kg copper and 474 mg/kg of lead. After five washings of the soil with saponin (30 g/L), the highest levels of zinc removal were 88%, and nickel removal was 76%. Copper removal (46%) was maximal with 2% rhamnolipids (pH 6.5). Multiple washings of the soil with 4% MEL (pH 5.6) provided low levels of removal (17% of the zinc and nickel and 36% of the copper). From the sediment, the highest level of zinc (33%) and lead removal (24%) were achieved with 30 g/L saponin (pH 5). Highest copper removal (84%) was achieved with 2% rhamnolipids (pH 6.5). Sequential extraction showed that the oxide fraction of zinc and organic fraction of copper were substantially reduced by the biosurfactants.

Slizovskiy et al. (2011) compared several surfactants including rhamnolipid for the removal of Zn, Cu, Pb, and Cd, and soil ecotoxicological reduction; 39%, 56%, 68%, and 43% of the respective metals were removed by the biosurfactant from an aged-field-contaminated soil. Bioaccumulation of the metals by two worm species (*Eisenia fetida* and *Lumbricus terrestris*) was reduced and biomass and survival increased, indicating reduced soil toxicity.

Dahrazma and Mulligan (2007) subsequently evaluated the performance of rhamnolipid, in a continuous flow configuration (CFC) for the removal of heavy metals (copper, zinc, and nickel) from sediments to simulate a flow-through remediation technique. Rhamnolipid solution was pumped at a constant rate through the sediment within a column. The effect of rhamnolipid concentration, additives, time, and flow rate was investigated. The heavy-metal removal was up to 37% of Cu, 13% of Zn, and 27% of Ni when rhamnolipid was applied. The addition of 1% NaOH to 0.5% rhamnolipid enhanced the removal of copper by up to four times compared with 0.5% rhamnolipid.

The size and morphology of rhamnolipid micelles were evaluated by Dahrazma et al. (2008). The SANS technique was used to perform the investigation. At high pH, large aggregates and micelles in the order of 17 Å were found. In acidic conditions, however, larger 500–600 Å diameter vesicles were formed. Therefore, there should not be any filtering effect in regard to soil flushing through pores that are typically in the order of 200 nm. Larger molecules such as exopolymers could thus cause plugging of the pores. Complexation of the micelles with metals did not have any significant effect on the size of the micelles.

A new approach for metal stabilization by biosurfactants was discovered (Massara et al., 2007). A study was conducted on the removal of Cr(III) by rhamnolipids from chromium-contaminated kaolinite. Results showed that the rhamnolipids have the capability to extract 25% of the more stable form of chromium, Cr(III), from the kaolinite, under optimal conditions. The removal of hexavalent chromium by rhamnolipids was also enhanced compared to the control by a factor of 2. The sequential extraction procedure results showed that rhamnolipids remove Cr(III) mainly from the carbonate, and oxide/hydroxide portions of the kaolinite. The rhamnolipids have also the capability of reducing almost completely the extracted Cr(VI) to Cr(III) over 24 days. The rhamnolipids, thus, could be beneficial for the removal of and the long-term conversion of Cr(VI) to Cr(III).

This work was continued to evaluate the use of rhamnolipid for the removal and reduction of hexavalent chromium from contaminated soil and water in batch experiments (Ara and Mulligan, 2008). The initial chromium concentration, rhamnolipid concentration, pH, and temperature affected the reduction efficiency. Complete reduction by rhamnolipid of initial Cr(VI) in water at optimum conditions (pH 6, 2% rhamnolipid concentration, 25°C) occurred at low concentration (10 ppm). For higher initial concentrations (400 ppm), 24 h were required to reduce Cr by 24.4%. In the case of soil, rhamnolipid alone could remove the soluble fraction of the chromium present in the soil. The extraction increased as the initial concentration increased in the soil but decreased slightly when the temperature increased above 30°C. A sequential extraction study was used on soil before and after washing to determine from what fraction the rhamnolipid removed the chromium. The exchangeable and carbonate fractions accounted for 24% and 10% of the total chromium, respectively. The oxide and hydroxide portions bound 44% of the total chromium in the soil. On the other hand, 10% and 12% of the chromium was associated with the organic and residual fractions. Rhamnolipid was able to remove most of the exchangeable (96%) and carbonate (90%) portions and some of the oxide and hydroxide portion (22%) but from the other fractions. This information is important in designing the appropriate conditions for soil washing.

The removal of another anionic contaminant, arsenic, was investigated from mining residues by rhamnolipids (Wang and Mulligan, 2009a). Only arsenic in the form of As(V) was extracted from the residues at high pH. Significant removal of Cu, Zn, and Pb simultaneously occurred and was positively correlated with arsenic removal. The arsenic mobilization was due either to organic complex formation or metal-bridging mechanisms.

Further work by Wang and Mulligan (2009b) focused on the development of mobilization isotherms of arsenic from the residues to predict the mobilization. Easily and moderately extractable arsenic could be removed but redox or methylation reactions did not occur to any significant effect. The rhamnolipid thus might be potentially useful for removal of As from mining tailings.

A plant-based biosurfactant, saponin, was evaluated for the removal of mixed contaminants, phenanthrene and cadmium, from soil (Song et al., 2008). Phenanthrene is removed by solubilization while cadmium is complexed by the carboxyl groups of saponin. The combined removal of both the hydrocarbon and metals were effective at levels of 87.7% and 76.2% for phenanthrene and cadmium, respectively. Further saponin studies were performed by Chen et al. (2008) to measure the removal of copper and nickel by saponin from kaolin. A concentration of 2000 mg/L of saponin could remove 85% of the nickel and 83% of the copper at pH 6.5. Metal desorption was in the following order: ethylenediaminetetraacetic acid (EDTA) > saponin ≫ SDS. A three-step washing mechanism of adsorption followed by the formation of ion pairs with adsorbed metal and then rearrangement to desorb the metal was postulated.

A biosurfactant from marine bacterium was isolated and evaluated for metal removal (Das et al., 2009). A concentration of five times the CMC enables the total removal of lead and cadmium. Transmission electron microscopy (TEM) analysis indicated metal binding on the micelle surface. Visible precipitation of the metal ions occurred, indicating potential for wastewater treatment.

Yuan et al. (2008) tested the removal of heavy metals by a tea saponin by ion flotation. The biosurfactant acted as both, the collector and the frother. The complexed metal ions adsorbed onto surfaces of the air bubbles. Lead (90%) was removed at higher levels than copper (81%) and cadmium (71%). Increasing ionic strength slightly decreased removal efficiencies. As in other studies, complexation via carboxylate groups with the divalent metal ions was determined.

Kilic et al. (2011) examined the removal of chromium from tannery sludge by saponin and hydrogen peroxide. The saponin removed 24% of the Cr, while the hydrogen peroxide was able to extract 70% and oxidize Cr(VI) to Cr(III). The organic content was the major inhibitor of the removal of Cr by the saponin and, therefore, pretreatment may be required.

Mercury has been recovered by foam fractionation using surfactin (Chen et al., 2011). Air is bubbled into the solution to separate the metal contaminant. Various parameters were evaluated. A surfactin concentration in the order of $10 \times$ CMC, low Hg concentration (2 mg/L), and high pH (8–9) were optimal. Results were superior to the chemical surfactants, SDS and Tween 80.

Various bacterial strains were screened for biosurfactant production from agro-based substrates and the ability to remove heavy metals (Hazra et al., 2010). *Pseudomonas aeruginosa* AB4 was isolated and its glycolipid biosurfactant was produced from renewable nonedible seed cakes. Preliminary tests also indicated that Pb and Cd could be chelated by the biosurfactant.

In Situ Soil Flushing Studies by Rhamnolipids

PAHs and PCP

To simulate in situ flushing conditions, soil column experiments were performed (Noordman et al., 1998). A concentration of 500 mg/L of rhamnolipids removed two to five times more phenanthrene than the control. Noordman et al. (2000), then, investigated adsorption of the biosurfactant to the soil as they must not adsorb strongly to the soil in in situ situations. The unitless retardation factor, R, used in models for transport estimation was determined: $R = 1 + (\rho/\varepsilon)k_d$, where ρ is the soil density, ε is the soil porosity and k_d is the soil–water partition coefficient. R was between 2 for naphthalene with silica and 700 for phenanthrene with octadecyl-derivatized silica. The addition of 500 mg/L of rhamnolipid decreased R by eight-fold. To limit interfacial hydrophobic adsorption of surfactant aggregates such as hemimicelles to the soil, biosurfactant concentrations higher than the CMC should be used. Adsorption was due to interfacial phenomena not partitioning into soil organic matter. The transport of the more hydrophobic compounds was facilitated while the less hydrophobic ones were retarded due to the sorption of biosurfactant admicelles (micelles adsorbed to the soil surface). The interactions could include ion exchange reactions with the anionic carboxylate portion of the rhamnolipid, complexation at the surface and hydrogen bonding of the rhamnose head groups (Somasundaran and Krishnakumar, 1997). Herman et al. (1995) determined that low concentrations of rhamnolipids below the CMC enhanced mineralization of entrapped hydrocarbons while concentrations above the CMC enhanced the mobility of the hydrocarbons.

Attempts are also being made to determine nonaqueous-phase liquid (NAPL) micelle-aqueous partition coefficients with rhamnolipids to understand equilibrium solubilization behavior in surfactant-enhanced soil remediation (McCray et al., 2001). Deviation from this ideal Raoult's law behavior occurred depending on the hydrophobicity of the compounds and the NAPL-phase mole fraction. Micelle–water partition coefficients were found to be nonlinear according to the NAPL-phase mole fraction. Enhancements by the biosurfactant were greater than predicted for hydrophobic and less than predicted for the more hydrophilic compounds. Empirical relationships were also developed for multicomponent NAPL and the biosurfactant. Correlations would then be incorporated into transport models in the future.

The effect of rhamnolipids on the partitioning of the PAHs, naphthalene, fluorene, phenanthrene, and pyrene from NAPLs has been examined by Garcia-Junco et al. (2003). Enhanced partitioning of most of the PAHs, with the exception of naphthalene, occurred even with humic acid–smectite clay complexes. The rhamnolipids sorbed onto the solids typical of those found in the subsurface and increased the amount of solid-phase PAHs. The equation $\frac{dC}{dt} = k(C_{eq} - C)$ was used, where C_{eq} is the equilibrium aqueous-phase concentration and C is the PAH concentration in aqueous and solid phases. It was determined that at biosurfactant concentrations above the CMC, k values were lower since there was competition for the PAH between the micelles and sorbed biosurfactants,

A rhamnolipid produced by *Pseudomonas putida* was evaluated for the desorption of phenanthrene from clay–loam soil (Poggi-Varaldo and Rinderknecht-Seijas, 2003). The biosurfactant (250 mg/L) improved desorption of the PAH linearly and therefore a linear k (k_l) was determined. Desorption with water (reference) gave a $k_{l,ref}$ = 139 mL/g and with the biosurfactant, k_l, = 268 mL/g. Therefore, there was an almost twofold improvement of the availability enhancement factor (AEF) by the biosurfactant.

The capability of a rhamnolipid in the form of foam was evaluated for the removal of pentachlorophenol (PCP) from soil (Mulligan and Eftekhari, 2003). The stability of the rhamnolipid foam was excellent. When the foam was injected into the contaminated soil (1000 mg/kg of PCP), 60% and 61% of the PCP was removed from a fine sand and sandy-silt soil, respectively. The foam can be injected into the soil at low pressures, which will avoid problems like soil heaving. The high-quality foams, generated in this study (99%), contain large bubbles with thin liquid films that can collapse easily and are less resistant to the soil, resulting in lower soil pressures.

Heavy Metals

Metal removal from a contaminated sandy soil (1710 ppm of Cd and 2010 ppm of Ni) was evaluated by a foam produced by 0.5% rhamnolipid solution, after 20 pore volumes (Mulligan and Wang, 2004). The biosurfactant foam removed 73.2% of Cd and 68.1% of Ni, compared to the biosurfactant liquid solution, where 61.7% Cd and 51.0% Ni were removed. This was superior to Triton X-100 foam, which removed 64.7% Cd and 57.3% Ni, and liquid Triton X-100, which removed 52.8% Cd and 45.2% Ni. Distilled water removed only 18% of both Cd and Ni. For a 90% foam quality, the average hydraulic conductivity was 4.1×10^{-4} cm/s, for 95%, it was 1.5×10^{-4} cm/s and for 99%, it was 2.9×10^{-3} cm/s. Increasing foam quality decreases substantially the hydraulic conductivity. All these values are lower than the conductivity of water at 0.02 cm/s. This higher viscosity will allow better control of the surfactant mobility during in situ use. Therefore, rhamnolipid foam may be an effective and nontoxic method of remediating heavy-metal, hydrocarbon-, or mixed-contaminated soils. Further efforts will be required to enable its use at field scale.

An evaluation was made of the capability of a rhamnolipid biosurfactant (JBR425) foam for the treatment of PAH-contaminated freshwater sediments with elevated levels of Pb, Zn, and Ni (Alavi and Mulligan, 2011). The biosurfactant foam was injected in the sediment column. The pressure gradient was monitored during the flushing tests to determine possible problems due to high pressure. Foam-quality rhamnolipid was varied between 85% and 99% with stabilities from 15 to 43 min. PAH and metal removal was then evaluated for sediment samples. Highest PAHs removal after 20 pore volumes obtained by a removal by biosurfactant foam (99% quality by a 0.5% rhamnolipid solution) was 44.6% of pyrene, 30% of benz(a)anthracene, and 37.8% of chrysene, while total removal efficiency (mobilization + volatilization) for the biosurfactant foam was 56.4% of pyrene, 41.2% of benz(a)anthracene, and 45.9% of chrysene. With biosurfactant liquid solution at the same pH as the aforementioned foam (pH 6.8), maximum removal (mobilization) was 31.4% of pyrene, 20.5% of benz(a)anthracene, and 27% of chrysene. No volatilization of PAHs was observed. The control (deionized water) did not remove any PAHs. The highest removal of metals was achieved with 0.5% rhamnolipid foam (99% quality, pH 10.0 was 53.3% of Ni, 56.8% of Pb, and 55.2% of Zn). Removal levels were reduced to 11%–17% for metals when a liquid 0.5% rhamnolipid solution was used instead of the foam or the control. The rhamnolipid foam could be a nontoxic and effective method of remediating PAH- and heavy-metal-contaminated soil/sediments. Further efforts will be required to optimize the performance of the foam.

Remediation of Oil-Contaminated Water by Rhamnolipids

The Amoco Cadiz spill off the Britanny coast in 1978 and the Exxon Valdez spill near Prince William Sound in 1989 are examples of significant coastline contamination. Biosurfactants can be useful for oil spills since they could be less toxic and more degradable than synthetic surfactants. Chakrabarty (1985) showed that an emulsifier produced by *P. aeruginosa* SB30 could disperse oil into fine droplets, which could enhance biodegradation. Chhatre et al. (1996) showed that four bacteria isolated from crude oil were able to degrade 70% of the Gulf and Bombay High Crude Oil. One of the isolates produced a rhamnolipid biosurfactant that enhanced biodegradation by emulsification of the crude oil.

The feasibility of using biosurfactants for dispersing oil slicks was studied (Holakoo and Mulligan, 2002). At 25°C and a salinity of 35‰, a solution of 2% rhamnolipids applied at a dispersant-to-oil ratio (DOR) of 1:2, immediately dispersed 65% of a crude oil. Applied at a DOR of 1:8, co-addition of 60% ethanol and 32% octanol with 8% rhamnolipids improved dispersion to 82%. Dispersion efficiency decreased in freshwater and at lower temperatures, but altering the formulation could improve efficiencies. A comparison of the dispersion behavior to the control showed that the rhamnolipids had excellent potential as nontoxic oil dispersing agents. Laboratory toxicity tests of the biosurfactant showed that IC_{50} values are low for marine flagellates, microalgae, and the bioluminescence of *Photobacter phosphoreum* (Lang and Wagner, 1987). Low concentrations of rhamnolipid (1 g/L) were able to convert a mousse oil into an oil-in-water emulsion. De-emulsification was performed to remove the oil. Subsequent rhamnolipid-assisted bioremediation of the remaining aqueous phase decreased the oil to undetectable levels (Nakata and Ishigami, 1999).

Therefore, despite the large amount of research on dispersants, there is very little on the use of biosurfactants as biodispersants despite their potential benefits, particularly for enhancing oil biodegradation and solubilization.

Simultaneous removal of metals and organic pollutants from water is challenging. Although ultrafiltration can remove high molecular weight molecules, it is not effective for removal of low molecular weight pollutants. Rhamnolipid biosurfactant (Elzeftawny and Mulligan, 2011) was utilized in micellar-enhanced ultrafiltration (MEUF) of heavy metals from contaminated waters. The effects of different major operating conditions on the MEUF system performance were investigated for copper, zinc, nickel, lead, and cadmium using two membranes. The optimal conditions were successfully applied to treat six contaminated wastewaters from metal refining industries using the two membranes (>99% rejection ratio). To efficiently choose the most influential factors to the MEUF system, optimization by the response surface methodology approach was utilized and data quality was examined. Optimization by the response surface methodology and validation experiments determined that the best operating conditions were a transmembrane pressure of 69 ± 2 kPa, biosurfactant-to-metal MRs of approximately 2:1, a temperature of 25°C ± 1°C, and pH of 6.9 ± 0.1. The rhamnolipid-enhanced ultrafiltration system was also shown to treat samples of six contaminated wastewaters from metal refining industries using membranes with molecular weight cutoffs (MWCO) of 10,000 and 30,000 Da. The resulting heavy-metal concentrations in the permeate were all significantly reduced to be in accordance with the federal Canadian regulations.

Micellar-enhanced ultrafiltration (MEUF) can be a more effective technique to remove contaminants as it makes use of the micellar properties of surfactant solutions to remove low molecular weight dissolved ions and/or organics. Compared to chemical surfactants, biosurfactants are less toxic. This is advantageous as they may create a secondary problem due to some leakage into the permeate. Therefore, the objective of this study was to evaluate the effect of a rhamnolipid biosurfactant on mixed contaminant removal from aqueous solutions (Ridha and Mulligan, 2011). The required quantity of rhamnolipid to remove the copper ions as a heavy-metal pollutant and benzene molecules as an organic pollutant separately was determined for different concentrations of pollutants. This quantity, indicated by the MR of biosurfactant to contaminant was 6.25 to obtain a 100% rejection for the copper ions and 1.33 to obtain the same rejection for benzene molecules. When copper and benzene were cocontaminants in the water, the MRs improved for benzene from 1.33 to 0.56 but remained the same for copper. In all cases, rhamnolipid proved excellent for the removal of contaminants and a rejection of 100% has been obtained for copper and benzene either separately or simultaneously, which indicates the potential of this approach for wastewater treatment.

EFFECT ON BIODEGRADATION AND CONTAMINANT REMOVAL BY SOPHOROLIPIDS

Although sophorolipids have been used to release bitumen from oil sands (Cooper and Paddock, 1984), few applications of these biosurfactants have been reported so far. High yields of the sophorolipid make this a potentially useful and economic biosurfactant. A crude preparation of biosurfactants from *Candida bombicola* was able to partially solubilize a North Dakota Beulah Zap lignite coal (Polman et al., 1994). Metal-contaminated soils and sediments have been treated with crude sophorolipids for the removal of heavy metals (Mulligan et al., 1999a, 2001b).

Oberbremer et al. (1990) added sophorolipids to a 10% soil and 1.35% hydrocarbon mixture. With the sophorolipids, 90% of the hydrocarbons were degraded in 79 h compared to 81% in 114 h without the biosurfactant. Schippers et al. (2000) studied the effect of sophorolipids on phenanthrene biodegradation and determined that the concentration of phenanthrene within 36 h decreased from 80 to 0.5 mg/L in the presence of the 500 mg/L of surfactant, compared to 2.3 mg/L without surfactant in a 10% soil suspension. The maximal degradation by *Sphingomonas yanoikuyae* was 1.3 mg/L-h with the sophorolipid compared to 0.8 mg/L-h without. The sophorolipids seem to enhance the phenanthrene concentration as shown by fluorescence measurements. In addition, toxicity of the sophorolipid was low for concentrations up to 1 g/L. The CMC of the sophorolipid increased to 10 mg/L in the presence of 10% soil suspensions from 4 mg/L in water, thus indicating adsorption of the surfactant onto the soil. Solubilization tests showed that 232 mg of phenanthrene in water and 80.7 mg in soil were solubilized by 1 g of sophorolipids. This is 10-fold higher than by other surfactants such as SDS. Therefore, these experiments have indicated that the sophorolipids enhance biodegradation of the phenanthrene through enhanced solubilization. More practical experience with more types of contaminants and more understanding of the mechanisms will be required.

SURFACTIN

Studies by Surfactin

Surfactin is another biosurfactant evaluated for environmental applications. Although it is a very effective biosurfactant, fewer studies have been performed with this biosurfactant in comparison with rhamnolipids. A strain of *Bacillus subtilis* O9 was isolated from contaminated sediments (Olivera et al., 2000). Surfactin was produced from sucrose. A crude form of surfactin was added to ship bilge waste to determine potential enhancement of biodegradation. Only 6.8% and 7.2% of the aliphatic and aromatic compounds in a nonsterile environment remained after 10 days as they were degraded more quickly in the presence of the biosurfactant. However, n-C17 pristane and n-C18 phytane degradation were not enhanced. More effective methods of production of surfactin are required to improve process feasibility.

A strain of *B. subtilis* was able to produce biosurfactant at 45°C at high NaCl concentrations (4%) and a wide pH range (4.5–10.5) (Makkar and Cameotra, 1997a,b). The biosurfactant was able to remove 62% of the oil in a sand pack saturated with kerosene. This indicates it could be used for in situ oil removal and cleaning sludge from sludge tanks.

Because of the presence of two negative charges, one on the aspartate and the other on the glutamate residues of surfactin, the binding of the metals, magnesium, manganese, calcium, barium, lithium, and rubidium, has been demonstrated (Thimon et al., 1992). Batch soil-washing experiments used surfactin from *B. subtilis* for the removal of heavy metals from a contaminated soil and sediments (Mulligan et al., 1999b). The sediments contained 110 mg/kg of copper and 3300 mg/kg of zinc. The contaminated soil had levels of copper, zinc, and cadmium of 550, 1200, and 2000 mg/kg, respectively. Water alone removed minimal amounts of copper and zinc (less than 1%). A series of five washings of the soil with 0.25% surfactin (1% NaOH) were required to remove 70% of the copper and 22% of the zinc. To determine the mechanism of metal removal by surfactin, the techniques of ultrafiltration, octanol–water partitioning and the measurement of zeta potential were used. Surfactin could remove the metals by sorption at the soil interface and metal complexation, followed by desorption of the metal through lowering of soil–water IFT and fluid forces, and finally complexation of the metal with the micelles (Figure 10.6).

The ability to produce lipopeptide biosurfactants by a *Bacillus* strain in situ in an oil reservoir was evaluated by Youssef et al. (2007). The amount of 90 mg/L was produced and is sufficient to mobilize entrapped oil in sandstone cores. Microorganisms were injected into the wells and showed enhanced oil recovery through in situ biosurfactant production. Limitation of nitrogen was necessary. Costs could be in the range of $10 per m³. The rates of biosurfactant production were more than required. This is a clear indication of the feasibility of field tests, which can have implications for remediation of contaminated sites.

Remediation of Heavy-Metal-Contaminated Water by Surfactin

Using MEUF, Mulligan et al. (1999b) studied the removal of various concentrations of metals from water by various concentrations of surfactin by a 50,000 Da molecular weight cutoff ultrafiltration membrane. Cadmium and zinc rejection ratios were close to 100% at pH values of 8.3 and 11, while copper rejection ratios were 85% at pH 6.7. The addition of 0.4% oil as a cocontaminant decreased slightly the retention of the metals by the membrane. The ultrafiltration membranes also indicated that metals were associated with the surfactin micelles as the metals remained in the retentate as shown in Figure 10.6. The ratio of metals to the surfactin was determined to be 1.2:1, which was only slightly higher than the theoretical value of 1 mol metal:1 mol surfactin due to the two charges on the surfactin molecule.

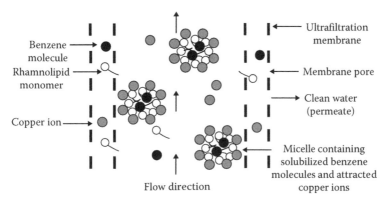

FIGURE 10.6 Simultaneous removal of Cu (II) ions and benzene molecules from aqueous solutions utilizing micellar-enhanced ultrafiltration. (Adapted from Ridha, Z.A.M. and Mulligan, C.N., Simultaneous removal of benzene and copper from contaminated water using micellar-enhanced ultrafiltration, *CSCE General Conference*, Ottawa, Ontario, Canada, June 14–17, 2011.)

Studies with Other Biosurfactants

Ex Situ Washing Studies by Other Biosurfactants

In addition to the biosurfactants previously discussed, other biosurfactants have been tested for environmental applications. A biosurfactant extracted from cactus named BOD-Balance™ was added in concentrations of 130–200 mg/L to a wastewater containing oil and grease from pet food (Nakhla et al., 2003). Oil and grease concentrations decreased from 66,300 to 10,200 mg/L and COD decreased from 59,175 to 35,000 mg/L within 2 months in a full-scale anaerobic digestion process. The economic feasibility of adding the biosurfactant, which costs $7 to $9/L, depends on the disposal of the sludge from the dissolved air flotation and the use of oil recycling. Ermolenko et al. (1997) used the strain *Mycobacterium flavescens* Ex 91 to produce Ekoil and found that this product could decontaminate oil-polluted water from a nuclear power station.

Hong et al. (2002) examined the removal of cadmium and zinc, by saponin, a plant-derived biosurfactant from three types of soils (Andosol soil, Cambisol, and Regosol). Highest removal was obtained from Regesol. Removals of 90%–100% for cadmium and 85%–98% for zinc were obtained. Saponin concentrations of 3% were optimal for metal removal within 6 h at a pH of 5.0–5.5. Sequential extraction tests showed that exchangeable and carbonate fractions were removed by the saponin, which is similar to surfactin and rhamnolipids tests for zinc performed by Mulligan et al. (2001). Heavy metals were recovered from the soil supernatants by precipitation by increasing the pH to 10.7 and the saponin was recycled for reuse; 80% of Cu, 86% of Cd, 90% of Pb, and 91% of Zn were recovered at pH 10.7.

Hong et al. (1998) have also evaluated a sodium salt of 2-(2-carboxyethyl)-3-decyl maleic anhydride derived from the dehydration of spiculisporic acid produced by *Penicillium spiculisporum*. Membranes with molecular weight cutoffs of 1000 and 3000 Da were studied for the removal of cadmium, copper, zinc, and nickel. Highest rejection rates were achieved when the molar biosurfactant and metal ion concentrations were equal.

Although biosurfactants have been used in a variety of environmental applications, little is known about the potential for biosurfactant production in contaminated soils by microorganisms. Bodour et al. (2003) determined that gram-negative strains were found in hydrocarbon or mixed contaminant soils, while gram-positive biosurfactant producers were found in heavy-metal and uncontaminated soils. A glycolipid produced by *Alcanivorax borkumiensis* enhanced the treatment of PCBs sorbed onto soil (Golyshin et al., 1999). In summary, much more effort is required to determine the role of biosurfactant-producing organisms in situ and to exploit their potential.

NEW BIOSURFACTANTS WITH POTENTIAL FOR ENVIRONMENTAL APPLICATIONS

Some efforts have been made to isolate and study new biosurfactants. A biosurfactant (JE-1058) was studied for the remediation of oil spills (Saeki et al., 2009). The biosurfactant was produced by *Gordonia sp.* isolated from soil and then evaluated by addition as an oil spill dispersant in seawater. It was found to stimulate the biodegradation of oil. It had low toxicity, and could disperse oil slicks and wash contaminated sea sand. Franzetti et al. (2009) further examined another strain, named BS29. Bioemulsans and cell-bound biosurfactants were produced. The bioemulsans were determined to be as effective as rhamnolipid for the removal of crude oil and PAHs from contaminated soils and thus they have potential as remediation agents. The removal of heavy metals and enhanced bioremediation seemed less effective but more evaluations are needed.

Nayak et al. (2009) identified a *Pseudoxanthomonas sp.* PNK-04 strain that produced rhamnolipids. The rhamnolipids were a mixture of mono- and di-rhamnose units. Emulsifying ability was high and showed enhanced biodegradation of 2-chlorobenzoic acid, 3-chlorobenzoic acid, and 1-methyl naphthalene through solubility enhancement compared to synthetic surfactants.

A biosurfactant-producing bacterium was isolated from oil-contaminated soil and water by Somayeh et al. (2008). The bacterium was identified as *P. aeruginosa* and showed a broth surface tension of less than 40 mN/m. Paraffin and glucose could be used as substrates and a wide range of pH (2–14) could be used as the media. The toxicity test and the emulsification of oil were used as the screening tests.

Wattanaphon et al. (2008) isolated a biosurfactant-producing bacterium from a fuel-oil-contaminated soil, identified as *Burkholderia cenocepacia* BSP3. The produced glycolipid showed a CMC higher than other biosurfactants (316 mg/L), a low surface tension (25 mN/m), and good emulsification ability. Methyl parathion, ethyl parathion, and trifluralin (three pesticides) could be solubilized by the biosurfactant, which could potentially be used for bioremediation of pesticide-contaminated soil.

Martins et al. (2009) showed that 99% of the PAHs up to three rings could be removed by the biosurfactant compared to 92% with the disperser. More than four rings were not effectively remediated. This biosurfactant had been previously produced by solid-state fermentation by the fungus *Aspergillus fumigatus* grown on rice bran and husk.

Pirollo et al. (2008) isolated a strain from hydrocarbon-contaminated soil. The isolate grows on diesel oil, kerosene, crude oil, and oily sludge and could potentially be used for remediation of hydrocarbon-contaminated sites. The culture was *P. aeruginosa* LB1 and reduced surface tensions to 32–36 mN/m. Emulsification ability showed good stability on heating over a wide range of pH (5–10).

Ilori et al. (2008) showed that biosurfactants were produced by the yeasts *Saccharomyces cerevisiae* and *Candida albicans* isolated from a polluted lagoon. Crude oil and diesel oil could be utilized as substrates by both strains.

A glycolipid biosurfactant was isolated from a thermophilic bacterium *Alcaligenes faecalis* (Bharali et al., 2011). Like other biosurfactants, the surface tension can be lowered to 32 mN/m and the CMC is low at 38 mg/L. However, it is very stable over a wide variety of conditions, pH (2–12), salinity (1%–6% NaCl), and up to 100°C. It also exhibited antimicrobial and antifungal properties. Biosurfactant could be produced in 2% diesel supplemented, but motor lubricating oil and crude oil were also good substrates, indicating the potential for use of this in bioremediation.

Although it has been shown that rhamnolipid was biodegradable by Mohan et al. (2006), there is little work on various biosurfactants in water and soil. Surfactin, iturin, fengycin, arthrofactin, glycolipid, and flavolipids biodegradability were evaluated (Lima et al., 2011). Overall, they had a higher biodegradation potential in the environment than SDS and thus are safer for use in environmental remediation applications.

Liu et al. (2012) isolated and characterized an isolate of *Bacillus amyloliqufaciens* BZ-6 able to treat an oily sludge from enhanced oil recovery; 88% of the oil was recovered and the surface tension was reduced to 29 mN/M. The biosurfactant produced consisted of four different fengycin A homologues and thus has potential for oil removal from sludges.

NEW TECHNOLOGIES

New developments have been made in the area of biosurfactants with regard to nanotechnology. Nanorods of NiO were produced by a water-in-oil microemulsion (Palanisamy, 2008) and a biosurfactant was subsequently added to heptane. The produced nanorods were 22 nm in diameter and 150–250 μm in length at pH 9.6. This could be due to the effect of pH on the biosurfactant morphology. The use of the biosurfactant is a more eco-friendly approach.

Reddy et al. (2009) showed that the synthesis of silver nanoparticles could be stabilized by surfactin. These nanoparticles have unique physical, chemical, magnetic, and structural properties. Various pH and temperature conditions were evaluated. The nanoparticles were stabilized for a period of 2 months by surfactin, which is a stabilizing agent that is renewable, has low toxicity, and is biodegradable, and is thus an environmentally friendly additive.

Rhamnolipid was evaluated for its effect on the electrokinetic and rheological behavior on nanozirconia particles (Biswas and Raichur, 2008). The biosurfactants adsorb increasingly onto the zirconia as the concentration increases. The biosurfactant was able to disperse the zirconia particles at pH 7 and above as shown by zeta-potential measurements, sedimentation, and viscosity tests. It can serve as an eco-friendly product for flocculation and the dispersion of high solid contents of microparticles.

Fatisson et al. (2010) evaluated various components on the stabilization of CMC-coated zero-nonvalent iron nanoparticles (nZVI). Stabilization is important to enhance the transport of the nZVI particles to the zone of contamination. The effects of fulvic acid and rhamnolipid were evaluated as these are naturally found in the groundwater and soil environment. The presence of the rhamnolipid led to the lowest rate of deposition of the particles on silica.

Fe/Ni particles were used to degrade PCE extracted by soil washing of various surfactants (Vipulanandan and Harendra, 2008). Within 45 min, CTAB extracted (280 mg/L of PCE), SDS (240 mg/L PCE) and UH biosurfactant (214 mg/L PCE) were completely degraded. More recent work by the same authors (Vipulanandan and Harendra, 2011) examined the in situ degradation of PCE with a surfactant–bimetallic nanoparticle colloidal solution. The colloidal solution was transported through a clayey soil in a soil column. The UH biosurfactant–Fe/Ni colloidal solution decreased PCE by 82%, compared to 77% by the CTAB Fe/Ni colloidal solution.

Laboratory experiments were conducted to investigate the effect of the presence of rhamnolipid on the production and stabilization of iron nanoparticles (Farshidy et al., 2011). In addition, the effect of rhamnolipid on the remediation of chromium (VI) from water using iron nanoparticles was tested. Iron nanoparticles were produced in the presence of different concentrations of rhamnolipid. Then, unmodified nanoparticles were treated with different concentrations of rhamnolipid and carboxymethyl cellulose. The TEM micrographs indicated that without adding rhamnolipid during the production process, the size of iron nanoparticles is high due to the formation of micron-sized clusters.

By adding low concentrations of rhamnolipid (90 mg/L), the diameter of the particles was reduced to less than 10 nm due to the coating and stabilization of nanoparticles by the rhamnolipid. This was confirmed by zeta-potential measurements on the modified iron nanoparticles. Furthermore, the effect of the presence of rhamnolipid on reductive remediation of hexavalent chromium, Cr (VI), to trivalent, Cr (III), was investigated and evaluated. By combining 0.034 g/L chromium (VI) with iron nanoparticles and rhamnolipi, the only positive effect was observed with concentrations of 0.08 g/L iron and 2% (w/w) of rhamnolipid. At this concentration, the remediation of chromium increased by 123% in 15 h compared with solutions containing only iron nanoparticles or only rhamnolipid. However, at lower rhamnolipid concentrations, the extent of remediation decreased.

Biosurfactants may also be beneficial for enhancing other remediation technologies. Zhu and Zhang (2008) investigated the enhanced uptake of pyrene in ryegrass by rhamnolipids. Low rhamnolipid concentrations (25.8 mg/L or $0.5 \times$ CMC) seemed to enhance the uptake, possibly by enhancing permeability of the root cells. The effect of tea saponin on *Zea mays* L. and *Saccharum officinarum* L. phytoremediation was studied (Xia et al., 2009). A tea saponin concentration of $0.2 \times$ CMC (0.01%) enhanced PCB and cadmium uptake from water and soil.

CONCLUSION AND FUTURE DIRECTIONS

Jeneil Biotech Corp (http://www.jeneilbiotech.com) has produced commercially biosurfactants such as rhamnolipids. Another company, AGAE Technologies, (http://www.agaetech.com/) also produces rhamnolipids from a unique strain of the bacterium *P. aeruginosa*. Various applications including remediation are emphasized such as in situ or ex situ soil remediation projects concerning hydrocarbons including PAHs. Elimination of heavy metals and pesticides from soil are also available applications. Remediation of industrial waste water from food processing, oil/gas operations, and other industries are also possible with our rhamnolipid biosurfactant.

The economics of producing the biosurfactants has limited commercial applications. Costs can be reduced by improving yields, rates, and recovery, using crude preparations and using cheap or waste substrates (Mulligan and Gibbs, 1993). Although more information is available concerning biosynthesis of rhamnolipids and surfactin, there is still a lack of information regarding the secretion of the biosurfactants, metabolic route, and primary cell metabolism (Peypoux et al., 1999). Research is thus required to accelerate the knowledge in this area as it could possibly enhance the applications of the surfactant. New forms of the biosurfactants could also become available.

According to Technical Insights, a division of Frost & Sullivan, Nonionic Surfactants, microbial surfactants have begun to enjoy a market (O'Connor, 2002). The most promising applications are oil spill (Figure 10.7) and oil-contaminated tanker cleanup, removal of crude oil from sludge, enhanced oil recovery (Figure 10.8), bioremediation of sites contaminated with hydrocarbons, other organic pollutants, and heavy metals. Contaminants such as PCB could also be bioremediated by biosurfactant addition, since natural surfactants such as humic acids were able to enhance the microbial degradation of PCBs (Fava, 2002). A summary of some of the studies involving the use of biosurfactants for biodegradation is shown in Table 10.2 and for washing or flushing is shown in Table 10.3. It can be seen that most of the

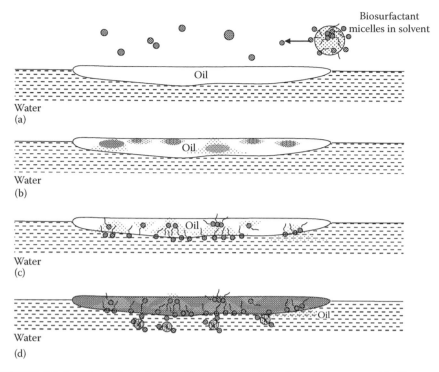

FIGURE 10.7 Dispersion of oil spill on water using dispersing agents. (a) Dispersant droplets applied to oil slick, (b) dispersant droplets coalesce with oil slick and diffuses into oil, (c) solvent delivers surfactants throughout oil and oil–water interface, and (d) dispersant-enriched oil readily disperses into droplets.

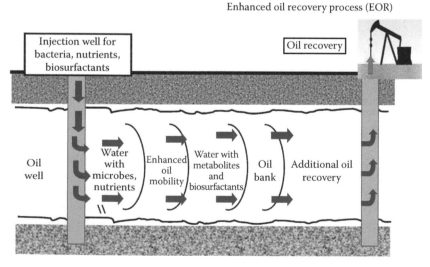

FIGURE 10.8 Enhanced oil recovery by biosurfactant injection. (Adapted from Marchant and Banat, 2012.)

TABLE 10.2
Summary of Biodegradation Studies Involving Biosurfactants

Biosurfactant	Medium	Microorganism	Contaminant	References
Rhamnolipid	Soil slurry	*P. aeruginosa* UG12	Hexachlorobiphenyl	Berg et al. (1999)
	Soil	*P. aeruginosa* UG12	Aliphatic and aromatic hydrocarbons	Sheibenbogen et al. (1994)
	Soil slurry	*P. aeruginosa* UG12	Phenanthrene	Providenti et al. (1995)
	Soil	*P. aeruginosa* UG2	Phenanthrene and hexadecane	Noordman et al. (1998, 2000, 2002)
	Soil	*P. aeruginosa* #64	Phenanthrene, fluoranthrene, pyrene, pentachlorophenol benzo[a]pyrene	Straube et al. (1999, 2003)
	Soil	*P. aeruginosa* ATCC 9027	Naphthalene and phenanthrene	Zhang et al. (1997)
	Soil	*P. aeruginosa* ATCC 9027	Octadecane	Zhang and Miller (1992)
	Soil	*P. aeruginosa* ATCC 9027	Phenanthrene and cadmium	Maslin and Maier (2000)
	Soil	*P. aeruginosa* ATCC 9027	Naphthalene and cadmium	Sandrin et al. (2000)
Mono-rhamnolipid	Oily sludge	*Consortium*	Alkanes	Rahman et al. (2003)
Di-rhamnolipid	Liquid	*Pseudomonas*	Toluene, ethyl benzene, butyl benzene	McCray et al. (2001)
Sophorolipid	Liquid	Mixed culture	14–16 C alkanes, pristane, phenyldecane, naphthalene	Oberbremer et al. (1990)
	Soil	*C. bombicola* ATCC 2214	Phenanthrene	Schippers et al. (2000)
Crude surfactin	Soil	*B. subtilis* ATCC 2423	Hexadecane and kerosene	Makkar and Cameotra (1997b)
	Soil	*B. subtilis* ATCC 2423	Endosulfan	Awashti et al. (1999)
	Seawater	*B. subtilis* O9	Aliphatic and aromatic hydrocarbons	Moran et al. (2000)
Alasan	Liquid	*Acinetobacter radioresistens* KA 53	Phenanthrene, fluoranthene, and pyrene	Barkay et al. (1999)
Rhamnolipid	Soil	Isolate	Diesel	Whang et al. (2008)

(continued)

TABLE 10.2 (continued)
Summary of Biodegradation Studies Involving Biosurfactants

Biosurfactant	Medium	Microorganism	Contaminant	References
Rhamnolipid	Liquid	Consortium	Diesel	Vasefy and Mulligan (2008)
	Liquid	*Pseudomonas* strain	Crude oil, diesel	Mehdi and Giti (2008)
Surfactin	Liquid	Isolates	Crude oil	Li et al. (2006)
Rhamnolipid	Soil	Consortium	Diesel	Jalali and Mulligan (2007)
Rhamnolipid	Soil	*Luteibacter* sp.	Petroleum and heavy metals	Zhang et al. (2011)
Biosurfactant	Sludge	*P. aeruginosa and rhodococcus* consortium	Oil	Cameotra and Singh (2008)
Rhamnolipid	Drill cuttings	Consortium	Oil	Yan et al. (2011)
Rhamnolipid	Wastewater	*P. aeruginosa*	Oil	Yin et al. (2008)
Biosurfactant	Water	Sludge	4-Chlorophenol	Usyal and Turkman (2007)
Rhamnolipid	Water	Consortium	PCE	Hamidi and Mulligan (2010)

Source: Adapted from Makkar and Rockne, 2003.
Note: All these studies showed a positive effect of the biosurfactant on biodegradation.

studies have involved rhamnolipids, although there are many that have been discovered (Table 10.1). Other biosurfactants may be more effective, however, and therefore more investigation is needed in the evaluation of other biosurfactants.

If the remediation can be performed in situ, production of the biosurfactants could also be in situ. This would be technically feasible and cost-effective. Less labor and transport would be required and the process would be ecologically acceptable. Implementation and survival of pure cultures on sites would be difficult since they would be outcompeted by indigenous microorganisms. However, strains of *Bacillus* and *Pseudomonas* at hydrocarbon-contaminated sites have been determined to be biosurfactant producers (Jennings and Tanner, 2000). Exploiting their presence may be the best strategy. As the bioavailability of the contaminant also plays an important factor in biodegradation, the role of biosurfactant production in the subsurface is not known but could significantly enhance the natural attenuation process (Yong and Mulligan, 2004). As the biosurfactants could influence the fate and transport of the contaminants in the subsurface studies are clearly needed in this area.

Dissolution of mixtures of multicomponent NAPLs by surfactant is not well-understood. Simple two- and three-component systems at most have been studied to gain a basic understanding. Only a few studies have also been completed regarding mixed organic and inorganic contamination. More research of more complex

TABLE 10.3
Summary of Soil Washing or Flushing Studies with Biosurfactants

Biosurfactant	Medium	Contaminant	References
Rhamnolipid	Soil	Oil	Harvey et al. (1990)
			Scheibenbogen et al. (1994)
			Van Dyke et al. (1993)
			Lafrance and Lapointe (1998)
			Urum et al. (2003)
		Hexadecane	Bai et al. (1998)
	Feathers, fur	Oil	Shulga et al. (2000)
	Sediment	Phenanthrene	Zhu et al. (2011)
	Soil	Phenanthrene	Pei et al. (2009)
Rhamnolipid, surfactin	Soil	Oil	Lai et al. (2009)
Rhamnolipid	Soil	Styrene	Guo and Mulligan (2006),
			Guo et al. (2011)
Lecithin		PCB	Occulti et al. (2007)
Rhamnolipid		HCB	Wan et al. (2011)
Rhamnolipid		PCE	Clifford et al. (2007)
Rhamnolipid, surfactin		PCE, TCE	Albin and Nambi (2009)
Rhamnolipid		Gazoil	Gozini et al. (2010)
Rhamnolipid, surfactin	Mining residues	Oil, heavy metals	Mulligan et al. (1999a,b)
Rhamnolipid	Soil	Pb, Zn, Cu	Neilson et al. (2003)
Rhamnolipid		Cu	Dahrazma and Mulligan
			(2004)
Rhamnolipid	Kaolinite	Cd, Pb	Juwarkar et al. (2007)
	Sepiolite, feldspar	Cd	Asci et al. (2008a)
		Zn	Asci et al. (2008b)
Biosurfactant	Feldspar	Pb	Kim and Vipulanandan (2006)
Rhamnolipid, MEL,	Kaolinite	Zn, Cu, Pb, Oil	Mulligan et al. (2007)
saponin	Soil, sediment	Cr	Massara et al. (2007)
Rhamnolipid	Water	Cr	Ara and Mulligan (2008)
	Soil, water	As	Wang and Mulligan (2009 a,b)
Saponin	Mining residues	Phenanthrene, Cd	Song et al. (2008)
	Soil	Cu, Ni	Chen et al. (2008)
	Kaolin	Cu, Cd, Pb	Yuan et al. (2008)
	Water	Cr	Kilic et al. (2011)
Surfactin	Tannery sludge	Hg	Chen et al. (2011)
Rhamnolipids	Liquid	Phenanthrene	Nordman et al. (1998, 2000)
Rhamnolipid	Soil	PAH, naphthalene, fluorine, phenanthrene, pyrene	Garcia-Junco et al. (2003)

(*continued*)

TABLE 10.3 (continued)
Summary of Soil Washing or Flushing Studies with Biosurfactants

Biosurfactant	Medium	Contaminant	References
Rhamnolipid	Soil	PCP	Mulligan and Eftekhari (2003)
	Soil	Cd, Ni	Mulligan and Wang (2004)
	Oil		Chakrabarty (1985)
Rhamnolipid	Sediments	PAH, Pb, Zn, Ni	Alavi and Mulligan (2011)
	Clay	PAHs	Shafeeq et al. (1989)
	Water	Oil	Holakoo and Mulligan (2002)
	Water	Cu, Zn, Ni, Pb, Cd	El Zeftawny and Mulligan (2011)
Glycolipid	Water	Cu, benzene	Ridha and Mulligan (2011)
	Soil	PCBs	Golyshin et al. (1999)

situations is needed, particularly in light of the potential for remediation of metal contamination by the biosurfactants.

More information is required concerning the interaction of the biosurfactants and the contaminant, relationship of biosurfactant structure and contaminant removal and the soil, scale-up and cost reduction for ex situ production, and understanding of the factors influencing the bioremediation of the compounds by enhanced bioavailability. Methods to enhance the economics of biosurfactant utilization will need to be developed further. For example, ultrafiltration (Mulligan and Gibbs, 1990) can be used to concentrate the biosurfactants for recovery and subsequent reuse, thus decreasing the amount of biosurfactant required.

The effects of soil components such as hydroxides, oxides, and organics on metal desorption by biosurfactants have been initiated in sediments, mining residues, and soil (Dahrazma and Mulligan, 2004; Mulligan and Dahr Azma, 2003; Mulligan et al. 1999). These types of studies provide clues into how difficult it will be to remove the metals from the soil by the biosurfactant. Further methods will need to be developed to enable one to predict and model the efficiency of the enhanced biodegradation, washing or flushing processes with biosurfactants under various hydrological and soil conditions.

Rhamnolipids and other biosurfactants including saponin have shown their potential for remediation of contaminated soil and water. More biosurfactants need to be investigated. Both organic and inorganic contaminants can be treated through desorption or biodegradation processes. The biosurfactants seem to enhance biodegradation by influencing the bioavailability of the contaminant through solubilization and emulsification of the contaminants. Due to their biodegradability and low toxicity, biosurfactants such as rhamnolipids are very promising for use in remediation technologies. In addition, there is the potential for in situ production, a distinct advantage over synthetic surfactants. This needs to be studied further. Further research regarding prediction of their behavior in the fate and transport of contaminants will be required. More investigation into the solubilization mechanism of both hydrocarbons and heavy metals by biosurfactants is required to enable model predictions for transport and remediation. New applications for the biosurfactants regarding nanoparticles are developing. Future research should focus on the stabilization of the nanoparticles by biosurfactants before addition during remediation procedures, decreasing production costs, in situ production, among other issues.

REFERENCES

Alavi, A. and Mulligan, C.N. 2011. Remediation of a heavy metal and PAH-contaminated sediment by a rhamnolipid foam. *Geo-Environmental Engineering 2011*, Takamatsu, Japan, May 21–22, 2011.

Al-Awadhi, N., Williamson, K.J., and Isok, J.D. 1994. Remediation of Kuwait's oil-contaminated soils. In *Hydrocarbon Contaminated Soils and Ground Water*, Vol. 3, Kostecki, P. and Calabrease, T. (eds.) Lewis Publishers, Chelsea, MI.

Albino, J.D. and Nambi, I.M. 2009. Effect of biosurfactants on the aqueous solubility of PCE and TCE. *Journal of Environmental Science and Health Part A* 44: 1565–1573.

Anonymous. 1984. Emulsan. *Chemical Week* 135, January, 58.

Ara, I. and Mulligan, C.N. 2008. Conversion of Cr(VI) in water and soil using rhamnolipid. *61st Canadian Geotechnical Conference*, Edmonton, Alberta, Canada, September 20–24, 2008.

Asci, Y., Nurbas, M., and Acikel, Y.S. 2007. Sorption of Cd(II) onto kaolinin as a soil component and desorption of Cd(II) from kaolin using rhamnolipid biosurfactant. *Journal of Hazardous Materials* B 139: 50–56.

Asci, Y., Nurbas, M., and Acikel, Y.S. 2008a. A comparative study for the sorption of Cd(II) by K-feldspar and sepiolite as soil components and the recovery of Cd(II) using rhamnolipid biosurfactant. *Journal of Environmental Management* 88: 383–392.

Asci, Y., Nurbas, M., and Acikel, Y.S. 2008b. Removal of zinc ions from a soil component Na-feldspar by a rhamnolipid biosurfactant. *Desalination* 233: 361–365.

Awashti, N., Kumar, A., Makkar, R., and Cameotra, S. 1999. Enhanced biodegradation of endosulfan, a chlorinated pesticide in presence of a biosurfactant. *Journal of Environmental Science and Health* B 34: 793–803.

Bai, G., Brusseau, M.L., and Miller, R.M. 1998. Influence of cation type, ionic strength and pH on solubilization and mobilization of residual hydrocarbon by a biosurfactant. *Journal of Contaminant Hydrology* 30: 265–279.

Barkay, T., Navon-Venezia, S., Ron, E., and Rosenberg. E. 1999. Enhancement of solubilization and biodegradation of polycyclic aromatic hydrocarbons by the emulsifier Alasan. *Applied and Environmental Microbiology* 65: 2697–2702.

Beal, R. and Betts, W.B. 2000. Role of rhamnolipid biosurfactants in the uptake and mineralization of hexadecane in *Pseudomonas aeruginosa*. *Journal of Applied Microbiology* 89: 158.

Bharali, P., Das, S., Konwar, B.K., and Thakur, A.J. 2011. Crude biosurfactant from thermophilic *Alcaligenes faecalis*: Feasibility in petro-spill bioremediation. *International Journal of Biodeteriation and Biodegradation* 65: 682–690.

Biswas, M. and Raichur, A.M. 2008. Electrokinetic and rheological properties of nano zirconia in the presence of rhamnolipid biosurfactant. *Journal of American Ceramic Society* 91: 3197–3201.

Bodour, A.A., Dress, K.P., and Maier, R.M. 2003. Distribution of biosurfactant-producing bacteria in undisturbed and contaminated arid southwestern soils. *Applied and Environmental Microbiology* 69: 3280–3287.

Burd, G. and Ward, O.P. 1996. Bacterial degradation of polycyclic aromatic hydrocarbons on agar plates: The role of biosurfactants. *Biotechnology Techniques* 10: 371–374.

Cameotra, S.S. and Singh, P. 2008. Bioremediation of oil sludge using crude biosurfactant. *International Biodeterioration and Biodegradation* 62: 274–280.

Chakrabarty, A.M. 1985. Genetically manipulated microorganisms and their products in the oil service industries. *Trends in Biotechnology* 3: 32–36.

Chang, J.-H., Qiang, Z., Huang, C.-P., and Ellis, A.V. 2009. Phenanthrene removal in unsaturated soils treated by electrokinetics with different surfactants-Triton X-100 and rhamnolipid. *Colloids and Surfaces A. Physicochemical Engineering Aspects* 348: 157–163.

Chhatre, S., Purohit, H., Shanker, R., and Khanna, P. 1996. Bacterial consortia for crude oil spill remediation. *Water Science and Technology* 34(10): 187–194.

Chuang, Y.-H., Liu, C.-H., Tzou, Y.-M., Chang, J.-S., Chiang, P.-N., and Wang, M.-K. 2010. Comparison and characterization of chemical surfactants and bio-surfactants intercalated with layered double hydroxides (LDHs) for removing naphthalene from contaminated aqueous solutions. *Colloids and Surfaces A. Physicochemical Engineering Aspects* 366: 170–177.

Chen, H.-R., Chen, C.-C., Reddy, A.S., Chen, C.-Y., Li, W.R., Tseng, M.-J., Liu, H.-T., Pan, W., Maity, J.P., and Atla, S.B. 2011. Removal of mercury by foam fractionation using surfactin, a biosurfactant. *International Journal of Molecular Sciences* 12: 8245–8258.

Chen, W.-J., Hsiao, L.-C., and Chen, K.K.-Y. 2008. Metal desorption from copper(II)/nickel(II)-spiked kaolin as a soil component using plant-derived saponic biosurfactant. *Process Biochemistry* 43: 488–498.

Churchill, S.A., Griffin, R.A., Jones, L.P., and Churchill, P.F. 1995. Biodegradation rate enhancement of hydrocarbons by an oleophilic fertilizer and a rhamnolipid biosurfactant. *Journal of Environmental Quality* 24: 19–28.

Clifford, J.S., Ionnidis, M.A., and Legge, R.L. 2007. Enhanced aqueous solubilization of tetrachloroethylene by a rhamnolipid biosurfactant. *Journal of Colloid and Interfacial Science* 305: 361–364.

Cooper, D.G. and Paddock, D.A. 1984. Production of a biosurfactant from *Torulopsis bombicola*. *Applied and Environmental Microbiology* 47: 173–176.

Dahrazma, B. and Mulligan, C.N., 2004. Extraction of copper from mining residues by rhamnolipids. *Practice Periodical on Hazardous Toxic and Radioactive Waste Management* 8(3): 166–172.

Dahrazma, B. and Mulligan, C.N. 2007. Investigation of the removal of heavy metals from sediments using rhamnolipid in a continuous flow configuration. *Chemosphere* 69: 705–711.

Dahrazma, B., Mulligan, C.N., and Nieh, M.P. 2008. Effects of additives on the structure of rhamnolipid (biosurfactant): A small-angle neutron scattering (SANS) study. *Journal of Colloid and Interfacial Science* 319: 590–593.

Daigger, G.T. and Crawford, G.V. 2005. Wastewater treatment plant of the future—Decision analysis approach for increased sustainability. In *2nd IWA Leading-Edge Conference on Water and Wastewater Treatment Technology, Water and Environment Management Series*. IWA Publishing, London, U.K. pp. 361–369.

Das, P., Mukherjee, S., and Sen, R. 2009. Biosurfactant of marine origin exhibiting heavy metal remediation properties. *Bioresource Technology* 100: 4887–4890.

Dean, S.M., Jin, Y., Cha, D.K., Wilson, S.V., and Radosevich, M. 2001. Phenanthrene degradation in soils co-inoculated with phenanthrene-degrading and biosurfactant-producing bacteria. *Journal of Environmental Quality* 30: 1126–1133.

Deschênes, L., Lafrance, P., Villeneuve, J.-P., and Samson, R. 1994. The impact of a biological and chemical anionic surfactants on the biodegradation and solubilization of PAHs in a creosote contaminated soil. Presented at the *Fourth Annual Symposium on Groundwater and Soil Remediation*, Calgary, Alberta, Canada, September 21–23, 1994.

El-Shafey, E.L. and Canepa, P. 2003. Remediation of a Cr (VI) contaminated soil: Soil washing followed by Cr (VI) reduction using a sorbent prepared from rice husk. *Journal de Physique IV (Proceedings)* 107(1): 415–418.

El Zeftawy, M.A.M. and Mulligan, C.N. 2011. Use of rhamnolipid to remove heavy metals from waste water by micellar-enhanced ultrafiltration (MEUF). *Separation and Purification Technology (SCI)* 77(1, 2): 120–127.

Ermolenko, Z.M., Kholodenko, V.P., Chugunov, V.A., Zhirkova, N.A., and Raulova, G.E. 1997. A *Mycobacteria* strain isolated from oil of Uktinskoe oil field identification and degradative properties. *Microbiology* 66: 542.

Falatko, D.M. 1991. Effects of biologically reduced surfactants on the mobility and biodegradation of petroleum hydrocarbons. MS thesis. Virginia Polytechnic Institute and State University, Blacksburg, VA.

Farshidy, M., Mulligan, C., and C. Bolton, K. 2011. *Stabilization of Iron Nanoparticles by Rhamnolipid for Remediation of Chromium Contaminated Water*, ICEPR, Ottawa, Ontario, Canada, August 17–19, 2011.

Fatisson, J., Ghoshal, S., and Tufenkji, N. 2010. Deposition of carboxymethylcellulose-coated zero-valent iron particles onto silica: Roles of solution chemistry and organic molecules. *Langmuir* 26: 12832–12840.

Fava, F. 2002. Natural surfactants enhance PCB bioremediation. *Industrial BioProcessing* 24: 2.

Fiebig, R., Schulze, D., Chung, J.C., and Lee, S.T. 1997. Biodegradation of polychlorinated biphenyls (PCBs) in presence of a bioemulsifier produced on sunflower oil. *Biodegradation* 8(2): 67–76.

Franzetti, A., Caredda, P., Ruggeri, C., La Colla, P., Tamburini, E., Papacchini, M., and Bestetti, G. 2009. Potential applications of surface active compounds by *Gordonia sp.* BS29 in soil remediation technologies. *Chemosphere* 75: 801–807.

Fraser, L. 2000. Innovations: Lipid lather removes metals. *Environmental Health Perspectives* 108: A320.

Garcia-Junco, M., De Olmedo, E., and Ortego-Calvo, J.-J. 2001. Bioavailability of solid and non-aqueous phase liquid-dissolved phenanthrene to the biosurfactant-producing bacterium *Pseudomonas aeruginosa* 19SJ. *Environmental Microbiology* 3: 561–569.

Garcia-Junco, M., Gomez-Lahoz, C., Niqui-Arroyo, J.-L., and Ortego-Calvo, J.-J. 2003. Biosurfactant- and biodegradation-enhanced partitioning of polycyclic aromatic hydrocarbons from nonaqueous-phase liquids. *Environmental Science and Technology* 37: 2988–2996.

Golyshin, P.M., Fredrickson, H.L., Giulano, L., Rothmel, R., Timmis, K.N., and Yakimov, M.N. 1999. Effect of novel biosurfactants on biodegradation of polychlorinated biphenyls by pure and mixed cultures. *Microbiologica*, 22: 257–278.

Gonzini, O., Plaza, A., Dipalma, L., and Lobo, M.C. 2010. Electrokinetic remediation of gasoil contaminated soil by rhamnolipid. *Journal of Applied Electrochemistry* 40: 1239–1248.

Guo, Y. and Mulligan, C.N. 2006. Combined treatment of styrene-contaminated soil by rhamnolipid washing followed by anaerobic treatment. In *Hazardous Materials in Soil and Atmosphere*, Hudson, R.C. (eds.) Nova Science Publishers, New York, pp. 1–38.

Guo, Y. and Mulligan ,C.N. 2007. Rhamnolipid-enhanced remediation of styrene-contaminated soil followed by anaerobic biodegradation. 60th Canadian Geotechnical Conference, Ottawa, October. 21–24, 2007.

Guo, Y., Mulligan, C.N., and Nieh, M.-P. 2011. An unusual morphological transformation of rhamnolipid aggregates induced by concentration and addition of styrene: A small angle neutron scattering (SANS) study. *Colloids and Surfaces A: Physicochemical and Engineering Aspects* 373(1–3): 42–50.

Hamidi, A. and Mulligan, C.N. 2010. Treatment of PCE-contaminated soil using biosurfactants and other additives. *11th International Environmental Specialty Conference*, Winnipeg, Manitoba, Canada, June 9–12, 2010.

Harenda, S. and Vipulanandan, C. 2008. Degradation of high concentrations of PCE solubilized in SDS and biosurfactant with Fe/Ni metallic particles. *Colloids and Surfaces A: Physicochemical Engineering Aspects* 322: 6–13.

Harvey, S., Elashi, I., Valdes, J.J., Kamely, D., and Chakrabarty, A.M. 1990. Enhanced removal of Exxon Valdez spilled oil from Alaskan gravel by a microbial surfactant. *Bio/Technology* 8: 228–230.

Hazra, C., Kundu, D., Ghosh, P., Joshi, S., Dandi, N., and Chaudhar, A. 2011. Screening and identification of *Pseudomonas aeruginosa* AB4 for improved production, characterization and application of a glycolipid biosurfactant using low-cost agro-based raw materials. *Journal of Chemical Technology and Biotechnology* 86: 185–198.

Herman, D.C., Artiola, J.F., and Miller, R.M. 1995. Removal of cadmium, lead and zinc from soil by a rhamnolipid biosurfactant. *Environmental Science and Technology* 29: 2280–2285.

Hester, A. 2001. IB market forecast. *Industrial Bioprocessing* 23(5): 3.

Holakoo, L. and Mulligan, C.N. 2002. On the capability of rhamnolipids for oil spill control of surface water. *Proceedings of the Annual Conference of the Canadian Society for Civil Engineering*, Montreal, Quebec, Canada, June 5–8, 2002.

Holden, T. 1989. *How to Select Hazardous Waste Treatment Technologies for Soils and Sludges*, Pollution Technology Review, No. 163, Noyes Data Corporation, Park Ridge, NJ.

Hong, J.-J., Yang, S.-M., Lee, C.-H., Choi, Y.-K., and Kajiuchi, T. 1998. Ultrafiltration of divalent metal cations from aqueous solutions using polycarboxylic acid type biosurfactant. *Journal of Colloid and Interfacial Science* 202: 63–73.

Hong, K.-J., Tokunaga, S., and Kajiuchi, T. 2002. Evaluation of remediation process with plant-derived biosurfactant for recovery of heavy metals from contaminated soils. *Chemosphere* 49(4): 379–387.

Ilori, M.O., Adebuseye, S.A., and Ojo, A. 2008. Isolation and characterization of hydrocarbon-degrading and biosurfactant-producing yeast strains obtained from a polluted lagoon water. *World Journal of Microbiology and Biotechnology* 24: 2539–2545.

Jalali, F. and Mulligan, C.N. 2007. Effect of biosurfactant production by indigenous soil microorganisms on bioremediation of a co-contaminated soil in batch experiments, *60th Canadian Geotechnical Conference*, Ottawa, Ontario, Canada, October 21–24, 2007.

Jennings, E.M. and Tanner, R.S. 2000. Biosurfactant-producing bacteria found in contaminated and uncontaminated soils. *Proceedings of the 2000 Conference on Hazardous Waste Research*, Denver, pp. 299–306.

Juwarkar, A.A. Nair, A., Dubey, K.V., Singh, S.K., and Devotta, S. 2007. Biosurfactant technology for remediation of cadmium and lead contaminated soils. *Chemosphere* 68: 1996–2002.

Kiliç, E., Font, J., Puig, R., Colak, S., and Celik, D. 2011. Chromium recovery from tannery sludge with saponin and oxidative remediation. *Journal of Hazardous Materials* 185(1): 456–462.

Kim, J. and Vipulanandan, C. 2006. Removal of lead from contaminated water and clay soil using a biosurfactant. *Journal of Environmental Engineering* 132(7): 777–786.

Lafrance, P. and Lapointe, M. 1998. Mobilization and co-transport of pyrene in the presence of *Pseudomonas aeruginosa* UG2 biosurfactants in sandy soil columns. *Ground Water Monitoring and Remediation* 18(4): 139–147.

LaGrega, M.D., Buckingham, P.L., and Evans, J.C. 2001. *Hazardous Waste Management*, McGraw Hill, Boston, MA.

Lai, C.-C., Huang, Y.-C., Wei, Y.-W., and Chang, J.-S. 2009. Biosurfactant-enhanced removal of total petroleum hydrocarbons from contaminated soil. *Journal of Hazardous Materials* 167: 609–614.

Lang, S. and Wagner, F. 1987. Structure and properties of biosurfactants. In *Biosurfactants and Biotechnology*, Kosaric, N., Cairns, W.L., and Gray, N.C.C. (eds.) Marcel Dekker, New York, pp. 21–45.

Li, Y-Y., Zheng, X-L., and Li, B. 2006. Influence of biosurfactant on the diesel oil remediation in soil-water system. *Journal of Environmental Science* 18: 587–590.

Lima, T.M.S., Procopio, L.C., Brandao, F.D., Carvalho, A.M.X., Totola, M.R., and Borges, A.C. 2011. Biodegradability of bacterial surfactants. *Biodegradation* 22: 585–592.

Liu, Z-F., Zeng, G.-M., Zhong, H., Yuan, Z-Z., Fu, H-Y., Zhou, M-F., Ma, X-L, Li, H., and Li, J-B. 2012. Effect of dirhamnolipid on the removal of phenol by laccase aqueous solution. *World Journal of Microbiology and Biotechnology* 28: 175–181.

Maier, R.M. and Soberon-Chavez, G. 2000. *Pseudomonas aeruginosa* rhamnolipids: Biosynthesis and potential applications. *Applied Microbiology and Biotechnology* 54: 625–633.

Makkar, R.S. and Cameotra, S.S. 1997a. Utilization of molasses for biosurfactant production by two Bacillus strains at thermophilic conditions. *Journal of the American Oil Chemists Society* 74: 887–889.

Makkar, R.S. and Cameotra, S.S. 1997b. Biosurfactant production by thermophilic *Bacillus subtilis* strain. *Journal of Industrial Microbiology and Biotechnology* 18: 37–42.

Makkar, R.S. and Rockne K.J. 2003. Comparison of synthetic surfactants and biosurfactants fin enhancing biodegradation of polycyclic aromatic hydrocarbons. *Environmental and Toxicological Chemistry* 22: 2280–2292.

Martins, V.G., Kalil, S.J., Alberto, J., and Costa, V. 2009. In situ bioremediation using biosurfactant produced by solid state fermentation. *World Journal of Microbiology and Biotechnology* 25: 843–851.

Marchant, R. and Banat, I.M. 2012. Microbial biosurfactants: Challenges and opportunities for future exploitation. *Trends in Biotechnology* 30: 558–565.

Maslin, P. and Maier, R.M. 2000. Rhamnolipid-enhanced mineralization of phenanthrene in organic-metal co-contaminated soils. *Bioremediation Journal* 4: 295–308.

Massara, H., Mulligan, C.N., and Hadjinicolaou, J. 2007. Effect of rhamnolipids on chromium contaminated soil. *Soil Sediment Contamínination: International Journal* 16: 1–14.

Mata-Sandoval, J.C., Karns, J., and Torrents, A. 2000. Effects of rhamnolipids produced by *Pseudomonas aeruginosa* UG2 on the solubilization of pesticides. *Environmental Science and Technology* 34: 4923–4930.

Mata-Sandoval, J.C., Karns, J., and Torrents, A. 2001. Influence of rhamnolipids and Triton X-100 on the biodegradation of three pesticides in aqueous and soil slurries. *Journal of Agricultural and Food Chemistry* 49: 3296–3303.

McCray, J.E., Bai, G., Maier, R.M., and Brusseau, M.L. 2001. Biosurfactant-enhanced solubilization of NAPL mixtures. *Journal of Contaminant Hydrology* 48: 45–68.

Mehdi, H. and Giti, E. 2008. Investigation of alkane biodegradation using the microtiter plate method and correlation between biofilm formation, biosurfactant production and crude oil biodegradation. *International Biodeterioration and Biodegradation* 62: 170–178.

Mench, M., Vangronsveld, J., Clijsters, H., Lepp, N.W., and Edwards, R. 2000. In situ metal immobilization and phytostabilization of contaminated soils. In *Phytoremediation of Contaminated Soil and Water*, Terry, N. and Bauelos, G. (eds.) Lewis Publishers, Boca Raton, FL, pp. 323–358.

Mohan, P.K., Nakhla, G., and Yanful, E.K. 2006. Biokinetics of biodegradation of surfactants under aerobic, anoxic and anaerobic conditions. *Water Research* 40(3): 533–540.

Moran, A., Olivera, N., Commedatore, M., Esteves, J., and Sineriz, P. 2000. Enhancement of hydrocarbon waste biodegradation by addition of a biosurfactant from *Bacillus subtilis* O9. *Biodegradation* 11: 65–71

Mosley, E. 2006. Developing a sustainability rating tool for wastewater systems. *WEFTEC Water Environment Federation* 21: 6283–6303.

Mulligan, C.N. and Dahr Azma, B., 2003. Use of selective sequential extraction for the remediation of contaminated sediments. *ASTM STP* 1442: 208–223; *Contaminated Sediments: Characterization, Evaluation, Mitigation/Restoration and Management Strategy Performance*, West Conshocken, PA, pp. 208–223.

Mulligan, C.N. and Eftekhari, F. 2003. Remediation with surfactant foam of PCP contaminated soil. *Engineering Geology* 70: 269–279.

Mulligan, C.N. and Gibbs, B.F. 1993. Factors influencing the economics of biosurfactants. In *Biosurfactants, Production, Properties, Applications*, Kosaric, N. (ed.) Marcel Dekker, New York, pp. 329–371.

Mulligan C.N., Oghenekevwe, C., Fukue, M., and Shimizu, Y. 2007. Biosurfactant enhanced remediation of a mixed contaminated soil and metal contaminated sediment. *7th Geoenvironmental Engineering Seminar*, Japan-Korea-France, Grenoble, France, May 19–24, 2007.

Mulligan, C.N. and Wang, S. 2004. Remediation of a heavy metal contaminated soil by a rhamnolipid foam. *Proceedings 4th BGA Geoenvironmental Engineering Conference*. Stratford-Upon-Avon, U.K., June 2004.

Mulligan, C.N., Yong, R.N., and Gibbs, B.F. 1999a. On the use of biosurfactants for the removal of heavy metals from oil-contaminated soil. *Environmental Progress* 18: 50–54.

Mulligan, C.N., Yong, R.N., and Gibbs, B.F. 1999b. Metal removal from contaminated soil and sediments by the biosurfactant surfactin. *Environmental Science and Technology* 33: 3812–3820.

Mulligan, C.N., Yong, R.N., and Gibbs, B.F., 2001a. Surfactant-enhanced remediation of contaminated soil: A review. *Engineering Geology* 60: 371–380.

Mulligan, C.N., Yong, R.N., and Gibbs, B.F. 2001b. Heavy metal removal from sediments by biosurfactants. *Journal of Hazardous Materials* 85: 111–125.

Nakata, K. and Ishigami, Y. 1999. A facile procedure of bioremediation for oily waste with rhamnolipid biosurfactant. *Journal of Environmental Science and Health* A34: 1129–1142.

Nakhla, G., Al-Sabawi, M., Bassi, A., and Liu, V. 2003. Anaerobic treatability of high oil and grease rendering wastewater. *Journal of Hazardous Material* 102: 243–255.

Nayak, A.S., Vijaykumar, M.H., and Karegoudar, T.B. 2009. Characterization of biosurfactant produced by Pseudoxanthomonas sp. PNK-04 and its application in bioremediation. *International Journal of Biodeterioration and Biodegradation* 63: 73–79.

Neilson, J.W., Artiola, J.F., and Maier, R.M. 2003. Characterization of lead removal from contaminated soils by non-toxic soil-washing agents. *Journal of Environmental Quality* 32: 899–908.

Nguyen, T.T., Youssef, N.H. McInerney, M.J., and Sabatini, D.A. 2008. Rhamnolipid biosurfactant mixtures for environmental remediation. *Water Research* 42: 1735–1743.

Noordman, W., Braussaeau, N., and Janssen, D. 1998. Effects of rhamnolipid on the dissolution, bioavailability and biodegradation of phenanthrene. *Environmental Science and Technology* 31: 2211–2217.

Noordman, W.H., Burseau, M.L., and Janssen, D.B. 2000. Adsorption of a multi-component rhamnolipid surfactant to soil. *Environmental Science and Technology* 34: 832–838.

Noordman, W.H., Wachter, J.J.J., de Boer, G.J., and Janssen, D.B. 2002. The enhancement by biosurfactants of hexadecane degradation by *Pseudomonas aeruginosa* varies with substrate availability. *Journal of Biotechnology* 94: 195–212.

Oberbremer, A., Muller-Hurtig, R., and Wagner, F. 1990. Effect of the addition of microbial surfactants on hydrocarbon degradation in a soil population in a stirred reactor. *Applied Microbiology and Biotechnology* 32: 485–489.

Occulti, F., Roda, G.C., Berselli, S., and Fava, F. 2007. Sustainable decontamination of an actual-site aged PCB-based washing followed by a photocatalytic treatment. *Biotechnology and Bioengineering* 99: 1525–1534.

Ochoa-Loza, F.J., Artiola, J.F., and Maier, R.M. 2001 Stability constants for the comxation of various metals with a rhamnolipid biosurfactant, *Journal of Environmental Quality*, 30, 479–485.

Ochoa-Loza, F. 1998. Physico-chemical factors affecting rhamnolipid biosurfactant application for removal of metal contaminants from soil. PhD dissertation, University of Arizona, Tucson, AZ.

Ochoa-Loza, F.J., Noordman, W.H., Jannsen, D.B., Brusseau, M.L., and Miller, R.M. 2007. Effects of clays, metal oxides, and organic matter on rhamnolipid sorption by soil. *Chemosphere* 66: 1634–1642.

O'Connor, L. 2002. Market forecast: Microbial biosurfactants *Industrial Bioprocessing* 24(3):10–11 .

Olivera, N.L., Commendatore, M.G., Moran, A.C., and Esteves, J.L. 2000. Biosurfactant-enhanced degradation of residual hydrocarbons from ship bilge wastes. *Journal of Industrial Microbiology and Biotechnology* 25: 70–73.

Palanisamy, P. 2008. Biosurfactant mediated synthesis of NiO nanorods. *Material Letters* 62: 743–746.

Papadopoulos, D., Pantazi, C., Savvides C., Harlambous, K.J., Papadopoulos A., and Loizidou, M. 1997. A study on heavy metal pollution in marine sediments and their removal from dredged material. *Journal of Environmental Health* A32(2): 347–360.

Park, K.S., Sims, R.C., and Dupont, R.R. 1990. Transformation of PAHs in soil systems. *Journal of Environmental Engineering* 116: 632–640.

Pei, X., Zhan, X., and Zhou, L. 2009. Effect of biosurfactant on the soption of phenanthrene onto original and H_2O_2 treated soils. *Journal of Environmental Science* 21: 1378–1385.

Peypoux, F., Bonmatin J.M., and Wallach, J. 1999. Recent trends in the biochemistry of surfactin. *Applied Microbiology and Biotechnology* 51: 553–563.

Pirollo, M.P.S., Mariano, A.P. Lovaglio, R.B. Costa, S.G.V.A.O., Walter, V., Hausmann, R., and Contiero, J. 2008. Biosurfactant synthesis by *Pseudomonas aeruginosa* LB1 isolated from a hydrocarbon-contaminated site. *Journal of Applied Microbiology* 105: 1484–1490.

Poggi-Varaldo, H.M. and Rinderknecht-Seijas, N. 2003. A differential availability enhancement factor for the evaluation of pollutant availability in soil treatments. *Acta Biotechnologica* 2–3: 271–280.

Polman, J.K., Miller, K.S., Stoner, D.L., and Brakenridge, C.R. 1994. Solubilization of bituminous and lignite coals by chemically and biologically synthesized surfactants. *Journal of Chemical Technology and Biotechnology* 61: 11–17.

Providenti, M.A., Fleming, C.A., Lee, H., and Trevors, J.T. 1995. Effect of addition of rhamnolipid biosurfactants or rhamnolipid producing *Pseudomonas aeruginosa* on phenanthrene mineralization in soil slurries. *FEMS Microbiological Ecology* 17(1): 15–26.

Rahman, K.S.M., Banat, I.M., Rahman, T.J., Thayumanavan, T., and Lakshmanaperumalsamy, P. 2002. Bioremediation of gasoline contaminated soil by a bacterial consortium amended with poultry litter, coir pith and rhamnolipid biosurfactant. *Bioresource Technology* 81: 25–32.

Rahman, K.S.M., Rahman, T.J., Kourkoutas, Y., Petsas, I., Marchant, R., and Banat, I.M. 2003. Enhanced bioremediation of n-alkane in petroleum sludge using bacterial consortium amended with rhamnolipid and micronutrients. *Bioresource Technology* 90: 159–168.

Reddy, A.S., Chen, C.-Y., Baker, S.C., Chen, C.-C., Jean, J.-S., Fan, C.-W., Chen, H.-R., and Wang, J.-C. 2009. Synthesis of silver nanoparticles using surfactin: A biosurfactant stabilizing agent. *Material Letters* 63: 1227–1230.

Ridha, Z.A.M. and Mulligan, C.N. 2011. Simultaneous removal of benzene and copper from contaminated water using micellar-enhanced ultrafiltration. *CSCE General Conference*, Ottawa, Ontario, Canada, June 14–17, 2011.

Robinson, K.G., Ghosh, M.M., and Shi, Z. 1996. Mineralization enhancement of non-aqueous phase and soil-bound PCB using biosurfactant. *Water Science and Technology* 34: 303–309.

Ron, E.Z. and Rosenberg, E. 2002. Biosurfactants and oil remediation. *Current Opinion in Biotechnology* 13: 249–252.

Saeki, M., Sasaki, H., Komatsu, K., Muira A., and Matsuda, H. 2009. Oil spill remediation by using the remediation agent JE1058BS that contains a biosurfactant produced by *Gordonia sp. Strain* JE1058. *Bioresource Technology* 100: 572–577.

Samson, R., Cseh, T., Hawari, J., Greer, C.W., and Zaloum, R. 1990. Biotechnologies appliquées à la restauration de sites contaminés avec d'application d'une technique physico chimique et biologique pour les sols contaminés par des BPC. *Science et Techniques de l'Eau* 23: 15–18.

Sandrin, T.R., Chech, A.M., and Maier, R.M. 2000. Rhamnolipid Biosurfactant Reduces Cadmium Toxicity during Naphthalene Biodegradation. *Applied and Environmental Microbiology* 66: 4585–4589.

Santa Anna, L.M., Siruabim, A.U., Gomes, A.C, Menezes, E.P., Gutarra, M.L., Freure, D.M.G, and Pereira, Jr., N. 2007. Use of biosurfactant in the removal of oil from contaminated soil. *Journal of Chemical Technology and Biotechnology* 82: 687–691.

Shafeeq, M., Kokub, D., Khalid, Z.M., Khan, A.M., and Malik, K.Am. 1989. Degradation of different hydrocarbons and production of biosurfactant by Pseudomonas aeruginosa isolated from coastal waters. *MIRCEN Journal of Applied Microbiology and Biotechnology* 5: 505.

Scheibenbogen, K., Zytner, R.G., Lee, H., and Trevors, J.T. 1994. Enhanced removal of selected hydrocarbons from soil by *Pseudomonas aeruginosa* UG2 biosurfactants and some chemical surfactants. *Journal of Chemical Technology and Biotechnology* 59: 53–59.

Schippers, C., Geβner, K., Muller, T., and Scheper, T. 2000. Microbial degradation of phenanthrene by addition of a sophorolipid mixture. *Journal of Biotechnology* 83: 189–198.

Shreve, G.S., Inguva, S., and Gunnan, S. 1995. Rhamnolipid biosurfactant enhancement of Hexadecane biodegradation by *Pseudomonas aeruginosa. Molecular and Marine Biology and Biotechnology* 4: 331–337.

Shulga, A., Karpenko, E., Vildanova-Martishin, R., Turovsky, A., and Soltys, M. 2000. Biosurfactant-enhanced remediation of oil contaminated environments. *Adsorption Science and Technology* 18(2): 171–176.

Slizovskiy, I.B., Klsey, J.W., and Hatzinger, P.B. 2011. Surfactant-facilitated remediation of metal-contaminated soils: efficacy and toxicological consequences to earthworms. *Environmental Toxicology and Chemistry* 30: 112–123.

Somasundaran, P. and Krishnakumar, S. 1997. Adsorption of surfactants and polymers at the solid liquid interface. *Colloids and Surfaces, A* 123–124: 491–513.

Somayeh, V., Abbas, A.S., and Nouhi, A.S. 2008. Study the role of isolated bacteria from oil contaminated soil in bioremediation. *Journal of Bacteriology* 1365: S678–S707.

Song, S., Zhu, L., and Zhou, W. 2008. Simultaneous removal of phenanthrene and cadmium from contaminated soils by saponin, a plant-derived biosurfactant. *Environmental Pollution* 156: 1368–1370.

Southam, G., Whitney, M., and Knickerboker, C. 2001. Structural characterization of the hydrocarbon degrading bacteria-oil interface: Implications for bioremediation. *International Biodeterioration and Biodegradation* 47: 197–201.

Straube, W.L., Nestler, C.C., Hansen, L.D., Ringleberg, D., Pritchard, P.J., and Jones-Meehan, J. 2003. Remediation of polyaromatic hydrocarbons (PAHs) through landfarming with biostimulation and bioaugmentation. *Acta Biotechnologica* 2–3: 179–196.

Straube, W.L., Jones-Meehan, J., Pritchard, P.J., and Jones W. 1999. Bench-scale optimization of bioaugmentation strategies for treatment of soils contaminated with high molecular weight hydrocarbons. *Resource, Conservation and Recycling* 27: 27–37.

Syldatk, C. and Wagner, F. 1987. Production of biosurfactants. In *Biosurfactants and Biotechnology*, Kosaric, N., Cairns, W.L., and Gray, N.C. (eds.) Vol. 25, Surfactant Science Series, Marcel Dekker, New York, pp. 89–120.

Tan, H., Champion, J.T., Artiola, J.F., Brusseau, M.L., and Miller, R.M. 1994. Complexation of cadmium by a rhamnolipid biosurfactant. *Environmental Science and Technology* 28: 2402–2406.

Thimon, L., Peypoux, F., and Michel, G. 1992. Interactions of surfactin, a biosurfactant from *Bacillus subtilis* with inorganic cations. *Biotechnology Letters* 14: 713–718.

Urum, K., Pekdemir, T., and Gopur, M. 2003. Optimum conditions for washing of crude oil contaminated soil with biosurfactant solutions. *Transactions of the Institute of Chemical Engineering* 81B: 203–209.

Uysal, A. and Turkman, A. 2007. Biodegradation of 4-CP in an activated sludge reactor: Effects of biosurfactant and the sludge age. *Journal of Hazardous Materials* 148: 151–157.

Van Dyke, M.I., Couture, P., Brauer, M., Lee, H., and Trevors, J.T. 1993. *Pseudomonas aeruginosa* UG2 rhamnolipid biosurfactants: Structural characterization and their use in removing hydrophobic compounds from soil. *Canadian Journal of Microbiology* 39: 1071–1078.

Vasefy, F. and Mulligan, C.N. 2008. Bioremediation of oil in a marine environment using rhamnolipid biosurfactant. *International Conference on Waste Engineering and Management*, Hong Kong, People's Republic of China, May 28–30, 2008.

Venghuis, T. and Werther, J. 1998. Flotation as an additional process step for the Washing of Soils Contaminated with heavy metals. *Proceedings of the 6th International FZK/TNO Conference*, ConSoil' 98, Edinburgh, Scotland, Thomas Telford Publishing, London, U.K., Vol. 1, 479–480, May 1998.

Vipulanandan, C. and Harenda, S. 2008. Remediation of PCE contaminated soil using nanoparticles. *Geotechnical* 117: 455–462.

Vipulanandan, C. and Harenda, S. 2011. Surfactant-bimetallic nanoparticle colloidal solutions to remediate PCE contaminated soils. *Geotechnical* 211: 875–884.

Vipulanandan, C. and Ren, X. 2000. Enhanced solubility and biodegradation of naphthalene with biosurfactant. *Journal of Environmental Engineering* 126: 629–634.

Wan, J., Chai, L., Lu, X., Lin, Y., and Zhang, S. 2011. Remediation of hexachlorobenzene contaminated soils by rhamnolipid enhanced soil washing coupled with activated carbon selective adsorption. *Journal of Hazardous Materials* 189: 458–464.

Wang, S. and Mulligan, C.N. 2009a. Rhamnolipid biosurfactant-enhanced soil flushing for the removal of arsenic and heavy metals from mine tailings. *Process Biochemistry* 44(3): 296–301.

Wang, S. and Mulligan, C.N. 2009b. Arsenic mobilization from mine tailings in the presence of a biosurfactant. *Applied Geochemistry* 24: 938–935.

Wattanaphon, H.T., Kerdsin, A., Thammacharaoen, C., Sangvanich, P., and Vangnai, A.S. 2008. A biosurfactant from *Burbholderia cenacepacia* BSP3 and its enhancement of pesticide solubilization. *Journal of Applied Microbiology* 105: 416–423.

Whang, L.-M., Liu, P.-W.G., Ma, C.-C, and Cheng, S.-S. 2008. Application of biosurfactants, rhamnolipid and surfactin for enhanced biodegradation of diesel-contaminated water and soil. *Journal of Hazardous Materials* 151: 155–163.

Xia, W.-B., Li., X., Gao, H., Huang, B.-R., Zhang, H.-Z., Liu, Y.-G., Zeng, G.-M., and Fan, T. 2009. Influence factors analysis of removing heavy metals from multiple metal-contaminated soils with different extractant. *Journal of Central Southern University Technology* 16: 108–111.

Yan, P., Lu, M., Guan, Y., Zhang, W., and Zhang, Z. 2011. Remediation of oil-based drill cuttings through a biosurfactant-based washing followed by a biodegradation treatment. *Bioresource Technology* 102: 10252–10259.

Yin, H., Qiang, J., Jia, Y., Ye, J., Peng, H., Qin, H., Zhang, N., and He, B. 2008. Characteristics of biosurfactant produced by *Pseudomonas aeruginosa* S6 from oil-containing wastewater. *Process Biochemistry* 44: 302–308.

Yong, R.N. and Mulligan, C.N. 2004. *Natural Attenuation of the Contaminants in Soil*, CRC Press, Boca Raton, FL.

Youssef, N., Simpson, D.R., Duncan, K.E., McInerney, M.J., Folmsbee, M., Fincher, T., and Knapp, R.M. 2007. In situ biosurfactant production by Bacillus strains injected into a limestone petroleum reservoir. *Applied and Environmental Microbiology* 73(4): 1239–1247.

Yu, H., Huang, G., Zhang, B., Zhang, X., and Cai, Y. 2010. Modeling biosurfactant-enhanced bioremediation processes for petroleum-contaminated sites. *Petroleum Science and Technology* 28: 1211–1221.

Yuan, X.Z., Meng, Y.T., Zwng, G.M., Fang, Y.Y., and Shi, J.G. 2008. Evaluation of tea-derived biosurfactant on removing heavy metal ions from dilute wastewater by ion flotation. *Colloid and Surfaces A: Physicochemical and Engineering Aspects* 317: 256–261.

Zhang, J., Li, J., Chen, L., and Thring, R.W. 2011. Remediation of refinery oily sludge using isolated strain and biosurfactant. *Water Resource and Environmental Protection (ISWREP), 2011 International Symposium*, Xian, China, May 20–22, 2011, Vol. 3, pp. 1649–1653.

Zhang, Y. and Miller. R.M. 1992. Enhanced octadecane dispersion and biodegradation by a *Pseudomonas* rhamnolipid surfactant (biosurfactant). *Applied and Environmental Microbiology* 58: 3276–3282.

Zhu, L.Z. and Zhang, M. 2008. Effect of rhamnolipids on the uptake of PAHs on ryegrass. *Environmental Pollution* 156: 46–52.

Zhu, S., Chen, Y., Li, G., and Liang, S. 2011. Effect of biosurfactant on desorption of sediment-sorbed phenanthrene. *IEEE Engineering in Medicine and Biology Society, 5th International Conference on Bioinformatics and Biomedical Engineering ICBBE 2011*, Wuhan, China, May 10–12, 2011.

11 Biosurfactant Complexation of Metals and Applications for Remediation

David E. Hogan, Tracey A. Veres-Schalnat,
Jeanne E. Pemberton, and Raina M. Maier

CONTENTS

INTRODUCTION

Humanity's ability to access and manipulate metals was the keystone to our advancement from the Stone Age to today's technologically advanced society. While this keystone drove technological innovation and societal development, it simultaneously resulted in unprecedented alterations of natural biogeochemical cycles and, arguably for the first time, affected the environment on a global scale. Indeed, paleopollution archaeology is a field of study devoted to evaluating this human impact by analyzing metal deposits in polar ice caps, bogs, and aquatic sediments to document and understand mining and smelting practices of ancient civilizations, such as the Roman Empire, as long as 2000 years ago (Nriagu, 1996). This analysis has shown that our utilization of metal resources is having local and global effects on human and environmental health. Practical, effective, and economical remediation technologies are needed to address large-scale metal contamination; microbially produced surfactants (biosurfactants) meet these requirements, and may be the basis for developing green remediation technologies. The goal of this chapter is to provide a brief introduction of metals and environmental metal contamination,* followed by an in-depth examination of metal interactions with biosurfactants. The chapter will conclude with a discussion of potential remediation techniques and technologies based on metal–biosurfactant interactions.

METALS AND METAL CONTAMINATION

In general, metals are lustrous solids (except Hg), usually ductile and malleable, capable of conducting both electricity and heat and forming alloys (Kotz et al., 2006). On the periodic table (Figure 11.1), they can be found to the left of the diagonal line drawn from B to At, excluding Ge, Sb, and H. Elements that display some physical characteristics of metals are defined as metalloids and include B, Si, Ge, As, Sb, and Te (Kotz et al., 2006). The two common terms used to describe metals, especially those of concern environmentally, are "trace elements"—elements in the Earth's crust at or below 100 mg kg^{-1}, for example, Zn, Ni, Cu, Co (Essington, 2004)—and "heavy metals"—elements with densities greater than 5 g cm^{-3}, for example, Cu, Co, Fe, Mn (Callender, 2003); these terms are used synonymously for the purpose of this chapter.

---

* For a more extensive review, see Nriagu (1990) and Callender (2003).

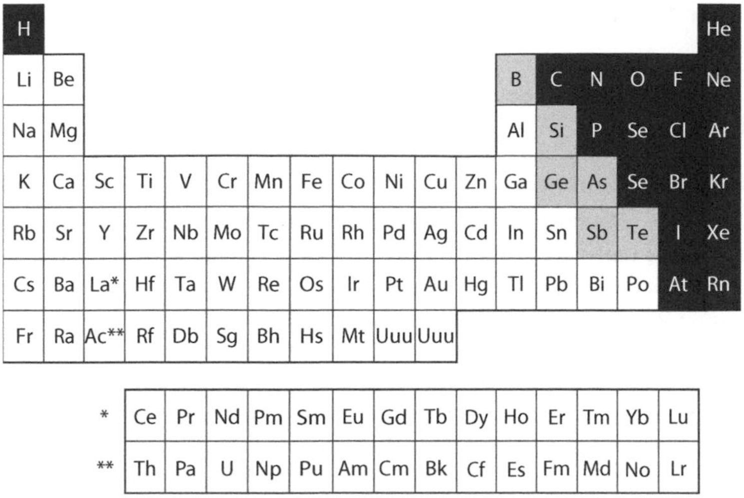

H																	He
Li	Be											B	C	N	O	F	Ne
Na	Mg											Al	Si	P	Se	Cl	Ar
K	Ca	Sc	Ti	V	Cr	Mn	Fe	Co	Ni	Cu	Zn	Ga	Ge	As	Se	Br	Kr
Rb	Sr	Y	Zr	Nb	Mo	Tc	Ru	Rh	Pd	Ag	Cd	In	Sn	Sb	Te	I	Xe
Cs	Ba	La*	Hf	Ta	W	Re	Os	Ir	Pt	Au	Hg	Tl	Pb	Bi	Po	At	Rn
Fr	Ra	Ac**	Rf	Db	Sg	Bh	Hs	Mt	Uuu	Uuu							

*	Ce	Pr	Nd	Pm	Sm	Eu	Gd	Tb	Dy	Ho	Er	Tm	Yb	Lu
**	Th	Pa	U	Np	Pu	Am	Cm	Bk	Cf	Es	Fm	Md	No	Lr

FIGURE 11.1   Metals (white), metalloids (grey), and non-metals (black) of the periodic table.

Trace metals generally do not occur as discrete minerals, but are rather "minor substituents in silicates and aluminosilicates (olivines, pyroxenes, amphiboles, micas, and feldspars); hydrous metal (iron, aluminum, and manganese) oxides; iron sulfides; calcium and magnesium carbonates; calcium, iron, and aluminum phosphates" (Essington, 2004). Anthropogenic activities, as discussed in the next section, have resulted in sites where soils are contaminated with metal concentrations well above background levels. The metals, as a result of processing, are no longer incorporated into mineral components, for example, silicates. As a result, they have increased bioavailability and the potential to exert toxicity. Such metal contaminants can become widely dispersed. For example, aerosols generated from metal processing or from metal-containing waste sites can contain elevated metal concentrations resulting in higher metal distribution via aerosol emission/deposition (Csavina et al., 2011). These aerosols can be dispersed over large areas, and since deposition is often a gradual process, elevated levels develop over time as metals accumulate.

Metal contamination cannot be mineralized to innocuous end products such as carbon dioxide and water like organic contaminants. Once in a system, metals can only be modified through changes in redox state, and thus speciation. For example, toxic species of selenium (Se^{6+}, Se^{4+}, and Se^0) can be methylated by bacteria to the volatile and less toxic species dimethylselenide [$(CH_3)_2Se$] or dimethyldiselenide [$(CH_3)_2Se_2$] (Maier et al., 2009). Thus, in order to treat a metal contamination event, the metal must either be physically removed or chemically modified to reduce toxicity and mobility. Due to the large area of most contamination events, the former solution is often infeasible or economically prohibitive; the latter solution is inevitably temporary as changes in environmental conditions due to natural processes or lack of management may result in the contamination returning to its initial or potentially worse condition. In the earlier example, microbial methylation of selenium accomplishes both chemical treatment via the generation of less toxic methylated species and physical removal via the generation of a volatile compound that is released to the atmosphere.

Sources of Metals*

Trace element contamination occurs both naturally and as a result of anthropogenic activities. The source and degree of contamination varies for atmospheric, aquatic, and soil environments.

Atmospheric Environments

Natural sources of atmospheric trace elements include volcanic activity, forest fires, wind-borne soil particles, seasalt spray, and biogenic, that is, biologically mediated volatilization and biological particulates. The level of trace elements emitted from each of these sources varies. Volcanic emanations account for 40%–50% of naturally emitted As, Cr, Cu, Ni, Pb, and Sb annually. Atmospheric particulates derived from soils can account for over 50% of Cr, Mn, and V. Seasalt spray and forest fires are relatively minor contributors to natural trace-element emissions with less than 10% of annual emissions for most elements. Biogenic contributions of metals in the atmosphere can account for over 50% of Se, Hg, and Mo annually, and 30%–50% of As, Cd, Cu, Mn, Pb, and Zn (Nriagu, 1989).

Anthropogenic atmospheric metal sources include fuel combustion (coal and oil), metal production (mining and smelting), secondary metal production (nonferrous and ferrous), refuse incineration, cement production, and wood combustion. Like naturally occurring sources, the amount and types of metals emitted from each source varies and depends on the source material for each process. For example, coal combustion represents a major contributor of Hg, Mo, Sn, Se, As, Cr, Mn, Sb, and Tl, while combustion of oil is a major contributor of V and Ni. Consumption of leaded gasoline, primarily in developing countries, represents the largest contributor to Pb emissions. Nonferrous metal production produces significant amounts of Pb, As, Cd, Cu, In, and Zn emissions, while ferrous metal production is the primary source of Cr and Mn (Nriagu and Pacyna, 1988; Pacyna and Pacyna, 2001). In 1990, anthropogenic emissions into the atmosphere exceeded natural emissions by a factor of 28 for Pb, 6 for Cd, and 3 for V and Zn.

Aquatic Environments

Natural contributions of trace elements to aquatic environments are primarily from settling of atmospheric particulates and fluvial movement of weathered material into bodies of water. The trace elements introduced vary by source. The primary anthropogenic sources of water contamination are domestic wastewater effluents (As, Cr, Cu, Mn, and Ni); sewage discharges (As, Mn, and Pb); coal consumption (As, Hg, and Se); ferrous and nonferrous metal production, that is, mining and smelting (Cd, Ni, Pb, Se, Cr, Mo, Sb, and Zn); and urban runoff (Nriagu, 1990; Nriagu and Pacyna, 1988). Atmospheric fallout from sources listed in the previous section is also a contributing factor. Anthropogenic metal inputs into aquatic environments are roughly twice those to the atmosphere (with the exception of Pb and Cd) (Callender, 2003).

---

* Those interested in quantitative information regarding metal contamination should review Niagru and Pacyna (1988) and Pacyna et al. (1995). From our literature review, it does not appear that there is a more recent global inventory of metal emissions than that supplied by these authors.

Soil Environments

The primary natural source of trace metals in soil environments is derived from pedogenesis, and metal content depends on the parent material. Like water, soils may also accumulate metals from atmospheric deposition. Deposited aerosols may contain background or elevated levels of metals; elevated levels can be generated from industrial or contaminated site point sources or from nonpoint sources, which might simply be a highly populated region. Such atmospheric contributions are important especially in soils found in rural and remote areas (Nriagu, 1990).

Anthropogenic inputs of trace elements to soils come from many different sources including fertilizers, pesticides, biosolids, metal mining and processing, and industrial wastes (Nriagu, 1990; Nriagu and Pacyna, 1988; Wuana and Okieimen, 2011). In 1990, antimony, Cu, Pb, and Zn anthropogenic soil emissions exceeded natural weathering flux threefold, and for Hg, ten-fold (Nriagu, 1990). Anthropogenic metal inputs to soil are several times greater than inputs into both air and water (Callender, 2003).

Summary

It is difficult to separate the atmosphere, hydrosphere, and pedosphere when examining the movement and cycling of trace elements considering their innate connectedness and interdependence. Overall, data reveal that anthropogenic trace element emissions are greater than natural emissions and humanity has now become the driving influence over the biogeochemical cycling of these elements globally (Callender, 2003; Nriagu, 1989, 1990; Nriagu and Pacyna, 1988). In fact, it has been shown that Pb, V, As, Sn, Zn, Cd, Hg, Sb, Cu, Ag, and Se emissions have caused perturbation of natural geochemical cycles at the global level, i.e., intercontinentally (Pacyna et al., 1995).

BIOSURFACTANT COMPLEXATION OF METALS

Biosurfactants are surface-active compounds derived from natural, biological sources; they are amphiphilic molecules consisting of two parts: a hydrophilic head group and a hydrophobic tail group. The hydrophilic group can be composed of a mono-, oligo- or polysaccharide, peptide, citric acid with two cadaverines, or a protein. The hydrophobic group is composed of saturated, unsaturated, and/or hydroxylated fatty acids or fatty alcohols (Bodour et al., 2004; Lang, 2002). Due to their amphiphilic nature, these molecules accumulate at interfaces. At the interface of immiscible fluids, surfactants increase the solubility and mobility of the hydrophobic or insoluble organic phase within the aqueous phase (Singh et al., 2007). Surfactants are characterized by their physical and chemical properties: ability to reduce surface tension, critical micelle concentration, hydrophile–lipophile balance, aggregate formation, charge, chemical structure, and source (Van Hamme et al., 2006).

The study of metal interactions with biosurfactants is a relatively new field of study. To the best of our knowledge, the first reports of biosurfactants interacting with metals involved the lipopeptide surfactin and some of the group I and II metals from a medical perspective (Thimon et al., 1992, 1993). Tan et al. (1994) was the first report of the interactions between a glycolipid, rhamnolipid, and a heavy metal, Cd, from an environmental perspective. A Web of Science search in 2012 showed 160 publications examining metal–biosurfactant interactions, an indication of the great interest in this area. The following sections review what is currently known about biosurfactant–metal interactions. It is not possible at this time to directly compare metal–surfactant complexation constants in all cases since different methodologies were used; however, we do so where possible.

RHAMNOLIPIDS

Rhamnolipids are glycolipidic biosurfactants first discovered as extracellular products of *Pseudomonas aeruginosa,* but they have recently been found to be produced by other species within and outside of *P. aeruginosa*'s phylum (Abdel-Mawgoud et al., 2010) (Figure 11.2). Rhamnolipids are characterized by a hydrophilic mono- or dirhamnose moiety (head group) and a hydrophobic lipid moiety consisting of one or two (in rare cases three) hydrocarbon chains (tail group); nearly 60 congeners of the rhamnolipid molecule have been reported, and congener distribution varies depending on bacterial strain and/or carbon source. The hydrophilic moiety contains a carboxylic acid group that makes rhamnolipid an anionic surfactant (Abdel-Mawgoud et al., 2010).

Metal Interactions with Rhamnolipid

Tan et al. (1994) reported a strong interaction between Cd^{2+} and monorhamnolipid. The monorhamnolipid was capable of complexing Cd^{2+} both rapidly (equilibrium within 15 min) and stably (for at least 27 h). The complexation reaction was independent of pH in a narrow range (6–7), buffer concentration, and temperature. The ability of rhamnolipid to complex Cd^{2+} was higher than reported values for bacterial cell and exopolymer complexation (previously described materials utilized for metal recovery). The stability constant of the monorhamnolipid–Cd^{2+} complex (log $K = 6.9$) was also stronger than reported stability constants for Cd^{2+} with other organic ligands that have carboxylate functional groups, including acetic acid, oxalic acid,

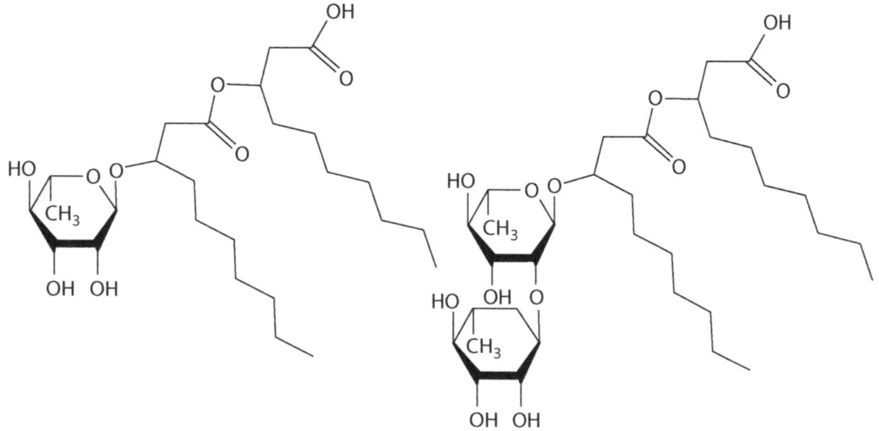

FIGURE 11.2   Mono- and dirhamnolipids (C10, C10) produced by *Pseudomonas aeruginosa.*

citric acid, and fulvic acid (ranging from log K = 1.2 to 4.5), all commonly found in soil (Ochoa-Loza et al., 2001). This study provided evidence that a biosurfactant could offer a versatile and powerful tool for dealing with heavy metal contamination.

Following this discovery, it was shown that monorhamnolipid forms strong complexes with many elements in addition to Cd with the following conditional stability constant sequence (from strongest, log K = 10.30, to weakest, 0.96): Al^{3+} > Cu^{2+} > Pb^{2+} > Cd^{2+} > Zn^{2+} > Fe^{3+} > Hg^{2+} > Ca^{2+} > Co^{2+} > Ni^{2+} > Mn^{2+} > Mg^{2+} > K^+ (Ochoa-Loza et al., 2001). This sequence shows that naturally occurring, abundant metal cations (Ca^{2+}, Mg^{2+}) have lower stability constants than trace elements of environmental concern (Pb^{2+}, Cd^{2+}, Hg^{2+}). These results, combined with the fact that monorhamnolipid complexed metals more strongly than other organic ligands that might compete in the soil environment, suggested that monorhamnolipid was a good candidate for metal removal applications.

Insight into the nature of the interactions of monorhamnolipid with metals has recently been elucidated at the molecular level using a combination of ^1H NMR, FTIR, and H/D exchange mass spectrometry (Schalnat, 2012). About 600 MHz ^1H NMR studies of monorhamnolipid solutions containing Pb^{2+} provide the surprising result that the carboxylate moiety is only weakly involved in the metal complexation. This assertion is supported by the insignificant chemical shift difference observed for the methylene protons immediately adjacent to the carboxylic acid in the absence and presence of metal cations in solution. For strong carboxylate–metal ion complexation, chemical shift changes of >2 ppm are typically observed (Bodor et al., 2002). In contrast, chemical shift changes for these methylene protons for the monorhamnolipids with Pb^{2+} are only 0.012–0.013 ppm, far smaller than the 2 ppm shift expected for carboxylate binding.

Given the large formation constants measured for these metal complexes as noted earlier, one must conclude that strong complexes are formed by binding of the metal cation to other parts of the monorhamnolipid molecule. Indeed, the small chemical shift changes observed are similar to those for metal-crown ether and carbohydrate–metal complexes (Ferrari et al., 2005; Karkhaneei et al., 2001; Pankiewicz et al., 2005; Rondeau et al., 2003; Rouhollahi et al., 1994). This similarity implies that strong binding might occur by the involvement of multiple atoms in the monorhamnolipid, possibly through the formation of a binding pocket involving the carboxylate and the rhamnose sugar. Indeed, an energy-minimized molecular mechanics model of the C10:C10 monorhamnolipid shown in Figure 11.3a documents hydrogen-bonding interactions between two of the rhamnose sugar hydroxyls and the carboxylic acid. This leads to the formation of an oxygen-rich cavity that might serve as a metal cation binding pocket in which shared coordination of metal cations can occur in much the same way as in a crown ether.

FTIR spectroscopy provides further evidence for metal complexation in a binding pocket involving both the carboxylate and the sugar hydroxyls. Fruitful spectral results can be found in two frequency regions of the spectrum. In one frequency region, the frequency difference between the $\nu_{as}(COO^-)$ and $\nu_s(COO^-)$ bands ($\Delta\nu$) of carboxylate species is sensitive to chemical environment and is useful for insight into metal cation binding. The value of $\Delta\nu$ is different for free carboxylates compared to those complexed to metal cations, thus providing an indicator of coordination that

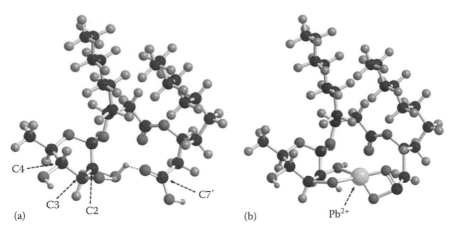

C4
C3
C2
(a)
C7'
(b)
Pb^{2+}

FIGURE 11.3 **(See color insert.)** (a) An energy-minimized molecular mechanics model of monorhamnolipid (C10, C10) showing the oxygen-rich cavity that may serve as a cation binding pocket. (b) A model showing how a Pb^{2+} ion might interact with the binding pocket of monorhamnolipid (C10, C10).

is sensitive to coordination type (Colthup et al., 1990; Mehrotra and Bohra, 1983; Palacios et al., 2004; Strathmann and Myneni, 2004). Due to symmetry differences between the free ionic species and monodentate metal carboxylate complexes, large increases in $\Delta\nu$ are observed. However, bidentate chelation, even weak bidentate coordination in which the interaction strength between the two carboxylate oxygen atoms and the metal cation is unequal, does not alter overall symmetry, and, therefore, does not significantly alter the spectral behavior from that of the ionic form, or gives rise to small decreases in $\Delta\nu$. For the free monorhamnolipid, $\Delta\nu$ is on the order of 160 cm^{-1}. For the Pb^{2+} complex, this $\Delta\nu$ decreases slightly to 154 cm^{-1}, consistent with bidendate or weak (i.e., asymmetric) bidentate binding.

Involvement of the sugar hydroxyls in the metal cation binding is also indicated in the FTIR spectra in the second useful spectral region. Sugar vibrational modes are known to undergo frequency shifts upon metal coordination (Tajmir-Riahi, 1985, 1989; Tian et al., 2000). For monorhamnolipid-Pb^{2+} complexes, the most significant spectral changes occur for the $[\delta(\text{COH}) + \nu(\text{C}-\text{O})]$ band, observed at 1233 cm^{-1} for the monorhamnolipid-Pb^{2+} complex but at 1191 cm^{-1} for the free monorhamnolipid, and the $[\delta_{ip}(\text{OH}) + \nu(\text{C}-\text{O})]$ band, observed at 1210 cm^{-1} for the monorhamnolipid-Pb^{2+} complex but at 1170 cm^{-1} for free monorhamnolipid. This shift to higher frequency is indicative of an increase in bond strength with a concomitant decrease in bond length upon Pb^{2+} coordination. Collectively, the data in these two spectral regions support involvement of both the carboxylate and the sugar in metal cation binding.

Further evidence for the existence of a binding pocket was sought through gas-phase hydrogen–deuterium exchange (HDX) ESI-MS experiments. This approach is predicated on changes in the rate of H/D exchange based on the acidity/basicity of labile hydrogen atoms in addition to the details of the three-dimensional structure of the molecule that dictates accessibility of these labile hydrogen atoms (Campbell et al., 1994, 1995; Green et al., 1995). It has been shown that, although HDX occurs in the gas phase, the exchange rates are directly related to the solution conformation and structure of a molecule. Thus, HDX was performed for the free monorhamnolipid and the monorhamnolipid-Pb^{2+} complex. Fully protonated free monorhamnolipid contains four exchangeable hydrogen atoms on the hydroxyls on carbons 2, 3, and 4 of the sugar and the oxygen of the carboxylic acid involving carbon 7' in addition to the ionizing proton which gives the ion its positive charge (Figure 11.3a). Exchange of four of these hydrogen atoms can be experimentally monitored (exchange of the carboxylic acid hydrogen is too rapid to be measurable). For the free monorhamnolipid, H/D exchange is complete for all four hydrogen atoms within 1 s. In contrast, for the complexed monorhamnolipid, which contains only three exchangeable hydrogen atoms, one of these hydrogen atoms, of the hydroxyl group on the rhamnose sugar C4 atom, exchanges at the same rate in the complex as in the free monorhamnolipid, suggesting that this hydroxyl is not involved in the metal binding. However, the hydrogen atoms of the hydroxyls on the rhamnose C3 and C2 atoms exchange more slowly by factors of 10 and 100, respectively. These results clearly indicate that these hydroxyl groups are highly inaccessible in the complex relative to the free monorhamnolipid, supporting the involvement of the sugar in a metal cation binding pocket. A model for what this metal binding in a pocket might look like is shown in Figure 11.3b.

Effects of Metals on Rhamnolipid Production

The strong monorhamnolipid–metal stability constants suggested that there may be a physiological reason for the interaction of this biosurfactant with metals. In fact, it has been shown that when *P. aeruginosa* IGB83 (which produces a mixture of mono- and dirhamnolipid) was exposed to subtoxic levels of Cd^{2+}, expression of one of the rhamnolipid genes, *rhlB*, was enhanced in mid-stationary phase and sustained through late-stationary phase (Neilson et al., 2010). The RhlB enzyme is responsible for catalyzing the addition of a second rhamnose sugar onto monorhamnolipid to form dirhamnolipid. As a result of this increased expression, there was an increase in the ratio of dirhamnolipid to monorhamnolipid produced. This is significant because the complexation constant for dirhamnolipid is several orders of magnitude higher than monorhamnolipid. Thus, it appears that the presence of Cd^{2+} during growth may increase production of dirhamnolipid as a detoxification mechanism. This is supported by an earlier study showing that the addition of monorhamnolipid to soils cocontaminated with phenanthrene and Cd^{2+} (levels high enough to exert toxicity) resulted in enhanced degradation of phenanthrene (Maslin and Maier, 2000).

Research has also linked rhamnolipid production to iron and magnesium status in the environment. Multiple studies have reported that iron limitation increases rhamnolipid production, while sufficient growth levels of iron suppress rhamnolipid production (Deziel et al., 2003; Glick et al., 2010; Guerrasantos et al., 1984; Ramana and Karanth, 1989). In contrast, magnesium limitation reduces rhamnolipid production, while sufficient levels of magnesium enhance production (Guerrasantos et al., 1986; Ramana and Karanth, 1989).

LIPOPEPTIDES

Lipopeptides are a class of biosurfactants with the general structure of a cyclic 7–10 amino acid peptide connected to a fatty acid chain. The best studied member of this class is the anionic surfactant surfactin, produced by *Bacillus subtilis*. Surfactin has a 7 amino acid peptide head group and a fatty acid chain of 13–16 carbons (Figure 11.4). Lesser studied members of this class include iturin (7 amino acid peptide with a C14–C17 fatty acid), fengycins (10 amino acid peptide with a C14–C18 fatty acid), and viscosin (9 amino acid peptide and a C10 fatty acid tail group) (Lang, 2002; Saini et al., 2008).

Metal Complexation by Lipopeptides

Early studies of surfactin interaction with cations focused on understanding their ability to disrupt erythrocyte membranes (Thimon et al., 1992). These studies showed that surfactin can complex Ca^{2+} and Mg^{2+} with constants of $1.5 \times 10^5\ M^{-1}$ and $1.9 \times 10^4\ M^{-1}$, respectively (Thimon et al., 1993). More recently, surfactin has been shown to interact strongly with Fe^{3+}, weakly with Cu^{2+}, and not at all with Nd^{3+}— biphasic liquid–liquid extraction percentages were >95%, ≈12%, and <5%, respectively (Dejugnat et al., 2011). This binding is associated with the negative charge generated by two carboxylic groups from the aspartate and glutamate residues of the peptide structure (Thimon et al., 1992). Spatial arrangement studies of surfactin suggest that the peptide portion of the molecule forms a "claw-like" structure with the aspartate and glutamate residues on opposite tips of the claw, creating a position for metals to bind (Bonmatin et al., 1994). As a result of the two charged groups, there are different association behaviors of surfactin with mono- and divalent cations. Divalent cations exhibit a single associated binding constant value that is higher

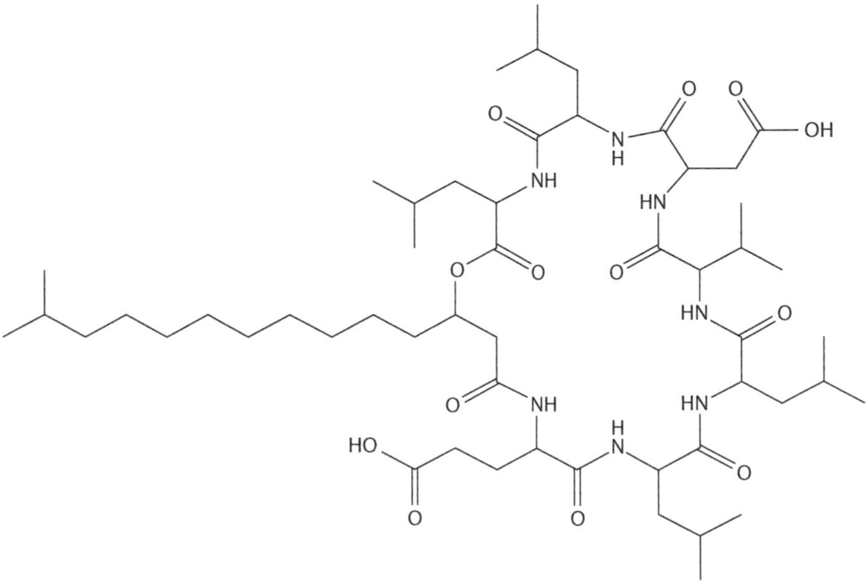

FIGURE 11.4   Surfactin produced by *Bacillus subtilis*.

than for monovalent cations. Monovalent cations have a choice of two binding sites on surfactin, each of which has different affinity for the cation. For example, for Rb^+ there is a site with high affinity (association constant of 71 M^{-1}) and a site with low affinity (association constant of 11 M^{-1}). Accordingly, surfactin shows a 1:1 molar ratio with Ca^{2+} and a 2:1 molar ratio with Rb^+ (Shen et al., 2011).

Other lipopeptides have also been shown to bind metals. Lichenysin is a lipopeptide produced by *Bacillus licheniformis* that structurally resembles surfactin. It has seven amino acids attached to a C13–C15 fatty acid tail. It differs from surfactin through the substitution of a glutaminyl residue for glutamic acid in the number one peptide position. This reduces the molecular charge from −2 in surfactin to −1 in lichenysin. Lichenysin has been reported to bind Ca^{2+} and Mg^{2+} fourfold and 10-fold, respectively, more strongly than surfactin (Grangemard et al., 2001). The affinity for Ca^{2+} over Mg^{2+} is only twofold higher for lichenysin in comparison to 10-fold for surfactin. This may result from the replacement of the more specific "claw" binding structure of surfactin with a less discriminate binding structure in lichenysin. Lichenysin also forms a 2:1 molar ratio with Ca^{+2} instead of the 1:1 reported for surfactin (Grangemard et al., 2001).

Viscosin is a lipopeptide produced by a variety of *Pseudomonas* sp. and is characterized by a fatty acid tail connected to two amino acids and a seven-member cyclic peptide. Viscosin has a conditional stability constant of 5.87 with Cd^{2+} (Saini et al., 2008).

Effects of Metals on Lipopeptides
It has been reported that solution cations (Mg^{2+}, Mn^{2+}, Ca^{2+}, Ba^{2+}, Li^+, Na^+, K^+, and Rb^+) can reduce the critical micelle concentration of surfactin from 4- to 12-fold, depending on the cation and concentration (Thimon et al., 1992). The shape of the aggregates is affected by both pH and cation concentration and type (mono- vs. divalent) (Shen et al., 2011; Thimon et al., 1992). Higher pH and lower cation concentration both result in smaller and rounder aggregates while lower pH and higher cation concentration result in aggregates that are rodlike to lamellar. These findings can be explained by the neutralization of charge in the surfactin head group by added cations. When the charge in the peptide ring is neutralized by a cation, the effective size of the head-group ring is reduced, allowing molecules to interact more closely. This results in a reduction of curvature in the aggregate, which favors larger rodlike and lamellar structures. A divalent cation can completely neutralize the charge of the surfactin head group, while a monovalent cation only partially neutralizes the charge and so the effect is not as great (the same concept applies to any charged surfactant molecule).

Surfactin effectively attaches to and integrates with lipid membranes. Its ability to do so is highly enhanced when combined with metals. Surfactin has been shown to produce ion-conducting pores in artificial lipid membranes when complexed with a metal cation. Further, surfactin has been shown to carry metal ions across a hydrocarbon barrier phase from one aqueous phase to another (Sheppard et al., 1991; Thimon et al., 1993). This ability can be attributed to its interaction with the metal cations; surfactin will not exhibit ionopore formation if not allowed to interact with cations first (Thimon et al., 1993). Recent data suggest that the complexation of surfactin with cations, especially divalent cations, reduces the hydrophilicity of the head group by neutralizing the charged residues in the peptide ring, allowing the surfactin–metal complex to integrate into hydrophobic structures (Shen et al., 2011). This ionophore activity has garnered a great deal of interest in the medical field due to its ability to cause hemolysis of erythrocytes, but has yet to be extensively explored for potential remediation applications.

Similar to rhamnolipids, metals have been found to influence surfactin production. Manganese added as $MnSO_4$ has been shown to increase the production of surfactin roughly 10-fold. In the same study, additions of Fe resulted in a reduction of surfactin produced on a per-biomass basis (Cooper et al., 1981). A more recent study of surfactin production by a thermophilic *B. subtilis* showed varying responses to a range of individual metal concentrations with the best productivity resulting from a mixture of added metals (Makkar and Cameotra, 2002). It is not yet clear whether the metal effects reported are a result of the metals alone or whether the presence/absence of other medium components also influenced surfactin production.

OTHER BIOSURFACTANTS

There are a variety of other biosurfactants produced but few are well-characterized in terms of their interactions with metals. The following section provides some examples of studies that have used these biosurfactants to complex metals in various applications even though, in some cases, the strength of the interaction is not characterized.

Siderolipids

Siderolipids (formerly flavolipids) are a recently described class of biosurfactants produced by *Pedobacter* sp. MTN11 (formerly *Flavobacterium* sp. MTN11) (Bodour et al., 2004) (Figure 11.5). The polar moiety of this anionic biosurfactant is composed of a citric acid and two cadaverine molecules, and the nonpolar moiety is composed of two branched chain acyl groups with 6–10 carbons each. It is reported to complex Cd with a conditional stability constant of log K = 3.6 (Bodour et al., 2004),

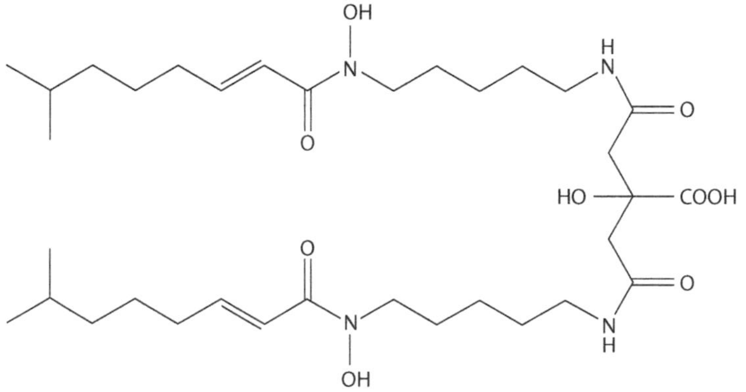

FIGURE 11.5   Siderolipid (U9, U9) produced by *Pedobacter* sp. MTN11.

which is weaker than the stability constant reported for rhamnolipid (6.9). The siderolipid–Cd complexation constant is comparable to those reported for natural organic acids: acetic acid, 1.2–3.2; oxalic acid, 4.1; and citric acid, 4.5. The structure of the siderolipid headgroup is similar to the siderophore aerobactin, which suggests that siderolipids may be iron chelators (Bodour et al., 2004).

Glycoglycerolipids

Glycoglycerolipids are produced by *Microbacterium* spp., *Micrococcus luteus*, and *Bacillus pumilus*. Figure 11.6 depicts a dimannosyl–glycerolipid produced by a strain of *M. luteus* isolated from the North Sea. These biosurfactants are an abundant constituent of plant and bacterial membranes. They are composed of carbohydrate unit(s), a glycerol moiety, and a variety of short or long-chained saturated or unsaturated fatty acids (Palme et al., 2010). A recent study showed that the partially purified glycoglycerolipids from four strains of *Microbacterium* sp. could be used to extract Zn^{2+} and Cd^{2+} from an industrial residue. The study reports varying efficiencies based on strain and carbon source (Aniszewski et al., 2010).

Saponins

Saponins are plant- and microorganism-derived biosurfactants. They are nonionic, acidic, high molecular weight glycosides characterized by a hydrophilic sugar moiety and hydrophobic tripertene or steroid aglycone group (Gusiatin and Klimiuk, 2012; Hong et al., 2000). Recent studies report their use as a soil washing agent of metal-contaminated materials.

Sophorolipids

Sophorolipids (Figure 11.7) are produced by various species of yeast including *Candida bombicola*, *Candida apicola*, *Wicherhamiella domercqiae*, *Pichia anomala*, and *Rhodotorula bogoriensis*. These molecules are characterized by a hydrophilic moiety composed of a disaccharide sophorose, a diglucose with a β-1,2 bond. The hydrophobic moiety is a terminal or subterminal hydroxylated fatty acid, β-glycosidically linked to the sophorose. Like rhamnolipids, sophorolipids

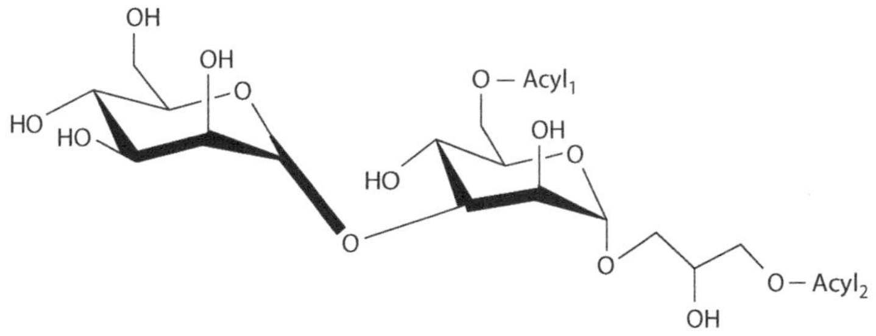

FIGURE 11.6 Dimannosyl–glycerolipid produced by the marine bacterium *Micrococcus luteus*. (With kind permission from Springer Science+Business Media: *Biosurfactants*, Selected microbial glycolipids: Production, modification and characterization, 672, 2010, 185–202, Palme, O., Moszyk, A., Iphoefer, D., and Lang, S.)

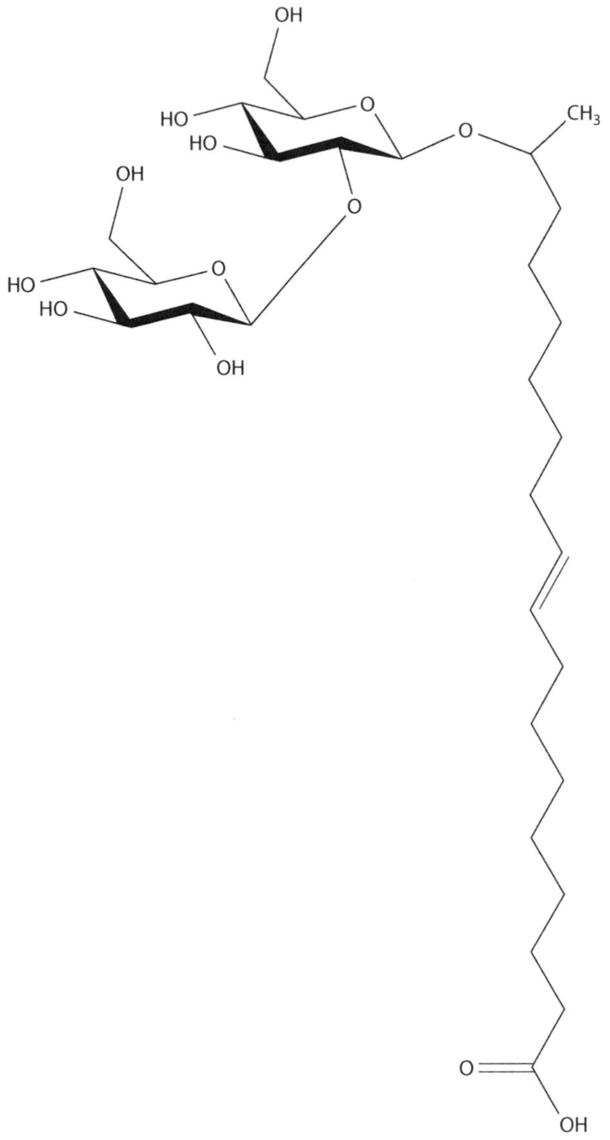

FIGURE 11.7   A common structure of sophorolipid.

are produced as a variety of congeners (Van Bogaert et al., 2011). A recent report shows that sophorolipids could be used to help synthesize silver nanoparticles (Kasture et al., 2008).

Spiculisporic Acids

Spiculisporic acids (S-acid) are microbially produced surfactants with two carboxcylic moieties and a lactone ring composing the hydrophilic head group; the hydrophobic portion is a decyl chain. The molecular structure can be modified through neutralization and saponification with NaOH to a tricarboxylic acid 2-(2-carboxyethyl)-3-decyl maleic anhydride (DCMA). This acid can further form mono-, di-, and tri-Na DCMA salts (DCMA-xNa where x is 1, 2, or 3, respectively) (Figure 11.8). The surfactant characteristics vary from form to form with the fully neutralized and saponified forms exhibiting the highest hydrophilicity and lowest surface activity (Ishigami et al., 2000). DCMA-xNa salts have been shown to disperse titanium dioxide and ferric oxide and sequester Ca^{2+}. The ability to sequester Ca^{2+} is strongest for the DCMA-3Na with decreasing sequestration from DCMA-3Na to DCMA-1Na (Ishigami et al., 1983). Later work showed that DCMA-3Na is capable of complexing various cations including Ca^{2+}, Cd^{2+}, Co^{2+}, Cu^{2+}, Mg^{2+}, Ni^{2+}, and Zn^{2+} (Choi et al., 1993; Hong et al., 1998). A binding affinity sequence of $Cd^{+2} > Cu^{+2} \approx Zn^{+2} > Ni^{+2}$ has been reported (Hong et al., 1998).

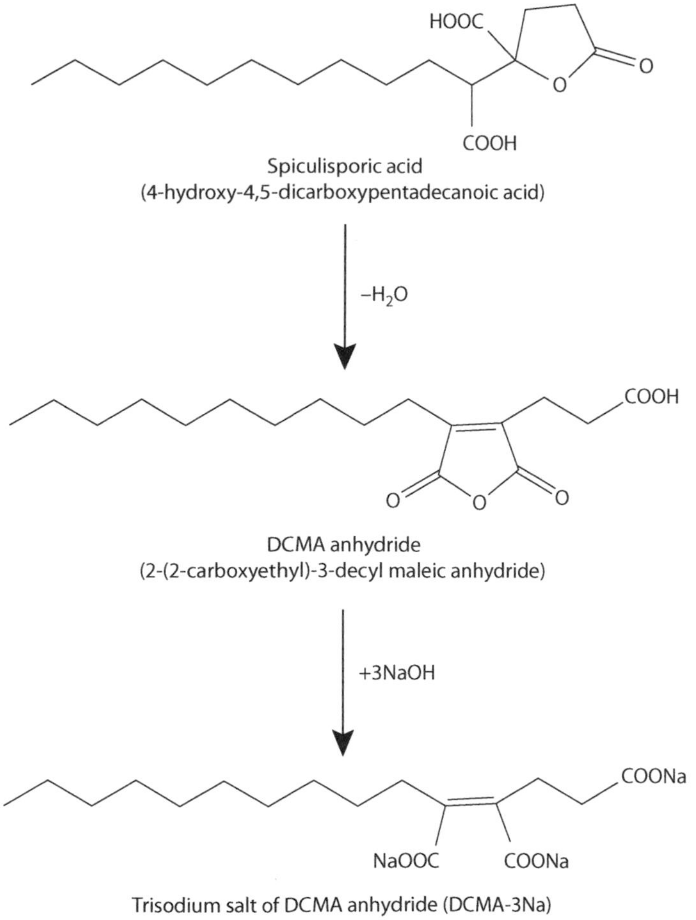

Spiculisporic acid
(4-hydroxy-4,5-dicarboxypentadecanoic acid)

$-H_2O$

DCMA anhydride
(2-(2-carboxyethyl)-3-decyl maleic anhydride)

$+3NaOH$

Trisodium salt of DCMA anhydride (DCMA-3Na)

FIGURE 11.8 The modification scheme of spiculisporic acid to DCMA Na-salts. (Reprinted from *J. Colloid Interface Sci.*, 173(1), Hong, J., Yang, S.M., Choi, Y.K., and Lee, C.H., Precipitation of tricarboxylic acid biosurfactant derived from spiculisporic acid with metal ions in aqueous solution, 92–103, doi: 10.1006/jcis.1995.1301, Copyright 1995, with permission from Elsevier.)

REMEDIATION APPLICATIONS

Biosurfactants are considered "green" materials with several advantages for use in bioremediation applications. They are naturally produced, that is, not fossil fuel derived, with potential to be produced in situ. Apart from the bioemulsifiers, they are of small molecular size with molecular weights generally less than 1500 (Miller, 1995). Biosurfactants accumulate at interfaces, which is beneficial for desorbing metal and organic contaminants, and they increase the apparent solubility of organic compounds, which is beneficial for physical and biodegradative removal. Biosurfactants are biodegradable and generally perceived as less toxic than synthetic analogs (Maslin and Maier, 2000; Van Hamme et al., 2006). A small number of studies have examined the biodegradation of rhamnolipid in soil. One showed that it took 8 weeks for rhamnolipid to biodegrade in a Brazito sandy loam (Maslin and Maier, 2000). A more recent study compared the biodegradation of EDTA (ethyl-enediamine tetraacetic acid: a common and strong metal chelant), citric acid, and rhamnolipid in three soils (Wen et al., 2009). Results showed that in the soils tested, the biodegradability of these materials increased in the order ETDA < rhamnolipid < citric acid. Rhamnolipid degradation was not complete in any of the soils in 20 days, the length of the experiment. Both studies show that rhamnolipids are biodegradable but the process is not rapid. This suggests that there would be an effective "window" of remediation time before their removal through biodegradation.

For the reasons provided earlier, biosurfactants have potential as remediation materials. In fact, much of the technology required to apply biosurfactants for reme-diation applications has already been developed for use with synthetic surfactants, meaning remediation procedures must only be adapted for use with biosurfactants. Their use has been constrained by lack of availability, which is primarily due to high costs of production. There is a need for research and avenues for commercialization to realize the potential for biosurfactants in remediation. The following sections will discuss research that has demonstrated the application of biosurfactants to various remediation technologies.

SOLID MEDIA

Soil Washing

Heavy metal remediation of soils, sediment, mine tailings, and industrial wastes has primarily focused on two strategies. The first is isolation and stabilization. Common approaches include excavation and landfilling; capping with impermeable materials; covering with soil; mixing or injecting inorganic or organic amendments (e.g., lim-ing agents, organic media; aluminosilicates, phosphates, metal oxides, etc.); and/or developing plant covers (phytostabilization). The goal of this strategy is to prevent the spread of contamination. The major disadvantage is that the metal contamina-tion is not removed. Thus, extended management, monitoring, and maintenance is required and there is generally a long-term risk of spreading contamination in the event of design failure (Lestan et al., 2008).

The second strategy is physical and chemical treatment applied to remove metal contamination. This technique generally involves washing soils with strong acids or metal-chelating compounds in order to mobilize metals, followed by a step that captures and reconcentrates the metals from solution. The use of strong acids has drawbacks because it leads to the disruption of the physical, chemical, and biological structure of soils, thus reducing possibilities for subsequent use as a soil medium (Malandrino et al., 2006; Neilson et al., 2003; Torres et al., 2012). Metal chelators can be effective and are less destructive. Many synthetic chelators, including some surfactants, have been studied; the most common of which is EDTA (Malandrino et al., 2006; Neilson et al., 2003; Torres et al., 2012). The need for an additive to facilitate metal removal has led to the notion of examining whether there are green additives, with reduced toxicity, that can be used in place of traditional chelators/surfactants.

There are several approaches to the mobilization step of soil washing. In situ soil washing is a viable technique for applications where the contaminated zone is underlain by a nonpermeable layer. This allows the washing solution to be leached through the contaminated zone and then pumped out and treated aboveground to remove metals (Lestan et al., 2008). Ex situ treatment basically involves removal and treatment of the contaminated soil. This can be done in a batch system, such as a soil slurry reactor, or can be done using heap or column leaching where the treatment solution is either gravimetrically percolated or pumped through the contaminated soil and then collected and treated. Finally, soils can be treated electrokinetically (generally ex situ) to remove metals by applying a direct current electrical field through a saturated contaminated soil. This causes the pore fluid to migrate by electro-osmosis and cationic metal ions to concentrate at the cathode (Lestan et al., 2008). For each of these approaches, in situ or ex situ soil washing or electrokinetic extraction, biosurfactants can be added to increase their efficacy in what is known as surfactant-enhanced soil washing (Torres et al., 2012).

The mechanism of biosurfactant-enhanced metal removal from solid surfaces is an interesting area of research. The sorption of metals onto mineral surfaces is controlled by many factors including ion exchange and association with Fe and Mn oxides, carbonates, and organic matter (precipitation/dissolution reactions). Two hypotheses have been offered to explain the mechanism behind surfactant-enhanced soil washing. First, direct complexation between the biosurfactant and solution phase metal effectively removes metal from solution and increases dissolution and desorption according to Le Chatelier's principle (Miller, 1995). The second is that biosurfactants can accumulate at the solid/liquid interface and interact with sorbed metals. Mulligan et al. (1999) designed a series of experiments to test these hypotheses using surfactin and a hydrocarbon-contaminated soil that was spiked with metals including Cu, Cd, Pb, and Zn. They conclude that interaction of surfactant with the soil allows direct interaction between the surfactant and the sorbed metal. The surfactant also reduces the interfacial tension at the liquid–solid interface, thereby making it easier to release the sorbed metal (Mulligan et al., 1999, 2001).

Soil Washing Efficacy of Biosurfactants on Artificial Contamination

Biosurfactants have been shown to be effective additives in soil washing technologies. Rhamnolipid, perhaps the best studied biosurfactant for surfactant-enhanced soil washing, was found to desorb metals from a sandy loam both when the metals were tested individually and in a mixture. An 80 mM monorhamnolipid (4%) solution removed 40% of Zn^{+2} and Pb^{+2} and nearly 60% of Cd^{+2} (Herman et al., 1995). A second study showed 99% removal of Pb, Ni, Cu, and Cd after flushing with a 1% glycolipid solution for 30 d (Parthasarathi and Sivakumaar, 2011). Yet a third study showed that 0.1% rhamnolipid removed 92% of Cd and 88% of Pb during a 36 h column leaching study, with no impact on culturable counts of bacteria or fungi (Juwarkar et al., 2007). Using a different approach, Wang and Mulligan (2004) examined whether a rhamnolipid foam could be used as a soil washing agent. Interestingly, a 0.5% rhamnolipid foam was 11% more effective at removing Cd and 17% more effective at removing Ni from a sandy soil than a 0.5% rhamnolipid solution. The authors hypothesized this was due to an increase in the homogeneity of the washing solution flow and surface contact when it was applied as a foam.

Saponin has also been studied as a soil washing agent. Saponin removed nearly 100% of Cd, Zn, and Cu from three soils with the removal efficiency of the soils being sandy loam > loam > silty clay. These batch studies showed that removal efficiency improved with multiple washings and that triplicate washes yielded optimal efficiency (Gusiatin and Klimiuk, 2012). Saponin was also shown to have high removal efficiencies for Cd (90%–100%) and Zn (85%–98%) from an Andosol, Cambisol, and Regosol when applied at a 3% concentration (Hong et al., 2002). Both Cu and Pb were also removed but with lower efficiencies (30%–60%).

Finally, surfactin has been demonstrated as a soil washing agent (Mulligan et al., 1999). Results of this study showed that metal removal by 0.25% surfactin solution was 25% for Cu and 6% for Zn. When the surfactin concentration in the washing solution was increased to 1%, the amount of metal removed decreased. Removal efficiency then increased again as surfactin was increased from 1% to 4%. As shown previously for rhamnolipids, multiple washes with surfactin improved metal recovery. For example, a 0.25% surfactin solution was used in five successive washes, removing 70% of Cu, 25% of Zn, and 15% of Cd.

Soil Washing Efficacy of Biosurfactants on Aged Contamination

While the earlier results sound promising, each of the aforementioned studies described used an artificially contaminated soil. Soil washing is not nearly as efficient in aged contaminated soils. When two historically contaminated soils were studied, one from a mining site and one from an abandoned army depot, rhamnolipid only removed a small fraction of the metals. Focusing on Pb which was present in both soils, after 10 washings under batch conditions, rhamnolipid removed 14.2% and 15.3% of total Pb from the soils, a relatively small amount. This was attributed to the association of the Pb with stable carbonate and oxide fractions in these soils. Rhamnolipid was effective in removing the soluble and exchangeable fraction of Pb from the soils but was much less effective at removing Pb bound to carbonates and amorphous iron oxides. This significantly reduced removal efficiency suggests that the longer a contaminant is in the soil, the more recalcitrant and associated with recalcitrant fractions of the soil that contaminant becomes (Neilson et al., 2003).

A historically contaminated acidic soil from the Palmerton Zinc Pile Superfund site was more successfully treated to remove Zn, Cu, Pb, and Cd (39%, 56%, 68%, and 43%, respectively) with rhamnolipid. The soil was reported to be less toxic to two species of earthworm after the washing (Slizovskiy et al., 2011). Taken together, these studies suggest that the contamination age of a soil along with its physical and chemical characteristics must be carefully considered to identify suitable extractants and extraction conditions. Further examination of historically contaminated soils is warranted, not only for biosurfactant application, but for the technology as a whole.

Sediments

Sediments have different characteristic properties than soil, for example, higher clay and organic matter content, and they too have been studied for surfactant-enhanced soil washing process efficacy. Mulligan et al. (2001) compared the ability of surfactin, rhamnolipid, and sophorolipid for removal of metals from a zinc- and copper-contaminated sediment. Results showed that rhamnolipid was most effective; a single washing of a 0.5% rhamnolipid solution removed 65% Cu and 18% Zn. A 4% sophorolipid solution removed 25% of Cu and 60% of Zn, while a 0.25% surfactin solution removed only 15% of Cu and 6% of Zn (Mulligan et al., 2001). In a second study that examined metal removal in continuous flow columns, 5% rhamnolipid was able to remove 37% of Cu, 7.5% of Zn, and 33% of Ni. Interestingly, the addition of 1% NaOH to 0.5% rhamnolipid increased the removal efficiency of Cu to levels near the 5% (10-fold higher) rhamnolipid treatment (Dahrazma and Mulligan, 2007). This increase in Cu recovery was due to the solubilization of organic matter by NaOH since Cu tends to partition into the organic fraction of soils and sediments. These results again suggest that the physicochemical characteristics of the system (sediment, soil, or water) must be carefully considered to allow extraction conditions to be optimized.

Mine Tailings

Mine tailings are the leftover waste material generated from mining processes. The materials, especially in older legacy mine tailings sites, can be highly contaminated with metals such as Pb and As. Surfactant-enhanced soil washing has recently been examined for use in the remediation of mine tailings. Aniszewski et al. (2010) examined the removal of Zn and Cd from a waste generated from mining zinc. They isolated four *Microbacterium* spp. with bioemulsifying activity and tested both cell-free supernatants and partially purified preparations for metal recovery. Cadmium removal ranged from 17% to 41% and Zn removal ranged from 14% to 68%, depending on the *Microbacterium* strain and the growth substrate used to produce the bioemulsifier.

Wang and Mulligan (2009a) used rhamnolipid to remove As from tailings, finding that the mobility of As increased with increasing rhamnolipid concentration and increasing pH. Since As is generally present as an anion, the authors discussed the potential mechanism of removal. The addition of rhamnolipid was found to lower the zeta potential of the tailings, suggesting that rhamnolipid was adsorbed to the tailings particle surfaces. It was hypothesized that this enhanced the mobilization of As anions due to increased charge density (negative) and repulsive electrostatic interactions (Wang and Mulligan, 2009a). In a second study, 70 pore volumes of a 0.1% rhamnolipid solution were applied to a column containing oxidized mine tailings with elevated levels of As, Cu, Pb, and Zn (Wang and Mulligan, 2009b).

Overall, the removal of these metals was low: 7% (As), 7% (Cu), 18% (Pb), and 5% (Zn). The mobilization of all these metals was found to be positively correlated with the removal of Fe.

Industrial Waste

Industrial wastes originating from sectors in aerospace to printed circuit boards to wastewater treatment can contain metal contaminants that range from low to very high in concentration. Several groups have examined the use of biosurfactants to recover metals from these waste streams. Gao et al. (2012) compared the removal of Pb, Ni, and Cr from an industrial water treatment plant sludge by saponin and sophorolipid. Both surfactants could aid in metal recovery with the amount of metal recovered related to increasing biosurfactant concentration and decreasing pH. Saponin, which removed 73.2%, 64.2%, and 56.1% of Pb, Ni, and Cr, respectively, performed better than sophorolipid (Gao et al., 2012). In a second study, saponin was found to remove 20%–45% of Cr, 50%–60% of Cu, and up to 100% of Pb from a municipal solid waste incinerator fly ash (Hong et al., 2000).

Summary

The variability in metal removal efficiency of the aforementioned studies is a testament to the challenges of using biosurfactants as soil washing agents. Comparing the studies is difficult because each utilized different soils and experimental treatments. Much of the variability is due to the nature of soils as dynamic and heterogeneous systems. Every soil characteristic that makes a soil unique, for example, soil pH, type, mineralogy, cation exchange capacity, organic matter content, porosity, contamination extent, contaminant age, etc., is a contributing factor that must be considered when developing a soil washing technique. For example, the mineralogy of the soil influences biosurfactant behavior: Ochoa-Loza et al. (2007) showed that monorhamnolipid sorption at low concentration follows the order hematite > kaolinite > MnO_2 ≈ illite ≈ Ca-montmorillonite > gibbsite ($Al(OH)_3$) > humic-acid-coated silica, while monorhamnolipid at high concentration sorbs in the order illite ≫ humic-acid-coated silica > Ca-montmorillonite >hematite > MnO_2 > gibbsite ≈ kaolinite. Adsorption is undesirable because it reduces the amount of solution phase surfactant available and further, can contribute to soil pore plugging.

As a second example, metal removal is optimal at a different pH for different surfactants. Saponin was most effective removing metals from kaolin at a pH of 6.5 (Chen et al., 2008). Rhamnolipid was most effective at a pH of 6.8 when tested individually on K-feldspar, sepiolite, quartz, and kaolin (Asci et al., 2007, 2008, 2010), but was more effective at a higher pH (10–11) when tested in soils (Wang and Mulligan, 2004), sediments (Mulligan et al., 2001), and mine tailings (Wang and Mulligan, 2009a). Surfactin was more effective at a pH of 10 than a pH of 8 (Mulligan et al., 2001).

As yet a third example, sorption behavior can also change as a function of surfactant type or congener mixture. For example, monorhamnolipid sorbed more strongly than a mix of the mono- and dirhamnolipid (Ochoa-Loza et al., 2007). Sorption has also been shown to be impacted by surfactant concentration, metal contamination (surfactants tend to sorb less when metal contamination is present), and ionic strength (Asci et al., 2007, 2008; Torrens et al., 1998). Clearly, surfactant-enhanced soil washing has the potential to be an effective remediation technique, but careful planning and design is required for every application if the process is to be successful.

PHYTOREMEDIATION

Phytoremediation technologies use plants to clean contaminated sites. Phytoextraction is a type of phytoremediation in which plants are used to concentrate and remove contaminants from soil (Lestan et al., 2008). Effective phytoextraction requires soil metals to be mobilized, absorbed by plant roots, translocated to shoot tissues, and stored, whereafter the biomass can be harvested and disposed of properly. A desirable plant for phytoremediation needs to be tolerant of elevated metal content, effective at accumulating metals in biomass, and productive with high biomass yields. Most of the plants already recognized as good candidates for phytoextraction are crop plants including sunflower (*Heliantus annuus*), corn (*Zea mays*), pea (*Pisum sativum*), and mustard (*Brassica juncae*) (Jordan et al., 2002). Chelates can be utilized to augment phytoremediation of contaminated soils. Chelate-assisted phytoextraction is based on the same principles as soil washing except that plants are used to remove the metals from the soil rather than a soil–water separation step. Thus, biosurfactants have the potential for use in chelate-assisted phytoextraction as long as they are compatible with plant growth, that is, not toxic. Indeed, many synthetic chelants, especially EDTA, have been shown to effectively increase concentrations of metals in plant tissues (Lestan et al., 2008).

Rhamnolipid has been examined in multiple studies to test its potential for chelate-assisted phytoextraction. Jordan et al. (2002) tested rhamnolipid with *Z. mays* and *Atriplex numilaria* (saltbrush) and the heavy metals Pb, Cu, and Zn. Rhamnolipid showed no effect on metal accumulation, yet was found in the plant tissues. It was hypothesized that the rhamnolipid micelles (where the metal is bound) are too big and thus excluded from the root, while rhamnolipid monomers are capable of entering the plant tissues. A second study examined the effect of rhamnolipid in a hydroponic system. This study found rhamnolipid did not enhance Cu uptake into *B. juncea* or *Lolium perenne* (ryegrass) (Johnson et al., 2009), even though rhamnolipid complexes with Cu strongly (log K = 9.27) (Ochoa-Loza et al., 2001). Exclusion of the soil–Cu sorption factor by this experiment supports the hypothesis that metal–micelle complexes are too big for root absorption. In a third study, rhamnolipid not only failed to enhance uptake of Zn and Cd, but it also showed toxic effects on *Z. mays* and *H. annuus* at concentrations as low as 0.2 mmol kg^{-1} of soil/week (Wen et al., 2010). Toxicity is also reported for *B. juncea* and *L. perenne* treated with 200 mg L^{-1} of rhamnolipid (Johnson et al., 2009).

Failure of rhamnolipid to act as an effective chelate-assisted phytoextraction amendment may be because experimental parameters were outside of the optimal conditions reported for surfactant-enhanced soil washing studies. For example, the pH range for the chelate-assisted phytoextraction studies was 6–7 while optimal conditions for rhamnolipid were shown to be near the pH of 10 for soils (Johnson et al., 2009; Wang and Mulligan,, 2009a; Wen et al., 2010). Further, the rhamnolipid concentration used in chelate-assisted phytoextraction studies was lower (4–5 mM) than those utilized for surfactant-enhanced soil washing studies (10–80 mM)

(Herman et al., 1995; Johnson et al., 2009; Jordan et al., 2002; Parthasarathi and Sivakumaar, 2011; Wang and Mulligan, 2004; Wen et al., 2010).

Two studies did show a benefit associated with the use of rhamnolipid in chelate-assisted phytoextraction. In the first, a moderate enhancement of Cu and Cd uptake by hydroponically grown *L. perenne* was achieved using 0.15 mM rhamnolipid (Gunawardana et al., 2010). Interestingly, uptake enhancement was significantly increased when rhamnolipid was combined with additional amendments including EDDS (an aminopolycarboxylic acid) or EDDS and citric acid. These combined treatments showed increased shoot Cu, Cd, and Pb concentrations (22- and 38-fold for Cu, 8- and 9-fold for Cd, and 2- and 3-fold for Pb, respectively). These treatments were also toxic to plant growth. The toxicity was attributed to both rhamnolipid, which showed some toxicity when used alone in metal-free controls, and the increased metal concentrations in the plant tissue. The rhamnolipid–EDDS–citric acid treatment was highly toxic to the plants so was applied only a few days prior to harvesting of the plants. Rhamnolipid–histidine and rhamnolipid–citric acid showed enhanced uptake of Cd, Cu, and Pb, but at lower levels and with less toxicity (Gunawardana et al., 2010).

In the second study, which examined the hydroponic growth of *Brassica napus* (canola) in Zn-limiting conditions, rhamnolipid was found to increase Zn absorption through nonmetabolically mediated pathways. Rhamnolipid–Zn complexes were found in the roots when examined by synchrotron XRF and XAS, and root K^+ efflux was increased, suggesting phytoxic effects of the treatment (Stacey et al., 2008). These studies suggest that physiological state of the plant may by a controlling factor in metal/metal-chelate uptake and accumulation. Indeed, studies have shown that root damage may be helpful for uptake of metal and/or metal–chelate complexes (Lestan et al., 2008). More research is required to better understand the mechanisms behind chelate-assisted phytoextraction and its optimization. Studies of additional biosurfactants and synergistic amendments are warranted as well.

REMEDIATION OF COCONTAMINATED SYSTEMS

Forty percent of the sites on the Environmental Protection Agency National Priority List are cocontaminated with heavy metals and organic compounds. Heavy metal toxicity in such sites can inhibit the biodegradation of organic contaminants by both aerobic and anaerobic consortia (Sandrin et al., 2000; Sandrin and Maier, 2003). Toxicity of metals to microorganisms occurs by a variety of mechanisms: competitive replacement of physiologically essential cations rendering enzymes nonfunctional, metal oxyanions may be substituted for similar essential oxyanions (e.g., arsenate for phosphate), and oxidative stress (Sandrin and Maier, 2003). Effective techniques for reducing metal toxicity rely either on the use of metal-tolerant bacteria or on the addition of materials to reduce metal bioavailability, for example, clays, calcium carbonate, phosphates, or chelating substances (Sandrin and Maier, 2003).

Biosurfactants can play a dual role in remediation of cocontaminated sites. They can complex metals thereby reducing their bioavailability and toxicity, but they can also increase the aqueous solubility of organic compounds resulting in enhanced biodegradation rates (Zhang and Miller, 1992). This was demonstrated in a study examining the effect of rhamnolipid on the degradation of naphthalene by a *Burkholderia* sp. in the presence of Cd. Results showed that a 1:1 molar ratio of rhamnolipid:Cd reduced cadmium toxicity, while increasing the molar ratio to 10:1 eliminated cadmium toxicity completely (Sandrin et al., 2000). In a second study, rhamnolipid was shown to enhance the degradation of phenanthrene by a soil consortium inhibited by the addition of Cd. In this study, phenanthrene mineralization reached the same levels as metal-free controls when several pulses of rhamnolipid were added. The pulsed addition technique was used to replenish rhamnolipid depleted through biodegradation (Maslin and Maier, 2000).

Biosurfactants have also been studied for simultaneous removal of organic and metal contamination from soil. In one study, saponin removed 76% of phenanthrene and 88% of Cd simultaneously (Song et al., 2008). Interestingly, removal levels for both phenanthrene and Cd were the same whether they were present individually or in combination. This suggests that the two contaminants do not compete with each other. This may be because Cd interacts with the head group of the surfactant and phenanthrene interacts with the micelle interior (Song et al., 2008). A second study examined removal of Cd and phenanthrene cocontaminants by four biosurfactants including surfactin; an iturin and fengycin mixture (lipopeptides produced by *Bacillus* sp.); arthrofactin (a lipopeptide produced by *Arthrobacter oxydans*); and flavolipid (now siderolipids) (Lima et al., 2011). Removal ranged from 79% to 87% and from 66.9% to 71.9% for phenanthrene and Cd, respectively, depending on the biosurfactant used. The use of an iodide ligand increased Cd removal from 74% to 99%. The iodide ligand is thought to form a neutral complex with Cd, which can then interact with the interior hydrophobic domain of the micelle, thereby increasing the amount of Cd associated with each micelle (Lima et al., 2011).

LIQUID MEDIA

There are a variety of approaches for the removal of metals from aqueous solutions. One is the use of sorbents, which can range from ion exchange resins to clays to microbial biomass. A recent report used rhamnolipid to assist in the process of sorption of Cu to clay. In this study, the efficiency of Cu sorption by an Na-montmorillonite clay was increased considerably when modified by rhamnolipid (Ozdemir and Yapar, 2009). A small amount of added rhamnolipid (2×10^{-6} M) acted to disperse the clay particles, thus increasing the total available surface area for Cu sorption. The sorbed metals were subsequently removed from the clay using a higher concentration rhamnolipid wash treatment (Ozdemir and Yapar, 2009).

Metals can also be removed directly from aqueous solution using biosurfactants. Once the aqueous solutions are treated with biosurfactants, the metal–surfactant complex must be removed from the solution. This can be accomplished in two ways: micellar-enhanced ultrafiltration and ion flotation. Micellar-enhanced ultrafiltration is a membrane-based separation process that uses anisotropic membranes with small size pores that do not allow surfactant micelles to pass through. Thus, the micelle-bound metal in solution is retained on the filter and removed from the bulk aqueous solution that passes through the filter (Fillipi et al., 1999). Pores sizes can range from 1,000 to 50,000 molecular-weight-cutoff (MWCO). Biosurfactants, because they are biodegradable, are particularly suited for this application; surfactant monomers can leak through the membrane leaving low levels of surfactant in the aqueous solution which can thereafter be biodegraded. The efficacy of this technique with biosurfactants has been examined. Micellar-enhanced ultrafiltration was tested with a spiculisporic acid derivative 2-(2-carboxyethyl)-3-decyl maleic anhydride (DCMA-3Na) (Hong et al., 1998). Results showed that DCMA-3Na removed 99%, 99%, and 93% of Cd, Cu, and Zn, respectively, when at equimolar concentrations using a 3000 MWCO membrane. DCMA-3Na exhibited a metal binding affinity of $Cd^{2+} > Cu^{2+} \sim Zn^{2+} > Ni^{2+}$ with Ni removal reaching only 65%.

A second study examined the use of rhamnolipid in micellar-enhanced ultrafiltration (El Zeftawy and Mulligan, 2011). Optimized conditions were determined for use of rhamnolipid to remove metals from six wastewater samples from the metal-refining industry. The wastewater samples contained Zn (60–130 mg L^{-1}); Cd (10–30 mg L^{-1}); and Pb, Cu, and Ni, which were all below 10 mg L^{-1}. The optimal conditions identified were as follows: membrane pressure, 69 ± 2 kPa; rhamnolipid: metal molar ratio, 2:1; temperature, 21 ± 1°C; and pH, 6.9 ± 0.1. Operating under these conditions, the level of all metals in the six wastewater samples was reduced to below 1.2 mg L^{-1} meeting Canadian Federal discharge limits (El Zeftawy and Mulligan, 2011).

A second approach to removing biosurfactant–metal complexes from solution is ion flotation, also known as dissolved air flotation or foam fractionation. This technique employs air sparging into a surfactant solution. The surfactant will absorb to the air bubbles generating a foam that can be harvested and removed from the solution. In the presence of metal ions, the surfactants can complex the ions and carry them into the foam (Chen et al., 2011). Using this technique, Chen et al. (2011) showed surfactin was able to remove 45% of the Hg from a 2 mg L^{-1} Hg solution. To achieve this removal, the optimal conditions were determined to be surfactin applied at 10 times the critical micelle concentration (10 × CMC) at a pH of 8 or 9. Overnight mixing prior to removal also helped increase Hg removal. In a second study, a 3:1 molar ratio of saponin to metal was used to remove lead, copper, and cadmium from wastewater with efficiencies of 90%, 81%, and 71%, respectively (Yuan et al., 2008). A third study used surfactants produced by *Candida* (likely sophorolipids) to achieve a removal of ≥98% of Fe (62 mg L^{-1}) and Mn (4 mg L^{-1}) from neutralized acid mine drainage using a 0.02% (200 mg L^{-1}) surfactant solution (Menezes et al., 2011).

Adsorbing colloid flotation is a method that combines the use of a colloidal sorbent material (e.g., clays and goethite) with the flotation technique (Zouboulis et al., 2004). This method essentially follows sorbent treatment of wastewater with surfactant flotation. Since the sorbent is present as a colloid, the surfactant foam will collect the metal ions adsorbed to the colloid particles by floating the sorbent. Similarly, compounds like Fe^{+3} can be used to coprecipitate metals for subsequent flotation. Using this technique, surfactin and lichenysin were used to collect either goethite (pH 4–7) or ferric hydroxide (pH 4) colloids that had sorbed Cr, resulting in the removal of Cr with almost 100% efficiency. Surfactin could also remove

~95% Zn when collecting ferric hydroxides (pH 6). Lichenysin, however, was ineffective for Zn removal (Zouboulis et al., 2004).

RECOVERY OF METAL FROM BIOSURFACTANT-METAL COMPLEXES

It is worth noting that once a biosurfactant–metal complex has been formed and used to remove the metal from a contaminated soil, sediment, or solution, the complex can be separated again through a simple pH adjustment. This allows recovery and recycling of both the metal and the biosurfactant. For anionic surfactants such as rhamnolipid or surfactin, the solution can be acidified (~pH 2) to precipitate the surfactant and release the metal ion. The biosurfactant can then be removed by centrifugation and recycled for reuse (Herman et al., 1995; Mulligan et al., 2001). For a nonionic surfactant, such as saponin, a pH adjustment to 11 will precipitate the complexed metals; the metals can then be removed by centrifugation and the biosurfactant collected for reuse (Gao et al., 2012).

CONCLUSION

As shown in this chapter, the body of literature on biosurfactant–metal interactions is large and continuing to grow. The potential for green, economical remediation technologies based on these interactions is enormous, but there remain several challenges that must be resolved before biosurfactants are viable alternatives for use in remediation technologies.

The first challenge is material cost. There is still no biosurfactant that can be competitively produced, in terms of cost, when compared to synthetic surfactants. The second challenge is in understanding the chemical properties of individual biosurfactant congeners so that they can be optimized for application in remediation technologies. For example, up to 60 rhamnolipid congeners are produced by wild-type bacteria, but these congeners can have very different chemical properties (e.g., CMC, interaction with metals). The ability to genetically manipulate bacteria to produce single congeners or alternatively, to chemically synthesize single congeners, has the potential to increase the efficacy of biosurfactants dramatically. The third challenge is to move research from the bench scale and into field testing. Optimally, this can be done by creating academic–industry partnerships, which can be implemented in two stages: first to demonstrate effectiveness, and second to scale-up production of biosurfactants for commercial use.

To close, it is clear humanity's impact on the biogeochemical cycling of metals is significant and resulting in increasing risks to public health due to metal exposures. As we continue to increase global consumption of new technologies, including cell phones and computers, the need for a variety of metals increases along with the resulting impacts of mining and metal consumption. As our demand for metals increases, we must find a way to reduce emissions and the impacts related to their use. The foundation for preserving the health and stability of our environment while encouraging innovation and economic growth lies in the development of green technologies and green chemicals; in the realm of metals, biosurfactants are one of the most promising possibilities available today.

REFERENCES

Abdel-Mawgoud, A. M., Lepine, F., and Deziel, E. (2010). Rhamnolipids: Diversity of structures, microbial origins and roles. *Applied Microbiology and Biotechnology, 86*(5), 1323–1336. doi: 10.1007/s00253-010-2498-2.

Aniszewski, E., Peixoto, R. S., Mota, F. F., Leite, S. G. F., and Rosado, A. S. (2010). Bioemulsifier production by microbacterium sp. strains isolated from mangrove and their application to remove cadmiun and zinc from hazardous industrial residue. *Brazilian Journal of Microbiology, 41*(1), 235–245. doi: 10.1590/S1517-83822010000100033.

Asci, Y., Nurbas, M., and Acikel, Y. S. (2007). Sorption of cd(II) onto kaolin as a soil component and desorption of cd(II) from kaolin using rhamnolipid biosurfactant. *Journal of Hazardous Materials, 139*(1), 50–56. doi: 10.1016/j.jhazmat.2006.06.004.

Asci, Y., Nurbas, M., and Acikel, Y. S. (2008). A comparative study for the sorption of cd(II) by soils with different clay contents and mineralogy and the recovery of cd(II) using rhamnolipid biosurfactant. *Journal of Hazardous Materials, 154*(1–3), 663–673. doi: 10.1016/j.jhazmat.2007.10.078.

Asci, Y., Nurbas, M., and Acikel, Y. S. (2010). Investigation of sorption/desorption equilibria of heavy metal ions on/from quartz using rhamnolipid biosurfactant. *Journal of Environmental Management, 91*(3), 724–731. doi: 10.1016/j.jenvman.2009.09.036.

Bodor, A., Banyai, I., and Toth, I. (2002). H-1- and C-13-NMR as tools to study aluminium coordination chemistry—Aqueous al(III)-citrate complexes. *Coordination Chemistry Reviews, 228*(2), 175–186. doi: 10.1016/S0010-8545(02)00039-5.

Bodour, A. A., Guerrero-Barajas, C., Jiorle, B. V., Malcomson, M. E., Paull, A. K., Somogyi, A., Trinh, L. N., Bates, R. B., and Maier, R. M. (2004). Structure and characterization of flavolipids, a novel class of biosurfactants produced by flavobacterium sp. strain MTN11. *Applied and Environmental Microbiology, 70*(1), 114–120. doi: 10.1128/AEM.70.1.114-120.2004.

Bonmatin, J. M., Genest, M., Labbe, H., and Ptak, M. (1994). Solution 3-dimensional structure of surfactin—A cyclic lipopeptide studied by H-1-nmr, distance geometry, and molecular-dynamics. *Biopolymers, 34*(7), 975–986. doi: 10.1002/bip.360340716.

Callender, E. (2003). 9.03—Heavy metals in the environment—Historical trends. In Heinrich D. H. and Karl K. T. (eds.), *Treatise on Geochemistry*, pp. 67–105. Oxford, U.K.: Pergamon. doi: 10.1016/B0-08-043751-6/09161-1.

Campbell, S., Rodgers, M. T., Marzluff, E. M., and Beauchamp, J. L. (1994). Structural and energetic constraints on gas-phase hydrogen-deuterium exchange-reactions of protonated peptides with D2o, Cd3od, Cd3co2d, and Nd3. *Journal of the American Chemical Society, 116*(21), 9765–9766. doi: 10.1021/ja00100a058.

Campbell, S., Rodgers, M. T., Marzluff, E. M., and Beauchamp, J. L. (1995). Deuterium exchange reactions as a probe of biomolecule structure. fundamental studies of cas phase H/D exchange reactions of protonated glycine oligomers with D2O, CD3OD, CD3CO2D, and ND3. *Journal of the American Chemical Society, 117*(51), 12840–12854. doi: 10.1021/ja00156a023.

Chen, H., Chen, C., Reddy, A. S., Chen, C., Li, W. R., Tseng, M., Liu, H. T., Pan, W., Maity, J. P., and Atla, S. B. (2011). Removal of mercury by foam fractionation using surfactin, a biosurfactant. *International Journal of Molecular Sciences, 12*(11), 8245–8258. doi: 10.3390/ijms12118245.

Chen, W., Hsia, L., and Chen, K. K. (2008). Metal desorption from copper(II)/nickel(II)-spiked kaolin as a soil component using plant-derived saponin biosurfactant. *Process Biochemistry, 43*(5), 488–498. doi: 10.1016/j.procbio.2007.11.017.

Choi, Y. K., Lee, C. H., Takizawa, Y., Gama, Y., and Ishigami, Y. (1993). Tricarboxylic acid biosurfactant derived from spiculisporic acid. *Journal of Japan Oil Chemists' Society, 42*(2), 95–99. doi: 10.5650/jos1956.42.95.

Colthup, N. B., Daly, L. H., and Wiberley, S. E. (1990). *Introduction to Infrared and Raman Spectroscopy*. Boston, MA: Academic Press.

Cooper, D. G., Macdonald, C. R., Duff, S. J. B., and Kosaric, N. (1981). Enhanced production of surfactin from *Bacillus subtilis* by continuous product removal and metal cation additions. *Applied and Environmental Microbiology, 42*(3), 408–412.

Csavina, J., Landazuri, A., Rheinheimer, P., Saez, A.E., Wonaschutz, A., Rine, K., Barbaris, B., Conant, W., and Betterton, E.A. (2011). Metal and metalloid contaminants in atmospheric aerosols from mining operations. *Water, Air, and Soil Pollution, 221*(1–4), 145–157.

Dahrazma, B. and Mulligan, C. N. (2007). Investigation of the removal of heavy metals from sediments using rhamnolipid in a continuous flow configuration. *Chemosphere, 69*(5), 705–711. doi: 10.1016/j.chemosphere.2007.05.037.

Dejugnat, C., Diat, O., and Zemb, T. (2011). Surfactin self-assembles into direct and reverse aggregates in equilibrium and performs selective metal cation extraction. *ChemPhysChem, 12*(11), 2138–2144. doi: 10.1002/cphc.201100094.

Deziel, E., Lepine, F., Milot, S., and Villemur, R. (2003). rhlA is required for the production of a novel biosurfactant promoting swarming motility in *Pseudomonas aeruginosa*: 3-(3-hydroxyalkanoyloxy)alkanoic acids (HAAs), the precursors of rhamnolipids. *Microbiology-Sgm, 149*, 2005–2013. doi: 10.1099/mic.0.26154-0.

El Zeftawy, M. A. M. and Mulligan, C. N. (2011). Use of rhamnolipid to remove heavy metals from wastewater by micellar-enhanced ultrafiltration (MEUF). *Separation and Purification Technology, 77*(1), 120–127. doi: 10.1016/j.seppur.2010.11.030.

Essington, M. E. (2004). *Soil and Water Chemistry: An Integrative Approach*. Boca Raton, FL: CRC Press.

Ferrari, E., Grandi, R., Lazzari, S., and Saladini, M. (2005). Hg(II)-coordination by sugar-acids: Role of the hydroxy groups. *Journal of Inorganic Biochemistry, 99*(12), 2381–2386. doi: 10.1016/j.jinorgbio.2005.09.005.

Fillipi, B. R., Brant, L. W., Scamehorn, J. F., and Christian, S. D. (1999). Use of micellar-enhanced ultrafiltration at low surfactant concentrations and with anionic-nonionic surfactant mixtures. *Journal of Colloid and Interface Science, 213*(1), 68–80. doi: 10.1006/jcis.1999.6092.

Gao, L., Kano, N., Sato, Y., Li, C., Zhang, S., and Imaizumi, H. (2012). Behavior and distribution of heavy metals including rare earth elements, thorium, and uranium in sludge from industry water treatment plant and recovery method of metals by biosurfactants application. *Bioinorganic Chemistry and Applications, 2012*, 173819. doi: 10.1155/2012/173819.

Glick, R., Gilmour, C., Tremblay, J., Satanower, S., Avidan, O., Deziel, E., Greenberg, E. P., Poole, K., and Banin, E. (2010). Increase in rhamnolipid synthesis under iron-limiting conditions influences surface motility and biofilm formation in *Pseudomonas aeruginosa*. *Journal of Bacteriology, 192*(12), 2973–2980. doi: 10.1128/JB.01601-09.

Grangemard, I., Wallach, J., Maget-Dana, R., and Peypoux, F. (2001). Lichenysin—A more efficient cation chelator than surfactin. *Applied Biochemistry and Biotechnology, 90*(3), 199–210. doi: 10.1385/ABAB:90:3:199.

Green, M. K., Gard, E., Bregar, J., and Lebrilla, C. B. (1995). H-D exchange kinetics of alcohols and protonated peptides—Effects of structure and proton affinity. *Journal of Mass Spectrometry, 30*(8), 1103–1110. doi: 10.1002/jms.1190300807.

Guerrasantos, L., Kappeli, O., and Fiechter, A. (1984). *Pseudomonas aeruginosa* biosurfactant production in continuous culture with glucose as carbon source. *Applied and Environmental Microbiology, 48*(2), 301–305.

Guerrasantos, L., Kappeli, O., and Fiechter, A. (1986). Dependence of *Pseudomonas aeruginosa* continuous culture biosurfactant production on nutritional and environmental-factors. *Applied Microbiology and Biotechnology, 24*(6), 443–448.

Gunawardana, B., Singhal, N., and Johnson, A. (2010). Amendments and their combined application for enhanced copper, cadmium, lead uptake by lolium perenne. *Plant and Soil, 329*(1–2), 283–294. doi: 10.1007/s11104-009-0153-4.

Gusiatin, Z. M. and Klimiuk, E. (2012). Metal (cu, cd and zn) removal and stabilization during multiple soil washing by saponin. *Chemosphere, 86*(4), 383–391. doi: 10.1016/j.chemosphere.2011.10.027.

Herman, D. C., Artiola, J. F., and Miller, R. M. (1995). Removal of cadmium, lead, and zinc fron soil by a rhamnolipid biosurfactant. *Environmental Science & Technology, 29*(9), 2280–2285. doi: 10.1021/es00009a019.

Hong, J., Yang, S. M., Choi, Y. K., and Lee, C. H. (1995). Precipitation of tricarboxylic acid biosurfactant derived from spiculisporic acid with metal ions in aqueous solution. *Journal of Colloid and Interface Science, 173*(1), 92–103. doi: 10.1006/jcis.1995.1301.

Hong, J., Yang, S. M., Lee, C. H., Choi, Y. K., and Kajiuchi, T. (1998). Ultrafiltration of divalent metal cations from aqueous solution using polycarboxylic acid type biosurfactant. *Journal of Colloid and Interface Science, 202*(1), 63–73. doi: 10.1006/jcis.1998.5446.

Hong, K. J., Tokunaga, S., Ishigami, Y., and Kajiuchi, T. (2000). Extraction of heavy metals from MSW incinerator fly ash using saponins. *Chemosphere, 41*(3), 345–352. doi: 10.1016/S0045-6535(99)00489-0.

Hong, K. J., Tokunaga, S., and Kajiuchi, T. (2002). Evaluation of remediation process with plant-derived biosurfactant for recovery of heavy metals from contaminated soils. *Chemosphere, 49*(4), 379–387. doi: 10.1016/S0045-6535(02)00321-1.

Ishigami, Y., Yamazaki, S., and Gama, Y. (1983). Surface active properties of biosoap from spiculisporic acid. *Journal of Colloid and Interface Science, 94*(1), 131–139. doi: 10.1016/0021-9797(83)90242-4.

Ishigami, Y., Zhang, Y. J., and Ji, F. X. (2000). Spiculisporic acid functional development of biosurfactant. *Chimica Oggi-Chemistry Today, 18*(7–8), 32–34.

Johnson, A., Gunawardana, B., and Singhal, N. (2009). Amendments for enhancing copper uptake by *Brassica juncea* and lolium perenne from solution. *International Journal of Phytoremediation, 11*(3), 215–234. doi: 10.1080/15226510802429633.

Jordan, F. L., Robin-Abbott, M., Maier, R. M., and Glenn, E. P. (2002). A comparison of chelator-facilitated metal uptake by a halophyte and a glycophyte. *Environmental Toxicology and Chemistry, 21*(12), 2698–2704. doi: 10.1897/1551-5028(2002)021<2698:ACOCFM>2.0.CO;2.

Juwarkar, A. A., Nair, A., Dubey, K. V., Singh, S. K., and Devotta, S. (2007). Biosurfactant technology for remediation of cadmium and lead contaminated soils. *Chemosphere, 68*(10), 1996–2002. doi: 10.1016/j.chemosphere.2007.02.027.

Karkhaneei, E., Zebarjadian, M. H., and Shamsipur, M. (2001). Complexation of Ba2+, Pb2+, Cd2+, and UO22+ ions with 18-crown-6 and dicyclohexyl-18-crown-6 in nitromethane and acetonitrile solutions by a competitive NMR technique using the li-7 nucleus as a probe. *Journal of Solution Chemistry, 30*(4), 323–333. doi: 10.1023/A:1010323106004.

Kasture, M. B., Patel, P., Prabhune, A. A., Ramana, C. V., Kulkarni, A. A., and Prasad, B. L. V. (2008). Synthesis of silver nanoparticles by sophorolipids: Effect of temperature and sophorolipid structure on the size of particles. *Journal of Chemical Sciences, 120*(6), 515–520. doi: 10.1007/s12039-008-0080-6.

Kotz, J. C., Treichel, P., and Weaver, G. C. (2006). *Chemistry & Chemical Reactivity*. Belmont, CA: Thomson Brooks/Cole.

Lang, S. (2002). Biological amphiphiles (microbial biosurfactants). *Current Opinion in Colloid & Interface Science, 7*(1–2), 12–20. doi: 10.1016/S1359-0294(02)00007-9.

Lestan, D., Luo, C., and Li, X. (2008). The use of chelating agents in the remediation of metal-contaminated soils: A review. *Environmental Pollution, 153*(1), 3–13. doi: 10.1016/j.envpol.2007.11.015.

Lima, T. M. S., Procopio, L. C., Brandao, F. D., Carvalho, A. M. X., Totola, M. R., and Borges, A. C. (2011). Simultaneous phenanthrene and cadmium removal from contaminated soil by a ligand/biosurfactant solution. *Biodegradation, 22*(5), 1007–1015. doi: 10.1007/s10532-011-9459-z.

Maier, R. M., Pepper, I. L., and Gerba, C. P. (2009). *Environmental Microbiology.* Amsterdam, the Netherlands; Boston, MA: Elsevier/Academic Press.

Makkar, R. S. and Cameotra, S. S. (2002). Effects of various nutritional supplements on biosurfactant production by a strain of *Bacillus subtilis* at 45 degrees C. *Journal of Surfactants and Detergents, 5*(1), 11–17. doi: 10.1007/s11743-002-0199-8.

Malandrino, M., Abollino, O., Giacomino, A., Aceto, M., and Mentasti, E. (2006). Adsorption of heavy metals on vermiculite: Influence of pH and organic ligands. *Journal of Colloid and Interface Science, 299*(2), 537–546. doi: 10.1016/j.jcis.2006.03.011.

Maslin, P. and Maier, R. M. (2000). Rhamnolipid-enhanced mineralization of phenanthrene in organic-metal co-contaminated soils. *Bioremediation Journal, 4*(4), 295–308.

Mehrotra, R. C. and Bohra, R. (1983). *Metal Carboxylates.* London, U.K.; New York: Academic Press.

Menezes, C. T. B., Barros, E. C., Rufino, R. D., Luna, J. M., and Sarubbo, L. A. (2011). Replacing synthetic with microbial surfactants as collectors in the treatment of aqueous effluent produced by acid mine drainage, using the dissolved air flotation technique. *Applied Biochemistry and Biotechnology, 163*(4), 540–546. doi: 10.1007/s12010-010-9060-7.

Miller, R. M. (1995). Biosurfactant-facilitated remediation of metal-contaminated soils. *Environmental Health Perspectives, 103*, 59–62. doi: 10.2307/3432014.

Mulligan, C. N., Yong, R. N., and Gibbs, B. F. (2001). Heavy metal removal from sediments by biosurfactants. *Journal of Hazardous Materials, 85*(1–2), 111–125. doi: 10.1016/S0304-3894(01)00224-2.

Mulligan, C. N., Yong, R. N., Gibbs, B. F., James, S., and Bennett, H. P. J. (1999). Metal removal from contaminated soil and sediments by the biosurfactant surfactin. *Environmental Science & Technology, 33*(21), 3812–3820. doi: 10.1021/es9813055.

Neilson, J. W., Artiola, J. F., and Maier, R. M. (2003). Characterization of lead removal from contaminated soils by nontoxic soil-washing agents. *Journal of Environmental Quality, 32*(3), 899–908.

Neilson, J. W., Zhang, L., Veres-Schalnat, T. A., Chandler, K. B., Neilson, C. H., Crispin, J. D., Pemberton, J. E., and Maier, R. M. (2010). Cadmium effects on transcriptional expression of rhlB/rhlC genes and congener distribution of monorhamnolipid and dirhamnolipid in *Pseudomonas aeruginosa* IGB83. *Applied Microbiology and Biotechnology, 88*(4), 953–963. doi: 10.1007/s00253-010-2808-8.

Nriagu, J. O. (1989). A global assessment of natural sources of atmospheric trace-metals. *Nature, 338*(6210), 47–49. doi: 10.1038/338047a0.

Nriagu, J. O. (1990). Global metal pollution: Poisoning the biosphere? *Environment: Science and Policy for Sustainable Development, 32*(7), 7–33. doi: 10.1080/00139157.1990.9929037.

Nriagu, J. O. (1996). A history of global metal pollution. *Science, 272*(5259), 223–223. doi: 10.1126/science.272.5259.223.

Nriagu, J. O. and Pacyna, J. M. (1988). Quantitative assessment of worldwide contamination of air, water and soils by trace-metals. *Nature, 333*(6169), 134–139. doi: 10.1038/333134a0.

Ochoa-Loza, F. J., Artiola, J. F., and Maier, R. M. (2001). Stability constants for the complexation of various metals with a rhamnolipid biosurfactant. *Journal of Environmental Quality, 30*(2), 479–485.

Ochoa-Loza, F. J., Noordman, W. H., Janssen, D. B., Brusseau, M. L., and Maier, R. M. (2007). Effect of clays, metal oxides, and organic matter on rhamnolipid biosurfactant sorption by soil. *Chemosphere, 66*(9), 1634–1642. doi: 10.1016/j.chemosphere.2006.07.068.

Ozdemir, G. and Yapar, S. (2009). Adsorption and desorption behavior of copper ions on na-montmorillonite: Effect of rhamnolipids and pH. *Journal of Hazardous Materials, 166*(2–3), 1307–1313. doi: 10.1016/j.jhazmat.2008.12.059.

Pacyna, J. M. and Pacyna, E. G. (2001). An assessment of global and regional emissions of trace metals to the atmosphere from anthropogenic sources worldwide. *Environmental Reviews, 9*(4), 269.

Pacyna, J. M., Scholtz, M. T., and Li, Y. F. (1995). Global budget of trace metal sources. *Environmental Reviews, 3*(2), 145–159.

Palacios, E. G., Juarez-Lopez, G., and Monhemius, A. J. (2004). Infrared spectroscopy of metal carboxylates—II analysis of fe(III), Ni and Zn carboxylate solutions. *Hydrometallurgy, 72*(1–2), 139–148. doi: 10.1016/S0304-386X(03)00137-3.

Palme, O., Moszyk, A., Iphoefer, D., and Lang, S. (2010). Selected microbial glycolipids: Production, modification and characterization. *Biosurfactants, 672*, 185–202.

Pankiewicz, R., Schroeder, G., Brzezinski, B., and Bartl, F. (2005). NMR, FT-IR and ESI-MS study of new lasalocid ester with 2-(hydroxymethyl)-12-crown-4 and its complexes with monovalent cations. *Journal of Molecular Structure, 749*(1–3), 128–137. doi: 10.1016/j.molstruc.2005.03.039.

Parthasarathi, R. and Sivakumaar, P. K. (2011). Biosurfactant mediated remediation process evaluation on a mixture of heavy metal spiked topsoil using soil column and batch washing methods. *Soil & Sediment Contamination, 20*(8), 892–907. doi: 10.1080/15320383.2011.620043.

Ramana, K. V. and Karanth, N. G. (1989). Factors affecting biosurfactant production using *Pseudomonas aeruginosa* cftr-6 under submerged conditions. *Journal of Chemical Technology and Biotechnology, 45*(4), 249–257.

Rondeau, P., Sers, S., Jhurry, D., and Cadet, F. (2003). Sugar interaction with metals in aqueous solution: Indirect determination from infrared and direct determination from nuclear magnetic resonance spectroscopy. *Applied Spectroscopy, 57*(4), 466–472. doi: 10.1366/00037020360626023.

Rouhollahi, A., Amini, M. K., and Shamsipur, M. (1994). Nmr-study of the stoichiometry, stability and exchange kinetics of complexation reaction between Pb2+ ion and 18-crown-6 in binary acetonitrile-water mixtures. *Journal of Solution Chemistry, 23*(1), 63–74. doi: 10.1007/BF00972608.

Saini, H. S., Barragan-Huerta, B. E., Lebron-Paler, A., Pemberton, J. E., Vazquez, R. R., Burns, A. M., Marron, M. T., Seliga, C. J., Gunatilaka, A. A., and Maier, R. M. (2008). Efficient purification of the biosurfactant viscosin from *Pseudomonas libanensis* strain M9-3 and its physicochemical and biological properties. *Journal of Natural Products, 71*(6), 1011–1015. doi: 10.1021/np800069u.

Sandrin, T. R., Chech, A. M., and Maier, R. M. (2000). A rhamnolipid biosurfactant reduces cadmium toxicity during naphthalene biodegradation. *Applied and Environmental Microbiology, 66*(10), 4585–4588. doi: 10.1128/AEM.66.10.4585-4588.2000.

Sandrin, T. R. and Maier, R. M. (2003). Impact of metals on the biodegradation of organic pollutants. *Environmental Health Perspectives, 111*(8), 1093–1101. doi: 10.1289/ehp.5840.

Schalnat, T. A. (2012). Metal complexation and interfacial behavior of the microbially-produced surfactant monorhamnolipid by *Pseudomonas aeruginosa* ATCC 9027. PhD. Tucson, AZ: University of Arizona.

Shen, H., Lin, T., Thomas, R. K., Taylor, D. J. F., and Penfold, J. (2011). Surfactin structures at interfaces and in solution: The effect of pH and cations. *Journal of Physical Chemistry B, 115*(15), 4427–4435. doi: 10.1021/jp109360h.

Sheppard, J. D., Jumarie, C., Cooper, D. G., and Laprade, R. (1991). Ionic channels induced by surfactin in planar lipid bilayer-membranes. *Biochimica Et Biophysica Acta, 1064*(1), 13–23. doi: 10.1016/0005-2736(91)90406-X.

Singh, A., Van Hamme, J. D., and Ward, O. P. (2007). Surfactants in microbiology and bio-technology: Part 2. Application aspects. *Biotechnology Advances, 25*(1), 99–121. doi: 10.1016/j.biotechadv.2006.10.004.

Slizovskiy, I. B., Kelsey, J. W., and Hatzinger, P. B. (2011). Surfactant-facilitated remediation of metal-contaminated soils efficacy and toxicological consequences to earthworms. *Environmental Toxicology and Chemistry, 30*(1), 112–123. doi: 10.1002/etc.357.

Song, S., Zhu, L., and Zhou, W. (2008). Simultaneous removal of phenanthrene and cad-mium from contaminated soils by saponin, a plant-derived biosurfactant. *Environmental Pollution, 156*(3), 1368–1370. doi: 10.1016/j.envpol.2008.06.018.

Stacey, S. P., McLaughlin, M. J., Cakmak, I., Hetitiarachchi, G. M., Scheckel, K. G., and Karkkainen, M. (2008). Root uptake of lipophilic zinc-rhamnolipid complexes. *Journal of Agricultural and Food Chemistry, 56*(6), 2112–2117. doi: 10.1021/jf0729311.

Strathmann, T. J. and Myneni, S. C. B. (2004). Speciation of aqueous ni(II)-carboxylate and ni(II)-fulvic acid solutions: Combined ATR-FTIR and XAFS analysis. *Geochimica et Cosmochimica Acta, 68*(17), 3441–3458. doi: 10.1016/j.gca.2004.01.012.

Tajmir-Riahi, H. (1985). Sugar complexes with alkaline earth metal ions synthesis, structure, and spectroscopic studies of mg(II), sr(II), and ba(II) complexes of d-glucur. *Journal of Inorganic Biochemistry, 24*(2), 127–136. doi: 10.1016/0162-0134(85)80004-0.

Tajmir-Riahi, H. (1989). Carbohydrate complexes with lead(II) ion. Interaction of pb(II) with beta-D-glucurono-6,3-lactone, D-glucono-1,5-lactone, and their acid anions and the effects of metal ion binding on the sugar hydrolysis. *Bulletin of the Chemical Society of Japan, 62*(4), 1281–1286. doi: 10.1246/bcsj.62.1281.

Tan, H., Champion, J. T., Artiola, J. F., Brusseau, M. L., and Miller, R. M. (1994). Complexation of cadmium by a rhamnolipid biosurfactant. *Environmental Science and Technology, 28*(13), 2402–2406. doi: 10.1021/es00062a027.

Thimon, L., Peypoux, F., and Michel, G. (1992). Interactions of surfactin, a biosurfactant from *Bacillus subtilis*, with inorganic cations. *Biotechnology Letters, 14*(8), 713–718. doi: 10.1007/BF01021648.

Thimon, L., Peypoux, F., Wallach, J., and Michel, G. (1993). Ionophorous and sequestering properties of surfactin, a biosurfactant from *Bacillus subtilis*. *Colloids and Surfaces B: Biointerfaces, 1*(1), 57–62. doi: 10.1016/0927-7765(93)80018-T.

Tian, W., Yang, L. M., Xu, Y. Z., Weng, S. F., and Wu, J. G. (2000). Sugar interaction with metal ions. FT-IR study on the structure of crystalline galactaric acid and its K+, NH4+, Ca2+, Ba2+, and La3+ complexes. *Carbohydrate Research, 324*(1), 45–52. doi: 10.1016/S0008-6215(99)00276-1.

Torrens, J. L., Herman, D. C., and Miller-Maier, R. M. (1998). Biosurfactant (rhamnolipid) sorption and the impact on rhamnolipid-facilitated removal of cadmium from various soils under saturated flow conditions. *Environmental Science and Technology, 32*(6), 776–781. doi: 10.1021/es970285o.

Torres, L. G., Lopez, R. B., and Beltran, M. (2012). Removal of as, cd, cu, ni, pb, and zn from a highly contaminated industrial soil using surfactant enhanced soil washing. *Physics and Chemistry of the Earth, 37–39*, 30–36. doi: 10.1016/j.pce.2011.02.003.

Van Bogaert, I. N. A., Zhang, J., and Soetaert, W. (2011). Microbial synthesis of sophorolipids. *Process Biochemistry, 46*(4), 821–833. doi: 10.1016/j.procbio.2011.01.010.

Van Hamme, J. D., Singh, A., and Ward, O. P. (2006). Physiological aspects—Part 1 in a series of papers devoted to surfactants in microbiology and biotechnology. *Biotechnology Advances, 24*(6), 604–620. doi: 10.1016/j.biotechadv.2006.08.001.

Wang, S. and Mulligan, C. N. (2004). Rhamnolipid foam enhanced remediation of cadmium and nickel contaminated soil. *Water Air and Soil Pollution, 157*(1–4), 315–330. doi: 10.1023/B:WATE.0000038904.91977.f0.

Wang, S. and Mulligan, C. N. (2009a). Arsenic mobilization from mine tailings in the presence of a biosurfactant. *Applied Geochemistry, 24*(5), 928–935. doi: 10.1016/j.apgeochem.2009.02.017.

Wang, S. and Mulligan, C. N. (2009b). Rhamnolipid biosurfactant-enhanced soil flushing for the removal of arsenic and heavy metals from mine tailings. *Process Biochemistry, 44*(3), 296–301. doi: 10.1016/j.procbio.2008.11.006.

Wen, J., McLaughlin, M. J., Stacey, S. P., and Kirby, J. K. (2010). Is rhamnolipid biosurfactant useful in cadmium phytoextraction? *Journal of Soils and Sediments, 10*(7), 1289–1299. doi: 10.1007/s11368-010-0229-z.

Wen, J., Stacey, S. P., McLaughlin, M. J., and Kirby, J. K. (2009). Biodegradation of rhamnolipid, EDTA and citric acid in cadmium and zinc contaminated soils. *Soil Biology and Biochemistry, 41*(10), 2214–2221. doi: 10.1016/j.soilbio.2009.08.006.

Wuana, R. A. and Okieimen, F. E. (2011). Heavy metals in contaminated soils: A review of sources, chemistry, risks and best available strategies for remediation. *ISRN Ecology ISRN Ecology, 2011*(4), 1–20.

Yuan, X. Z., Meng, Y. T., Zeng, G. M., Fang, Y. Y., and Shi, J. G. (2008). Evaluation of tea-derived biosurfactant on removing heavy metal ions from dilute wastewater by ion flotation. *Colloids and Surfaces A-Physicochemical and Engineering Aspects, 317*(1–3), 256–261. doi: 10.1016/j.colsurfa.2007.10.024.

Zhang, Y. M. and Miller, R. M. (1992). Enhanced octadecane dispersion and biodegradation by a pseudomonas rhamnolipid surfactant (biosurfactant). *Applied and Environmental Microbiology, 58*(10), 3276–3282.

Zouboulis, A. I., Matis, K. A., Lazaridis, N. K., and Golyshin, P. N. (2004). The use of biosurfactants in flotation: Application for the removal of metal ions. *Minerals Engineering, 17*(1), 105–105. doi: 10.1016/j.mineng.2003.11.017.

12 Biosurfactants
Future Trends and Challenges

Catherine N. Mulligan, Sanjay K. Sharma, and Ackmez Mudhoo

CONTENTS

INTRODUCTION

Often, activities of industries in general will come into conflict with the goals of a sustainable environment. This is based on the assumption that (1) nonrenewable natural resources (materials and energy) are required to fuel the engine of *industry*, (2) nonrenewable source materials are used in the production or manufacture of goods, and (3) the smokestack emissions and discharge of liquid and solid wastes from these industries are harmful to both human health and the environment (Figure 12.1).

The basic concept of *industrial ecology* is to apply a systems approach to protect the environment and conserve environmental resources as components of industrial production and development (Yong et al., 2006). Through integration of industry with the environment, *industrial ecology* focuses on (1) renewable and nonrenewable environmental resource consumption and conservation with regard to raw materials for industrial activities; (2) efficient industrial production through technology

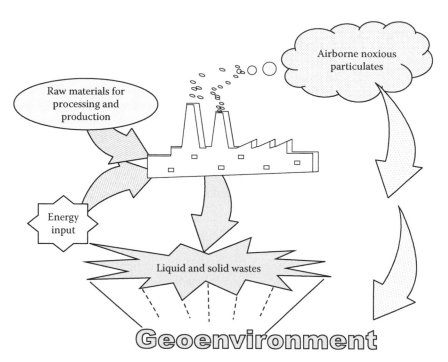

FIGURE 12.1 Interactions between industries and environment.

and resource conservation through recycling, recovery, reduction, and reuse of waste products; and (3) environmentally sensitive management of emissions and disposal of waste products from industrial activities (Figure 12.2). Ideally, industrial ecology is a holistic approach to industrial production of goods, that is, it takes into account the goals of environment and resources sustainability whilst meeting its goals of production of goods and other life support systems to the benefit of consumers.

An example is the production of chemical surfactants such as detergents and cleaners that are mainly produced from various petrochemicals (Desai and Banat, 1997). Biologically produced surfactants are potentially viewed as more environmentally friendly than synthetic surfactants due to their biodegradability based on the concept of industrial ecology. Research on biosurfactants has been extensive due to their potential applications in the cosmetic, food, pharmaceutical, and environmental fields. They are biodegradable, effective, and exhibit many interesting properties. Attempts are underway to decrease costs through media optimization, use of inexpensive or waste substrates, or process optimization. Although there have been various biosurfactants that have been characterized, the most studied are glycolipids, lipopeptides, trehalose lipids, sophorolipids, and mannosylerythritrol lipids (MELS).

In addition, in comparison to synthetic surfactants, biosurfactants can be produced in a more sustainable manner. The raw materials should be renewable and nonpolluting. Any secondary products must be nonpolluting and of beneficial use. To enhance the commercialization of the biosurfactants, production and purification conditions with inexpensive substrates must be optimized. This chapter is an effort

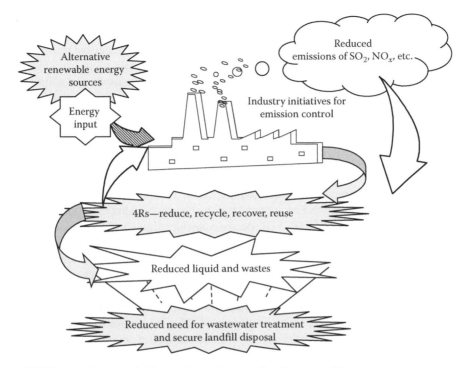

FIGURE 12.2 Industry initiatives for reduction of environmental impacts.

to review the state of the art of biosurfactants and their applications and to determine approaches to overcome the challenges.

TYPES OF BIOSURFACTANTS

RHAMNOLIPIDS

Rhamnolipids are produced by *Pseudomonas aeruginosa*, a gram-negative opportunistic pathogen. Different homologues and congeners of rhamnolipids with varying ratios of the components are produced depending on the production conditions. However, Marchant and Banat (2012) indicated that manipulation of the ratio of the two main components is difficult due to the synthetic pathway of producing the dirhamnolipid from the monorhamnolipid. More knowledge is needed in this area. The focus of many researchers is to find and isolate microorganisms capable of overproducing rhamnolipids or tailoring some nonpathogenic microorganisms to reduce the production complications and cost.

Rhamnolipids have outstanding properties including high surface activity, antimicrobial activity, degradability, renewability, low toxicity, and stability in emulsions with hydrocarbons. They are excellent candidates for use in bioremediation, petroleum, pharmaceutical, cosmetic, and food industries and thus have many benefits for the public. Although they have been demonstrated in many applications, their current main market is in the environmental and petroleum industries.

Nevertheless, research has shown that this valuable material can be used as a fungicide, and for decomposing agricultural wastes, and where there is a large potential market in the agricultural sector. Other applications are in the medical and pharmaceutical fields, suggesting the use of these active agents for vast applications from fighting and antiproliferative activity against human breast cancer; to be used as an additive in nanoparticle production; or functioning as an antimicrobial agent. In the cosmetic and food industries, emulsifying and antimicrobial activity make them ideal additives for use as emulsifiers in food production processes and in moisturizing creams.

Heyd et al. (2008) examined methods to analyze and purify rhamnolipids. They indicated that the methods range from colorimetric to chromatographic. The most precise for analysis of rhamnolipids is HPLC coupled with MS. However, colorimetric tests are ideal for screening strains and media for production. The development of methods for quantization of rhamnolipids that are rapid will be beneficial. Although some work has been performed to develop FTIR-ATR spectroscopy for culture broth analysis, complex rhamnolipid analysis is difficult. Continuous removal of biosurfactants from culture media is essential for decreasing costs related to product recovery. Foam fractionation, adsorption, or membrane methods are under development. Currently, chromatography followed by precipitation is the main recovery method used.

For rhamnolipids to compete with synthetic surfactants currently in the market, the cost of the production needs to be decreased. This is possible through the selection of low-cost substrates, optimization of the fermentation environment, and isolation and creation of new strains that could produce rhamnolipids at greater rates, along with a higher ratio of the desirable congeners. Renewable substrates will also enhance process sustainability.

Sophorolipids

Sophorolipids are mainly produced by the yeasts *Candida bombicola* or *C. apicola*. They are a mixture of eight major components with another 15 minor components (Van Bogaert et al., 2007). Sophorose (that may or may not be acetylated) is attached to a fatty acid chain of 16–18 carbons. Although some control on the composition of the sophorolipids has been reported due to strain isolation and media conditions, changing the backbone of the structure has not been achieved. The main component is a lactonic compound with 17 hydroxy-octadecanoic acid. Glycolipids from *C. bombicola* have interesting properties. They are highly biodegradable, low foaming, and of low toxicity with potential use for detergent formulations. Surfactant–enzyme combinations have also shown promise. Surfactin–subtilisin A is an example of this (Onaizi et al., 2009).

This is a clear indication that sophorolipid has a very vast application area and is becoming technologically important, particularly in Japan and South Korea. In some specific industrial sectors, such as cosmetic and pharmaceutical, biosurfactants have a high potential and will probably play a major role in a short period of time. Sophorolipid is a choice molecule because of highly valued surface properties, high degree of biodegradability, and environmental compatibility. Yields have been greater than 100 g/L, making it more economic to produce and isolate. Sophorolipids have potential for the cosmetics industry due to their various properties including stimulation of dermal fibroblast metabolism, and hygroscopic. Future prospects for sophorolipid-based products include several types of facial cosmetics, lotions, beauty washes, and hair products. However, the different forms of sophorolipid have shown different properties. Van Bogaert et al. (2011) have indicated that the acidic form has a lower CMC and higher surface adsorption, whereas the lactonic form is more surface-active. Therefore, either a pure form or a controlled mixture of components will need to be achieved.

Still, the industrial use of the sophorolipid is limited because of the costs involved in the production process, and other fermentation conditions must be optimized to achieve cheaper production technology. The production process of native sophorolipids from glucose and vegetable oil substrates is already optimized and yields are much higher than 100 g/L. Thus, optimization of the production process leading to a less complex mixture of sophorolipids is the key factor to reduce costs. The genetic and metabolic engineering using recombinant techniques could be a better solution for the manipulation of biosurfactant production and properties. Genetic manipulation of *C. bombicola* has shown some promise in changing chain length and degree of saturation of the tail and acetylation of the carbohydrate head. Overcoming these limitations is necessary for enhancing sophorolipid commercialization.

MELS

Mannosylerythritol lipids (MELS) are classified as the four forms of 4-o-β—mannopyranosyl-D-erythritol attached to two chains of fatty acyl esters(MEL-A, B, C, and D) (Arutchelvi and Doble, 2011). They are produced by yeast and basdiomycota (i.e., *Pseudozyma antarctica* and *Ustilago maydis*). Water solubility of MEL-A is low, limiting some applications. However, yields of 165 g/L have been obtained with MEL-A as the predominant form. Similar to sophorolipids, MELS are produced by cells in the resting phase, which allows high yields (Arutchelvi and Doble, 2011). These high yields on wastes or by-products may allow enhanced commercial viability. Through lipase catalysis of hydrolysis, a nonacylated product (MEL-D) can be produced. MEL-D was a lamellar-forming glycolipid (Fukuoka et al., 2011).

MARINE SOURCES OF BIOSURFACTANTS

Although biosurfactants have been the subject of intense investigation during the past few decades, relatively small numbers of microorganisms and research output have focused on their production from marine microorganisms. Nevertheless, some marine microbial communities including *Acinetobacter, Arthrobacter, Pseudomonas, Halomonas, Bacillus, Rhodococcus, Enterobacter, Azotobacter, Corynebacteium,* and *Lactobacillus* have been explored for the production of surface-active molecules, both biosurfactants and bioemulsifiers. Such biological surfactants have important potential application in different industries and the marine ecosystems can provide an excellent opportunity to select unique and diverse producing microorganisms and chemical products. Effective screening and purification techniques are essential in order to explore and discover unique and effective biosurfactants and bioemulsifiers able to be produced using cost-effective technology processes and at acceptable yields and quality. Promising recent biotechnological approaches coupled with highlighting of the importance of the marine resource for such novel compounds and their environmental credentials are expected to support both search and potential industrial application of biosurfactants in many industrial applications.

LIPOPEPTIDES

Surfactin has very interesting surfactant properties. Potential medical applications are related to anti-inflammatory, antiviral, antibiotic, and anti-adhesive activities. However, the economics are not competitive due to poor yields and the requirement for expensive and complex substrates. Portilla-Rivera et al. (2009) have postulated that biosurfactant costs can be as low as $0.50/L from molasses sugarcane. Low-cost purification methods are also needed as downstream costs can account for 60% of the cost (Mukherjee et al., 2006). Purity of the product is also a major consideration; 98% pure surfactin is sold for $10 per mg but can be reduced to $2–$4/kg for tank cleaning or oil recovery applications (Bognolo, 1999). Although more information is available concerning biosynthesis of surfactin, there is still a lack of information regarding its secretion, metabolic route, primary cell metabolism, and physicochemical properties of the biosurfactant. Research is thus required to enhance the applications of the surfactant. New forms of surfactin and other lipopeptides could also become available.

TREHALOSE BIOSURFACTANTS

Trehalose lipids have been extensively studied over the past several decades and thus there now has been a significant amount of research regarding their production, chemical structures and properties. Despite their favorable physicochemical and biological activities, which determine their potential applications in environmental and industrial biotechnologies, these affordable biosurfactants have not been commercialized extensively. This is mainly due to the fact that trehalose lipids are mostly cell-bound and are produced from nonrenewable carbon sources. There is still a lack of research on the development of efficient bioprocesses, using low-cost and renewable resources and the optimization of the culture conditions, including cost-effective recovery processes. Although that different attractive aspects of trehalose lipids have been recently documented, such as their biomedical and therapeutic properties, future fundamental research should be focused toward the development of novel genetically engineered hyperproducing strains coupled with economization of the biosurfactant production process. The efficient combination of these options could open up perspectives for higher yields and successful large-scale economically profitable production of these unique biomolecules.

APPLICATIONS OF BIOSURFACTANTS

The market for biosurfactants was 2210 million USD in 2011 and will grow from 2011 to 2018 at a rate of 3.5% (Transparency Market Research) (Sekhon et al., 2012). Producers of soaps, toothpastes, washing powders, detergents are increasingly becoming more aware and interested in biosurfactants due to their excellent properties of biodegradability, low toxicity, and surface activity. Some of the applications are detailed in Table 12.1 and in the following sections.

TABLE 12.1
Various Applications of Biosurfactants

Industry	Applications
Petroleum	Enhanced oil recovery, de-emulsification
Environmental	Bioremediation, soil washing, and flushing
Food	Emulsification and de-emulsification
Pharmaceutical	Antibacterial, antifungal, antiviral agents
Agriculture	Biocontrol of parasites, microorganisms
Cosmetics	Health and beauty agents

Source: Adapted from Kapadia Sanket, G. and Yagnik, B.N., *Asian J. Exp. Biol Sci.*, 4, 1, 2013.

FOOD ADDITIVES

Increasing demand for natural ingredients and green chemistry stimulates the exploitation of biosurfactants. These molecules show properties highly useful in food processing as ingredients, as cleaning agents, or high-value products obtained from food waste substrates. Food and cosmetic applications related to the promotion of emulsion stability have the most potential. For food uses, *Lactobacillus* or yeasts would be safer sources of biosurfactants (Nitschke and Costa, 2007).

To improve biosurfactant utilization by the food industry, more data about their toxicity including in vivo tests and more research to evaluate not only biosurfactant properties but also their interaction with food components are needed. Furthermore, there is a lack of information about their contribution to the sensory properties of food in which they are incorporated or in contact. The use of wastes or by-products can reduce costs and valorize the residues. However, simple substrates that need few preparation steps and standardization should be preferred to turn the process cost-effective. The discovery of new applications will contribute to the valorization of these molecules, increasing even more their demand. Finally, the improvement of microbial strain productivity, safety, and stability is imperative to the success of biosurfactants.

ENVIRONMENTAL APPLICATIONS

Rhamnolipids and other biosurfactants including saponin have shown their potential for remediation of contaminated soil and water for organic and inorganic contaminants. Due to the binding of contaminants to soil, biosurfactants have been proposed to enhance bioavailability and mobilization of contaminants. More biosurfactants need to be investigated. The biosurfactants seem to enhance biodegradation by influencing the bioavailability of the contaminant through solubilization, increasing the hydrophobicity of the cell surface and emulsification of the contaminants. Some cases have resulted in inhibition of the biodegradation, which is not well-understood. Modeling of the biodegradation is not reliable and experimental testing is required.

For soil washing applications, solubilization and mobilization are the two main mechanisms as indicated by the effectiveness of the biosurfactants mentioned earlier and below the CMC (Franzetti et al., 2010). Mulligan (2009) has reviewed the application for many types of hydrocarbon- and metal-contaminated soils. Arsenic and chromium have recently been added to list.

Due to their biodegradability and low toxicity, biosurfactants such as rhamnolipids are very promising for use in remediation technologies. In addition, there is the potential for in situ production, a distinct advantage over synthetic surfactants. This needs to be studied further. Some preliminary work by Jalali and Mulligan (2010) and others for microbial enhanced oil recovery (MEOR) showed that the indigenous bacteria in contaminated soil were able to produce biosurfactants once sufficient nutrients were available. Further research regarding prediction of their behavior in the fate and transport of contaminants will be required. More investigation into the solubilization mechanism of both hydrocarbons and heavy metals by biosurfactants is required to enable model predictions for transport and remediation. New applications for the biosurfactants regarding nanoparticles are developing. Future research should focus on the stabilization of the nanoparticles by biosurfactants before addition during remediation procedures, decreasing production costs, in situ production, among other issues will be required.

The most promising industrial application is enhanced oil recovery as only 40%–45% of the oil in the reservoirs can currently be recovered (Banat et al., 2010). However, most studies have been at lab scale to simulated oil reservoir conditions. Three strategies exist regarding the use of biosurfactants. Biosurfactants can be produced off-site for subsequent addition; biosurfactant can be produced by added microorganisms or stimulation of in situ production of biosurfactants. Recent research has focused on the isolation of new strains that simulate the extreme condition of oil reservoirs (Agarwal and Sharma, 2009). Some field trials involved the addition of microorganisms for MEOR (Sen, 2008). However, the lack of control well tests makes it difficult to conclude regarding the effectiveness of the microbial addition. Other oil applications include removal of oil from tank bottom sludge (Perfumo et al., 2010b) and oil dispersion.

Research on biosurfactant–metal interactions is large and continues to grow. The potential for green, economical remediation technologies based on these interactions is enormous, but there remain several challenges that must be resolved before biosurfactants are viable alternatives for use in remediation technologies. The first challenge is the product cost. Biosurfactants are not currently very competitive with synthetic surfactants. The second challenge is in understanding the chemical properties of individual biosurfactant congeners so that they can be optimized for application in remediation technologies. The ability to manipulate bacteria to produce congeners is appropriate for specific applications. The third challenge is the lack of full-scale applications. Research needs to be performed in pilot demonstration tests in the field before full-scale applications can be performed to demonstrate effectiveness. A protocol as indicated in Figure 12.3 should be followed. Although there has been substantial evidence of the benefits of biosurfactants on bioremediation, inhibition has also been observed. Therefore, more research on the complex interactions between the pollutants, biosurfactants, and microorganisms is needed.

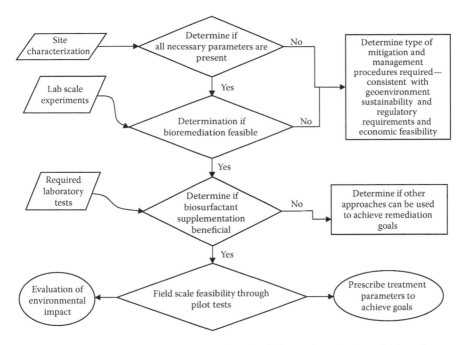

FIGURE 12.3 General protocol to determine feasibility and application of biosurfactants during remediation of contaminated sites.

BIOMEDICAL APPLICATIONS

Medical applications of biosurfactants are related to the disruption of surface properties affecting adhesion of microorganisms and disruption of cell membranes causing cell lysis. Healing of wounds (Piljac et al., 2008) is another promising application for biosurfactants.

Biosurfactants exhibit antimicrobial, antiviral, and antifungal properties. Lipopeptides are the most commonly studied. MELS are also promising. Inhibition of biofilm formation and swarming motility can enhance the reduction of infection in medical devices and food processing environments. Chemical surfactants do not specifically disrupt biofilms and, therefore, addition of biosurfactants may help inhibit biofilms formed by washing at low temperatures. Other properties are anti-inflammatory and antitumor activities.

Marchant and Banat (2012) examined replacement of chemical surfactants by biosurfactants. They have indicated that sophorolipids exhibit sufficient yields, whereas rhamnolipids and MELS require further development. Surfactin-mediated gold nanoparticles (Reddy et al., 2009) is a new potential application in the biomedical field. Tailoring of the biosurfactant to specific applications will be key for future development. Biomedical applications are more likely to lead to economic feasibility. Nanotechnology developments will also enhance these applications. Further studies will be needed on microbial interactions, biofilm formation, and motility.

PRODUCTION COSTS

Mulligan and Gibbs (1993) analyzed the economics of biosurfactant production. Many of their conclusions are still relevant today. The aspects included

- Choosing inexpensive raw materials
- Increasing yields and production rates
- Optimizing the fermentation step
- Reducing product recovery costs
- Producing biosurfactants for suitable applications

The choice of raw materials for production influences substantially the final product cost. Bulk, readily available, or waste material use is advantageous. Some substrates are listed in Table 12.2. Agro-industrial by-products have been the most frequently studied, whereas oil-based substrates such as motor oil and frying oil require higher downstream costs. Other wastes include molasses, whey, animal fats and oils, starchy wastes, cassava wastewater, and refinery wastes (Saharan et al., 2011). Glycerol derived from biodiesel production has potential as a cost-efficient substrate (Zheng et al., 2008). If waste materials are considered, low-cost hydrolysis methods for lignocellulosics are necessary. Product yields based on cost must be determined. *Bacillus, Pseudomonas,* and *Candida* are the most studied for large-scale production (Mukherjee et al., 2006). Other media components including the source of nitrogen, magnesium, manganese, phosphorus, sulfur, and iron have also significant impacts on biosurfactant yields. Substrates with appropriate amounts of these components will decrease the cost of additional nutrient requirements.

Enhanced recovery of the biosurfactants must also be achieved through increasing production yields or product recovery techniques. Screening for overproducers, and media and fermentor design and manipulation to control biosynthesis are highly important for yield enhancement. Targeted or random mutagenesis has not been extremely effective in enhancing product yields. The genetics of rhamnolipid production is quite complex. In addition, rhamnolipids are mixtures of monorhamnolipid and dirhamnolipid congeners.

TABLE 12.2
Waste Substrates for Microbial Surfactants

Substrate	Reference
Casava flour	Nitschke and Pastore (2003)
Sugar beet peels	Onbasli (2009)
Sweet potato peels	Makkar and Cameotra (2002)
Sweet sorghum peels	Makkar and Cameotra (2002)
Rice and wheat bran husk	Barrios-Gonzalez et al. (1988)
Sugarcane bagasse	Krieger et al. (2010)
Cashew apple juice pomace	Rocha et al. (2007)
Soybean oil	Lima et al. (2009)
Whey	Dubey and Juwarkar (2004)
Coffee wastes, saw dust, distillery wastes, tea waste, agricultural wastes	Pandey et al. (2000)

Fermentor design has not evolved substantially and is still mainly of stirred tank design. Approaches such as surface response methodology and other statistical techniques have been recently employed to enable optimization of the various factors that influence biosurfactant production (Kronemberger et al., 2008; Sivapathasekaran et al., 2010). Some of the major environmental factors include aeration, agitation, pH, and temperature. Using continuous production will enable optimal conditions for biosurfactant production to be maintained. Such conditions include nutrient or aeration control. New control devices are also being developed that will assist optimization.

Some research has been performed on solid-state fermentation and increasing oxygen transfer and mixing efficiency. Continuous removal to reduce product inhibition and enhance removal efficiency through foam collection, ion exchange resins, and other ultrafiltration and liquid membrane processes needs to be optimized. This approach reduces solvent requirements and increases product concentrations. Recycling of effluents has not been widely practiced but can reduce nutrient, cell, and water requirements. Capital costs may also be reduced. Process scale-up is still lacking but should follow the approach in Figure 12.4.

Purification requirements also have a substantial influence on product costs. It has been estimated as 60% of the total costs (Saharan et al., 2011). For biomedical applications, pharmaceutical grades will be required. However, for environmental applications, crude biosurfactants may provide adequate effectiveness and can substantially reduce product costs. Little research has been performed on crude products. In addition, the recovery of the biosurfactants must utilize less toxic and inexpensive solvents.

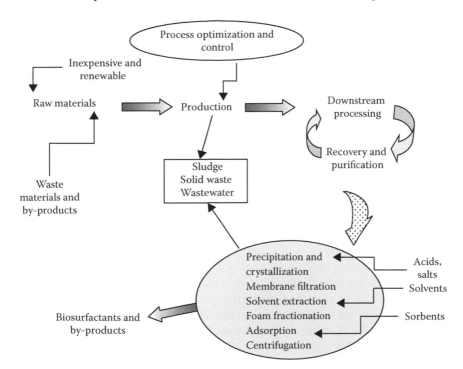

FIGURE 12.4 Processes and waste products generated in biosurfactant production.

TABLE 12.3
Industrial Production of Biosurfactants

Company	Products and Applications
Agae Technologies (USA)	R95, an HPLC grade rhamnolipid, most likely from glycerol
Jeneil Biotechnology (USA)	Various concentrations of rhamnolipid (2%–99% pure), ZONIX rhamnolipid as biofongicide, RECO for oil recovery from storage tanks
Fraunhofer IGB (Germany)	Glycolipids, cellobiose lipids, MELS
Cognis (USA and Germany)	Glycolipids, cellobiose lipids, MELS
Ecover (Belgium)	Sophorolipids
Saraya Co. Ltd. (Japan)	Sophorolipids from palm oil
MG Intobio (South Korea)	Sophorolipids for soaps for acne
Groupe Soliance (France)	Sophorolipids from rapeseed for skin care
Paradigm Biomedical (USA)	Pharmaceuticals from rhamnolipids

Sources: Adapted from Marchant, R. and Banat, I.M., *Biotechnol. Lett.*, 34, 1597, 2012; Sekhon, K.K. et al., *Petrol. Environ. Biotechnol.*, 3, 11, 2012.

Surfactant effectiveness will have a substantial influence on future commercialization. Mulligan and Gibbs (1993) indicated that the CMC of most biosurfactants is in the range of 1–2000 mg/L. If a surfactant costs $1/kg with a CMC of 8 mM, then a biosurfactant of CMC 0.01 mM and costs $20/kg would be less expensive to use.

Sophorolipids and MELS can be produced at high concentrations (Kitamoto et al., 2001), whereas rhamnolipids are produced in the range of 10–20 g/L. Despite this, rhamnolipids are produced by Jeneil Biotechnology, Milwaukee, USA (www.jeneilbiotech.com). AGAE Technologies (www.agaetech.com) also produces rhamnolipids and recently announced that their proprietary rhamnolipid biosurfactant has application for the prevention and cure of various superbug infections. A summary is shown in Table 12.3.

Various properties can influence economics, such as biodegradability, toxicity and temperature, salt, and pH stability. Adsorption during surfactant flushing in enhanced oil recovery or environmental applications will decrease surfactant effectiveness.

CONCLUSION

The research in the field of biosurfactants is advancing rapidly and it encompasses fields as diverse as medicine, surface science, organic chemistry, molecular biology, etc. Thus, there is a remarkable increase in the applications of biosurfactants and hence the need for its increased production. Traditional chemical synthesis methods raise concerns about toxicity and environmental impact. This favors the switch to more sustainable synthesis options. Biosurfactants that are produced from environmentally renewable resources are preferred as they use low-temperature, cost-effective methods, retain sustainability, and generate minimum waste. Methods for recovery of biosurfactants must be modified to reduce the problem of environmentally hazardous and costly organic solvents. Inexpensive raw materials, fermentation processes, and recovery methods must all be optimized for more technical and economic efficiency. The research on biosurfactants for extended applications has been continuing. Also, as the data on the phase behavior of different biosurfactants capable of forming micro-emulsion/reverse-micro-emulsion systems is not very well known, standard experimental protocols need to be established. The use of biosurfactants can lead to the greener synthesis of nanoparticles. Sophorolipids and MELS biosurfactants are also highly promising due to their safety and high yields. Co-isolation of other by-products such as enzymes may also increase economic feasibility of biosurfactants production.

REFERENCES

Agarwal, P. and Sharma, D.K. 2009. Studies on the production of biosurfactant for the microbial enhanced oil recovery by using bacteria isolated from oil contaminated wet soil. *Petroleum Science and Technology* 27: 1880–1893.

Arutchelvi, J. and Doble, M. 2011. Mannosylerythritol lipids: Microbial production and their applications. In *Biosurfactants from Genes to Applications,* Soberon-Chavez, G. (ed.) Vol. 20, Springer-Verlag, New York, pp. 145–177.

Banat, I.M., Franzetti, A., Gandolfi, I., Bestetti, G., Martinotti, M.G., Fracchia, L., Smyth, T.J., and Marchant, R. 2010. Microbial biosurfactants production, applications and future potential. *Applied Microbiology and Biotechnology* 87: 427–444.

Barrios-Gonzalez, J., Tomassini, A., Viniegra-Gonzalez, G., and Lopez, L. 1988. Penicillin production by solid state fermentation. *Biotechnology Letters* 10: 793–799.

Bognolo, C. 1999. Biosurfactants as emulsifying agents for hydrocarbons. *Colloids and Surfaces A Physico Engineering Aspects* 152: 41–52.

Desai, J.D. and Banat, I.M. 1997. Microbial production of surfactant and their commercial potential. *Microbiology and Molecular Biology Reviews* 61: 47–64.

Dubey, K. and Juwarkar, A. 2004. Determination of genetic basis for biosurfactant production in distillery and curd whey wastes utilizing *Pseudomonas aeruginosa* strain B52. *Indian Journal of Biotechnology* 3: 74–81.

Franzetti, A., Caredda, P., Ruggeri, C., La Colla, P., Tamburini, E., Papacchini, M., and Bestetti, G. 2009.Potential applications of surface active compounds by *Gordonia* sp. BS29 in soil remediation technologies. *Chemosphere* 75: 801–807.

Fukuoka, T., Yanagihara, T., Imura, T., Morita, T., Sakai, H., Abe, M., and Kitamoto, D. 2011. Enzymatic synthesis of a novel glycolipid biosurfactant, mannosylerythritol, lipid D and its aqueous behaviour. *Carbohydrate Research* 346: 266–271.

Heyd, M., Kohnert, A., Tan, T.-H., Nusser, M., Kirschhofer, F., Brenner-Weis, G., Franzreb, M., and Berenmeier, S. 2008. Development and trends of biosurfactant analysis and purification using rhamnolipids as an example. *Analytical and Bioanalytical Chemistry* 391: 1579–1590.

Jalali, F. and Mulligan, C.N. 2007. Effect of biosurfactant production by indigenous soil microorganisms on bioremediation of a co-contaminated soil in batch experiments. *60th Canadian Geotechnical Conference*. Ottawa, Canada. October 21–24.

KapadiaSanket, G. and Yagnik, B.N. 2013. Current trend and potential for microbial biosurfactants. *Asian Journal of Experimental and Biological Sciences* 4: 1–8.

Kitamoto, D., Ikegami, T., Suzuki, G. T., Sasaki, A., Takeyama, Y., Idemoto, Y., Koura, N., and Yanagishita, H. 2001. Microbial conversion of n-alkanes into glycolipid biosurfactants, mannosylerythritol lipids by *Pseudozymaantartica. Biotechnology Letters* 23: 1709–1714.

Krieger, N., Doumit, C., and David, A.M. 2010. Production of microbial biosurfactants by solid-state cultivation. *Advances in Experimental Medicine and Biology* 672: 203–210.

Kronemberger, F.D., Anna, L., Fernandes, A., de Menezes, R.R., Borges, C.P., and Freire, D.M.G. 2008. Oxygen controlled biosurfactant production in a bench scale bioreactor. *Applied Biochemistry and Biotechnology* 147: 22–45.

Lima de, C.J.B., Ribeiro, E.J., Servulo, E.F.C., Resende, M.M., and Cardoso, V.L. 2009. Biosurfactant production by *Pseudomonas aeruginosa* grown in residual soybean oil. *Applied Biochemistry and Biotechnology* 152: 156–168.

Makkar, R.S. and Cameotra, S.S. 2002. An update on use of unconventional substrates for biosurfactants production and their new applications. *Applied Microbiology and Biotechnology* 58: 428–434.

Marchant, R. and Banat, I.M. 2012. Biosurfactants: A sustainable replacement for chemical surfactants? *Biotechnology Letters* 34: 1597–1605.

Mukherjee, S., Das, P., and Sen, R. 2006. Towards commercial production of microbial surfactants. *Trends in Biotechnology* 24: 509–515.

Mulligan, C.N. 2009. Recent advances in the environmental applications of biosurfactants. *Current Opinion in Colloid and Interfacial Science* 14: 372–378.

Mulligan, C.N. and Gibbs, B.F. 1993. Factors influencing the economics of biosurfactants. In *Biosurfactants, Production, Properties and Applications,* Kosaric, N. (ed.) Marcel Dekker Inc., New York, Chapter 13, pp. 329–372.

Nitschke, M. and Costa, S.G.V.A.O. 2007. Biosurfactants in food industry. *Trends in Food Science and Technology* 18: 252–259.

Nitschke, M. and Pastore, G.M. 2003. Cassava flour wastewater as a substrate for biosurfactant production. *Applied Biochemistry and Biotechnology* 106: 295–302.

Onaizi, S.A., He, L., and Middelberg, A.P.J. 2009. Rapid screening of surfactant and biosurfactant surface cleaning performance. *Colloids and Surfaces B* 72: 68–74.

Onbasli, A.B. 2009. Biosurfactant production in sugar beet molasses by some *Pseudomonas* spp. *Journal of Environmental Biology* 30: 161–163.

Pandey, A., Soccol, C.R., and Mitchell, D. 2000. New development in solid state fermentation: I-bioprocesses and products. *Process Biochemistry* 35: 1153–1169.

Perfumo, A., Rancich, I., and Banat, I.M. 2010b. Possibilities and challenges for biosurfactants in petroleum industry. In *Biosurfactant's Advances in Experimental Medicine and Biology*, Sen, R. (ed.) Vol. 672, Springer, Berlin, Germany, pp. 135–157.

Piljac, A., Stipcevic, T., Piljac-Zegarac, J., and Piljac, G. 2008. Successful treatment of chronic decubitus ulcer with 0.1% dirhamnolipid ointment. *Journal of Cutaneous Medicine and Surgery* 12: 142–146.

Portilla-Rivera, O.M., Teliez-Luis, S.J., Ramirez de Leon, J.A., and Vasquez, M. 2009. Production of microbial transglutaminase on media made from sugar cane molasses and glycerol. *Food Technology and Biotechnology* 47: 19–26.

Reddy, A.S., Chen, C.Y., Chen, C.C., Jean, J.S., Fan, C.W., Chen, H.R., Wang, J.C., and Nimje, V.R. 2009. Synthesis of gold nanoparticles via an environmentally benign route using a biosurfactant. *Journal of Nanoscience and Nanotechnology* 9: 6693–6699.

Rocha, M., Souza, M., and Benedicto, S. 2007. Production of biosurfactant by grown on cashew apple juice. *Biochemistry and Biotechnology* 53:136–140.

Saharan, B.S., Sahu, R.K., and Sharma, D. 2011. A review on biosurfactants: Fermentation, current development and perspectives. *Genetic Engineering and Biotechnology Journal* 2011, accepted November 7, 2011.

Sekhon, K.K., Khanna, S., and Cameotra, S.S. 2012. Biosurfactant production and potential correlation with esterase activity. *Petroleum and Environmental Biotechnology* 3: 11–10.

Sen, R. 2008. Biotechnology in petroleum recovery: The microbial EOR. *Progress in Energy Combustion* 34: 714–724.

Sivapathasekaran, C., Mukherjee, S., Ray, A., Gupta, A., and Sen, R. 2010. Artificial neural network modeling and genetic algorithm based medium optimization for the improved production of marine biosurfactant. *Bioresource Technology* 101: 2884–2887.

Soberon-Chavez, G. (ed.) 2011. *Biosurfactants, from Genes to Application*. Springer, Heidelberg, Germany.

Van Bogaert, I.N.A., Saerens, K., De Muynck, C., Develter, D., Soetaert, W., and Vandamme, E.J. 2007. Microbial production and application of sophorolipids. *Applied Microbiology and Biotechnology* 76: 23–34.

Yong, R.N., Mulligan, C.N., and Fukue, M. 2006. *Geoenvironmental Sustainability*. CRC Press, Boca Raton, FL.

Zheng, Y.G., Chen, X.L., and Shen, Y.C. 2008. Commodity chemicals derived from glycerol, an important biorefinery feedstock. *Chemical Reviews* 108: 5253–5277.

Index

T - #0358 - 071024 - C4 - 234/156/16 - PB - 9780367378899 - Gloss Lamination